W9-CUB-430

Toward a Dialogue
of Understandings

Toward a Dialogue of Understandings

Loren Eiseley
and the Critique of Science

Mary Ellen Pitts

Lehigh
University
Press

Bethlehem: Lehigh University Press
London: Associated University Presses

Associated University Presses
440 Forsgate Drive
Cranbury, NJ 08512

Associated University Presses
25 Sicilian Avenue
London WC1A 2QH, England

Associated University Presses
P.O. Box 338, Port Credit
Mississauga, Ontario
Canada L5G 4L8

The paper used in this publication meets the requirements
of the American National Standard for Permanence of Paper
for Printed Library Materials Z39.48-1984.

Library of Congress Cataloging-in-Publication Data
Pitts, Mary Ellen.
  Toward a dialogue of understandings : Loren Eiseley and the critique of science / Mary Ellen Pitts.
    p.   cm.
  Includes bibliographical references and index.
  ISBN 0-934223-37-8 (alk. paper)
  1. Eiseley, Loren C., 1907–1977—Knowledge—Science.
  2. Literature and science—United States—History—20th century.
  3. Science—Historiography.   4. Science in literature.   I. Title.
PS3555.I78Z83   1995
818.5409—dc20   94-46994
                                                                       CIP

Contents

Eiseley's Texts and Abbreviations Used

Scholarship

Darwin's Century (1958)	DC

Collections of Popular Essays

The Immense Journey (1957)	IJ
The Firmament of Time (1960)	FT
The Unexpected Universe (1969)	UU
The Invisible Pyramid (1970)	IP
The Night Country (1971)	NC
The Man Who Saw Through Time (1973), originally published in 1962 as *Francis Bacon and the Modern Dilemma*	MWSTT (FBMD)
The Star Thrower (1978)	ST
Darwin and the Mysterious Mr. X (1979; posthumous)	DMX

Autobiography

All the Strange Hours (1975)	ASH

Poetry

Notes of an Alchemist (1972)	NA
The Innocent Assassins (1973)	IA
Another Kind of Autumn (1977)	AKA
All the Night Wings (1979; posthumous)	ANW

Notebooks

The Lost Notebooks of Loren Eiseley (Edited by Kenneth Heuer; posthumous)	LN

Preface

My interest in Loren Eiseley began long ago when I read "The Long Loneliness." I was haunted by his exploration both of scientific evidence of communication among porpoises, highly intelligent beings lacking the ability to communicate beyond their own generation through writing, and of *other* understandings of the relationship of human life to the natural world. When I read his entire oeuvre, I responded, as do most readers, with admiration for his sensitivity, his poetic prose, and his combination of literary and scientific interests. Still, something more about Eiseley's texts intrigued me. I came to see Eiseley's quest, not simply for bringing together science and literature in what some have called a "new genre" (in one sense a continuation of what the eighteenth- and nineteenth-century British naturalists and Thoreau had done) but for an epistemological shift—a movement toward an epistemological dialogue that involves both the analytical method of science and the insights associated with intuitive knowing, with the synthesizing insights of literature.

As I delved more deeply into Eiseley's epistemological quest, I began to see how his project anticipated the work of Thomas S. Kuhn and Gerald Holton, who posit, respectively, the role of the established paradigm and of the thematic and emotional involvement of the scientist in the history of science, and of E. H. Gombrich, who posits the role of making before matching, of schemata and correction, in the psychology of human perception, especially as applied to the history of art.

These realizations—Eiseley's proximity to those who questioned the "objectivity" on which notions of the scientific method are based, his desire to recognize in science the "supplement" (in great science, the necessity) of poetic knowledge, and his questioning of the very possibility of exact representation—led me to see in Eiseley's oeuvre an anticipation of the epistemological skepticism and reexamination that have emerged in recent philosophy and in literary theory. Indeed, as an anthropologist whose love for the written word was evident from his high-school declaration that he wanted "to be a nature writer" and from his college literary activities,

Eiseley was clearly knowledgeable about the role of language and
rhetoric in shaping human understanding. In his texts he equates
scientific symbols with literary symbols and the creative act in
science with that of art ("The Illusion," *ST,* 275).

Eiseley was, I came to realize, actually defamiliarizing (and anticipating contemporary approaches to literary and philosophical
texts by questioning hidden assumptions, decentering, finding
points of faltering, undermining hierarchical patterns, questioning
any system that totalizes meaning or life or signification, examining
the often unwitting role of metaphor) certain basic attitudes toward
science that we have inherited from the nineteenth century, especially the notion of "nature." Michel Foucault has contended that
"man" himself, seen as separate from the rest of the world, is an
eighteenth-century construct [1973, xxiii]. In studying the history
of evolutionary thinking for the seminal *Darwin's Century* (1958),
Eiseley came to see that, as a result of scientific studies, nature
has become externalized, particularized, mechanized, separated
from the human and fragmented, reduced to conflict without consideration of cooperation, confined to reductionist and positivist
study. The results for humankind as part of the *biota*—Eiseley's
concern as a writer—are far-reaching.

Eiseley himself began his career as a writer for nonspecialist
audiences with a study in appreciation of Sir Francis Bacon's contribution to the emergence of the modern world and its privileging
of "observation"; yet Eiseley's emerging emphasis even in this
study is on Bacon's vision of "the uses of life" (which Eiseley
interprets in terms of all life, not just human life) to which the
scientific method could lead. And in *Darwin's Century,* as he painstakingly traced the use of the Baconian method in the discovery
of evolution, Eiseley became aware of the contributions of a "more
gracious, humane tradition" that is traceable to John Ray and Gilbert White, eighteenth-century parson-naturalists, and to "literary
observers" such as Thoreau and Hudson, so that two separate
"streams" were formed that "have at times mingled, influenced and
affected each other but . . . have remained in some degree apart
in method and in outlook" (*DC,* 13).

Of both streams Eiseley found himself the inheritor. The second,
"the more gracious, humane tradition," helped move him, even as
he was preparing the manuscript for *Darwin's Century,* away from
purely scholarly writing and toward the literary essay. It was this
stream that compelled him to reexamine, for the nonspecialist audience, the excesses of experimentalism, and to seek an epistemology that incorporates both Baconian observation and experiment

and humane insight—an epistemology powerfully reminiscent of Martin Heidegger's nostalgic existentialism, of his "dwelling poetically" (1975c, 211–29). Jacques Derrida has suggested that although we will probably never separate ourselves from Plato, we can at least reexamine the assumptions that we have unwittingly inherited from the Platonic tradition. Eiseley suggests that, as a scientific civilization, we cannot and should not separate ourselves from experimentalism, but we can reexamine the basic assumptions of the civilization that this method has created. And it is in this reexamination, this defamiliarization of scientific thought, especially as applied to the "natural" world, that Eiseley's unique impact as a thinker and a literary figure emerges. His reexamination of science and of the linguistic and rhetorical constructs that underlie the way we understand science and the world is what concerns me here. I have specified Eiseley's approach to the "natural" because his primary concern is with the biological world and its interrelationship with the cultural world that humankind has constructed. "Nature," for Eiseley, emerges as a metonym for a view of the physical world, of the *biota,* and of humankind that must be reexamined if life is to survive.

*   *   *

Note on the use of nonsexist language:

In keeping with the time when he wrote and with anthropological convention, Eiseley uses language that is often sexist (though he attacks Victorian and other prejudices against women). Whenever possible, I have quoted parts of sentences and referred to *humankind* and to *humans* to avoid the outdated use of *man.* My own text also alternates *she* and *he* in the generic.

Acknowledgments

For the final setting-forth of this book, my debts are many. To the graduate professors in the English Department at the University of Florida who stimulated my interest in interdisciplinary studies and in literary theory when I was just beginning to work on Eiseley's texts, I am grateful. Greg Ulmer, Alistair Duckworth, Sid Homan, and Robert D'Amico supported my early attempts at interdisciplinarity and stretched my mental capacities. At The University of Memphis (then Memphis State University), former chair William H. O'Donnell supported my application for a Faculty Development Assignment that enabled me to complete this book. My colleagues in the Society for Literature and Science, especially Bill McRae, Fred White (who read the manuscript and made important suggestions), and Ed Block, offered valuable advice. At Lehigh University Press, Philip Metzger smoothed the way for this project and advised me throughout the review process. Editors at the Associated University Presses who guided the manuscript were always helpful. Faults in the manuscript, of course, are mine alone.

My extended family network was were there when I needed hearing and endured real difficulties with me: Virginia Ruth and Hudge, Pat and Fred, Pat and Charles, Charmaine and Bill, Edythe, Mary Sue, Bruce, and Joan helped in intangible but vital ways.

Acknowledgments are due to Doubleday Publishing Company for permission to quote from *Darwin's Century;* to Random House, Inc., for permission to quote from *The Immense Journey* and *The Star Thrower;* to Simon & Schuster-Scribner's for permission to quote from *The Firmament of Time, The Invisible Pyramid, The Night Country,* and *All the Strange Hours,* and from "The Maya" from *Another Kind of Autumn;* to Harcourt Brace & Company for permission to quote from *The Unexpected Universe;* to Little, Brown and Company for permission to quote from *The Lost Notebooks of Loren Eiseley* under the doctrine of fair use. Ed Block granted permission to quote from a paper presented to the Society for Literature and Science in 1991, and John Clifford granted permission to quote from a paper presented in 1986 to the Conference on College Communication and Composition, later published in

Literary Nonfiction, edited by Chris Anderson (Carbondale: Southern Illinois University Press, 1989: 247–61).

Portions of this material appeared in similar form in earlier publications. I thank the University of Georgia Press for permission to use conceptual material from my essay "Reflective Scientists and the Critique of Mechanistic Metaphor," published in *The Literature of Science: Perspectives on Popular Scientific Writing,* edited by Murdo William McRae (Athens: University of Georgia Press, 1993: 249–72); *Centennial Review* for permission to quote conceptual material from my article titled "Toward a Dialectic of the Open End: The Scientist as Writer and the Revolution Against Measurement," *Centennial Review* 38.1 (Winter 1994): 179–204; and *Modern Language Studies* for permission to reprint the explanation of the holographic paradigm that appeared in my article titled "The Holographic Paradigm: A New Model for the Study of Literature and Science," *Modern Language Studies* 20.4 (1990): 80–89.

Toward a Dialogue
of Understandings

1

The Scientist as Writer: Eiseley and the Ironic Imperative

The rational and the intuitive are complementary modes of functioning of the human mind. Rational thinking is linear, focused, and analytic. It belongs to the realm of the intellect, whose function it is to discriminate, measure, and categorize. Thus rational knowledge tends to be fragmented. Intuitive knowledge, on the other hand, is based on a direct, nonintellectual experience of reality arising in an expanded state of awareness. It tends to be synthesizing, holistic, and nonlinear.
—Fritjof Capra, *The Turning Point*

The special value of science . . . lies not in what it makes of the world, but in what it makes of the knower.
—Quoted by Loren Eiseley, *The Firmament of Time*

In the last two decades a number of practicing scientists—members of the scientific community with outstanding credentials—have turned to writing about science itself, about the role of the individual in relation to science. As William Zinsser has noted, literary nonfiction has often outsold fiction in recent years, displacing forms, such as novels and poems, that were defined as "literary" in the nineteenth century (1985, 55). Carl Sagan's *Cosmos* rivals the Bible in sales, Loren Eiseley's essays are still widely read, and Richard Selzer's *Mortal Lessons* and Oliver Sacks's *Awakenings* have both been reprinted with new introductions. That these examples of literary nonfiction are written by scientists who explore the human side of science and the role of science in today's world is indicative of what I believe to be a singularly important trend in contemporary writing. In his essay "Do Scientists Need a Philosophy?" Gerald Holton observes that scientists today seem less concerned than early twentieth-century scientists with the kind of philosophical exploration that characterized the writings of Einstein and others—the quest for a unified *Weltbild* that underlay Einstein's work at the deepest level and undoubtedly was intertwined with the *themata* that helped to shape his work (1987, 170). Holton argues that such philosophical concerns are replaced today by a turn toward considerations of ethical problems raised by science (177). I suggest that the surprisingly popular writings of scientists in the last two decades reflect such an ethical concern—an imperative to write about the role of science in society, about the ethos that science itself constructs, about the role of culture in constructing scientific understandings, about the ethical role of science and technology and of the human beings who *do* science and technology. The awareness that we participate in shaping the new evolution, cultural evolution, even as our technology can affect physical evolution, has led reflective scientists in a broad spectrum of activities to speculate on the relationship of science and the individual in a world that consists, not of separate parts to be identified, analyzed, and set aside, but of an interlocked web of ongoing processes.

One such writer was Loren Eiseley, whom one biographer credits with discovering "a new prose idiom for science and for literature" (Carlisle 1977, 112) and another credits with inventing "a new genre—an imaginative synthesis of literature and science—one that enlarged the power and range of the personal essay" (Angyal 1983, 39). In an early essay, Eiseley writes of a hybrid genre that "contains overtones of thought which is not science, nor intended to be, and yet without which science itself would be the poorer" (1957, 482). Eiseley's texts draw on science, personal reflection, autobiographical elements, and "personal experiences capable of being shared by every perceptive human being," experiences that "are part of that indefinable country which lies between the realm of natural objects and the human spirit which moves among them" (480). Although I will more specifically address the question of Eiseley's genre in a later chapter, the point here is that he is almost always cited as anticipating the kind of writing currently practiced by Stephen Jay Gould, Richard Selzer, Lewis Thomas, Carl Sagan, Fritjof Capra, David Bohm, Oliver Sacks, Rupert Sheldrake, and Ilya Prigogine—all of whom not only share many of Eiseley's thematic and ethical concerns but continue to write to question and to reexamine science and the assumptions that underlie it.

Similarly, in recent years scholars across the disciplines have discussed the "interpretive turn," the "textual turn," or the "narrative turn" in philosophy, social sciences, the humanities, and science.[1] To understand this "turn," however, we need to situate it in relation to the hegemony of the scientific epistemology. Through-out most of the nineteenth century and as institutionalized science has developed in the twentieth century, science has been considered the model of knowledge. To be a "science" is to be a valid, coherent, and worthwhile activity. The social sciences long assumed the scientific model for their activities—to the extent that some practitioners found comfort in insisting that psychology and economics *are* sciences in the Baconian experimental tradition. In literary studies, Northrop Frye provided the quintessential statement of the need for a "science" of literary studies; distinguished from the reading and interpretation of literary texts, his science of literary studies was "to advance our knowledge about such activities, a knowledge divorced from practice, just as linguistics is divorced from prescribing standards" (Seamon 1989, 297). Frye's emphasis is on an "impersonal body of consolidating knowledge" (cited in Seamon 1989, 297). Similarly, Roland Barthes in his early texts such as *Elements of Semiology* (before *S/Z*, in which he openly embraced the plural and fragmented text, concluding that the text "is not a thing," but a "galaxy of signifiers" to which there are multiple paths of access [1968, 5–6]), joined others in calling for a structuralist science of literature parallel to structuralist linguistic studies.[2] At this point in his career Barthes argues that his goal is "not the decipherment of a work's meaning but the reconstruction of the rules and constraints of that meaning's elaboration" (1964, 218).

The "interpretive" or "textual" turn, however, requires a realization of the role of interpretation, of the perspective that narration provides, within a field of knowledge. This turn places any given text within its cultural and historical milieu. The turn is related to the questioning of "representation" that began in the early twentieth century in a variety of fields and has ranged in recent critical theory from E. H. Gombrich's investigations into the psychology of human representation to Jacques Derrida's concern with the instability of the text of philosophy. Given strong momentum by Michel Foucault, the interpretive turn requires a realization that any knowledge is contextual, that history or literature or economic theory both shapes and is shaped by its cultural context. Foucault contends, "Historians want to write histories of biology in the eighteenth century; but they do not realize that biology did not exist then, and that the pattern of knowledge which has been familiar to us for a hundred and fifty years is not valid for a previous period. . . . All that existed [in the eighteenth century] was living beings, which were viewed through a grid of knowledge constituted by *natural history*" (1973, 127–28).

Thus in writing history one is also writing *about* history; in studying art or literature one is writing *about* the position of the work in its cultural context. Robert D'Amico effectively states the significance of the "interpretive turn" in terms of historicism, which is, he argues, "a position about the limits of knowledge, how human understanding is always a 'captive of its historical situation'" (1989, x). Historicism follows a "clearly sceptical strategy" that takes "evaluations, universal claims, or standards" as "constitutive of the world, not representative of the world" (xii). Emphasizing the "textuality" of philosophy, D'Amico contends that philosophers must understand the text in terms of both its historical context and the present reflection. Thus the historicist reading "assumes that there will be and must be indeterminacy," for "no understanding of the world is a direct and unmediated contact with reality" (xiii).

The textual model—as explored by structuralists and poststructuralists, by reader-response theorists, by Marxist theorists, and by feminist critics—suggests (1) the indeterminacy of the text, which Roland Barthes parallels to the Einsteinian "relativity of the frames of reference" (1977b, 156); (2) the importance of the cultural frame within which a text is produced and read; (3) the text as a space of interactivity between the author's codes and the reader's response; and (4) the interaction between observer and observed—not only in anthropological studies or in noncontrolled medical experiments, but also in writing history or evaluating the contributions of nineteenth-century biologists to scientific thought. Accepting skepticism about "certain knowledge of the past," the textual or interpretive turn is a turn to the text in context, "to reconstructions or interpretations" (D'Amico 1989, 143).

The scientist-writers with whom I am concerned participate in this "textual" or "interpretive" turn. As reflective scientists they have sought to expose the assumptions that lie beneath the revered scientific establishment—to reexamine the nineteenth-century notion of scientific "objectivity" and to place it in its context of Newtonian assumptions about the world. As Shirley Staton explains,

> The older, nineteenth-century, commonsense view had depicted an objective world which people knew through direct experience of their senses. In that dualistic empirical model, the words people used to describe the world corresponded to objects "out there." . . . Finding

meaning depended upon determining the right correspondences between mind and matter. (1987, 3)

In *The Structure of Scientific Revolutions,* Thomas S. Kuhn emphasizes the extent to which the established frame or method—what he calls "normal science"—forms a paradigm, a shared way of looking at the world or a "disciplinary matrix" (1969, 182) that influences the scientist's interpretation of data. Thus he posits that the individual scientist is caught within scientific conventions that are part of the culture. Gerald Holton, in *Thematic Origins of Scientific Thought,* stresses the role of *themata* and of the pull of emotion in scientific "observation." Gombrich's *Art and Illusion,* as its subtitle, *A Study in the Psychology of Pictorial Representation,* indicates, is a major synthesis of the effects of "style" on seeing; although he focuses on art, Gombrich also addresses everyday perception, science, and language. Even before these writers, concern was growing for what Roger S. Jones has called "idolatry" toward science (1982, 14), as exemplified in Jacques Barzun's earlier comment that there has emerged "a public unenlightened by the achievements of modern science," which only "gapes at its wonders, incapable of critical judgment" (1981, 335).

The work of these writers in the history of science suggests a growing tendency toward "textualizing" or "historicizing" science itself, a trend that grows out of the early twentieth-century shift in *episteme* (Foucault's term) from being to language (in the broad sense of both speaking and writing), awareness of the mediating role of language in all human activity. This shift to the linguistic *episteme* underlies the "interpretive turn" and characterizes much of twentieth-century philosophy, linguistics, and philosophy of science. Emphasis on the separate reality of the thing has been displaced in favor of the interactive understanding that we derive from language. From the view of the relationship of a sign and its referent as essential and emblematic, as in hieroglyphics (Irwin 1983, 7), to a recognition of the arbitrariness with which sign and referent are connected is the move verbalized by Ferdinand de Saussure in the foundational work of structural linguistics, his *Course in General Linguistics* (published by his students in 1916 from his lecture notes). Phonetic writing, with (ideally) a single symbol for a single sound, is linked to the relativism of truth (truth as assigned in the process of observation), and to the arbitrariness of the relationship between signifier and signified.[3] If nature itself partakes of indeterminacy and evolves probabilistically, then language, too, comes to be seen as indeterminate and probabilistic.[4] Significantly, at the same time this arbitrariness was being explained, Albert Einstein observed the arbitrariness—the "'purely fictitious character'"—of the "'fundamentals of scientific theory'" (cited in Holton 1987, 14). For Einstein, the relationship between sense experience and concept "'is not analogous to that of soup to beef, but rather to that of check number to overcoat'" (cited in Holton 1987, 14). Virtually all fields of study have been touched by the shift in emphasis toward system, process, and relationships of difference and similarity, as well as by the new emphasis on language as a medium of knowledge.

In this textualization of science, or subjecting science itself to interpretation within its cultural setting, these writers do not see science as *just* text, but as including both textuality (i.e., being subject to interpretation) and reality. As D'Amico suggests, the historicist does not argue for the ultimate aporia of all interpretation. Interpretive pursuits are not dismissed as meaningless; they are certainly "possible and coherent but marked by indeterminacy over the translation of basic concepts" (1989, 143–44). These scientists who write share with Eiseley a participation in the linguistic *episteme*—a concern for the role of language itself in human understanding, and specifically with the essential metaphoricity of language, as well as their attempt through literary nonfiction to resituate language within the understanding of science. In recognizing the basic metaphoricity of language, these writers explicitly recognize the problematic of metaphor—its inevitability, its enigmatic cognitive role, its relation to what Kuhn calls the "mutual accommodation between experience and language" (1981, 418). They recognize the equivocality of language and the effect that metaphor has on human thinking. Eiseley sees analogy, or "root metaphor," as the means by which "science succeeds in extending its domain" ("How the World," 1980, 20). Gould writes that we can trace a writer's "primary commitments by recording his metaphors" (1987a, 45). All of these writers participate in a reexamination of the linguistic—hence, cultural—codes through which science is communicated. In this reexamination, they question these codes, propose redefinition, undermine codes, or create alternative codes—sometimes in the form of alternative root metaphors—in a self-conscious attempt to situate science in the culture that produces it and to reintroduce the individual human being into the institution of science. They demystify (Gould's term is *debunk*) old metaphors and sometimes remotivate them. In a sense, their reexamination is popularization, because these writers are concerned with popular awareness of the role of science. But this reexamination is more: it is a thinking-through, a self-reflexiveness on the part of the scientist in the quest to understand the relationship of science and nature, of science and culture.

The Ironic Imperative

The self-reflexiveness and the linguistic reflectiveness of these writers not only marks a critical imperative that foregrounds the role of language in constructing any reality, including scientific reality, but they exemplify, I believe, a maturation of scientific discourse that can be understood in terms of Hayden White's complex schema paralleling Piaget's stages in the development of cognition with stages in the development of discourse. White contends that discourse "serves as a model of the metalogical operations by which consciousness" deals with its own environment (1978, 5). Discourse itself, he continues, moves "through all of the structures of relating self to other which remain implicit as different ways of knowing in the fully matured consciousness" (11). For White, the four basic tropes—metaphor, metonymy, synecdoche, and irony—parallel these structures of knowing. Metaphor, metonymy, and synecdoche form the "prelogical" means—indeed, they are "paradigms, provided by language itself"—by which consciousness "*pre*figure[s] the field" before treating data, in a "poetic act [that] is indistinguishable from the linguistic act in which the field is made ready for interpretation" (1987, 30). These tropes, then, are "linguistic protocols" that form "the languages of identity (Metaphor), extrinsicality (Metonymy), and intrinsicality (Synecdoche)" (36).

These first three tropes, however, White labels as "naive" because they reflect an assumption that language can indeed grasp "the nature of things" figuratively. Irony, however, is not naive but (1) "dialectical" because it employs metaphor self-consciously and negatively, (2) aporetic because it suggests disbelief in the statement itself, and (3) "metatropological" because it suggests "the self-conscious awareness of the possible misuse of figurative language" (1987, 36–37). Irony, White finds, is the mode of figuration dominant in late nineteenth-century philosophy of history, for in the ironic mode the viewer experiences a self-critical understanding of the world (37). And, I submit, it is the ironic mode of self-reflective questioning of the discourse itself that twentieth-century thinkers in science, literature, and philosophy, as well as history, inherited.

If the twentieth century inherited both an ironic, self-reflective, linguistically *reflexive* (turning back to itself) tendency and a cosmos in which indeterminacy and probability increasingly displaced causality and determinism, then the scientist who writes is clearly a part of the cultural web. My point is that the scientist who writes—be he popularizer (as Eiseley is often categorized) or debunker (as Eiseley was and Stephen Jay Gould avowedly is) or visionary (as Fritjof Capra becomes in *The Turning Point*)—becomes a *textual* and *intertextual* part of the evolving culture; her texts become part of the "always already" of a culture that does not necessarily move "forward" or toward "truth," but assuredly does not stand still. Discourse, White argues, "is quintessentially a *meditative* enterprise," which is "always as much *about* the nature of interpretation itself as it is *about* the subject matter" (1978, 4).

These writers share not only a concern for reflection on the role of science and technology, and of the human beings who *do* science and technology in an evolving world, but they share also a concern with the role of language itself in human understanding, particularly with the essential figurativeness of language; and they share, too, the attempt through literary nonfiction to resituate language within the understanding of science. I am concerned with the impact of the rhetoric of the scientist/writer who brings insights before an audience that consists not only of general readers but also of specialists so diverse that their communication must take place through a broader medium than the scholarly essay. In view of their impact today, the texts of the serious reflective scientist should not be marginalized, either as mere "popularization" or as devoid of literary themes, purposes, and techniques. Moreover, Eiseley writes that the popularizer "was often a very significant figure in the earlier centuries of science. His work might plant the germ of new ideas in other, more systematic minds, and the actual diffusion of his books, as represented by numbers of editions and translations, *can throw light upon the ideas which were beginning to intrigue the public imagination*" (1961, 30; emphasis added). Anticipation—planting the germ of an idea—and "intrigue" of "the public imagination" both valorize the texts of the scientist who writes and emphasize these texts as *textual* and *intertextual* parts of the "always already" of culture. As a member of both "literary" and "scientific" discourse communities, the scientist who writes literary nonfiction brings cultural interdisciplinarity to her texts. Albert Einstein was strongly committed to sharing his understandings across perceived "boundaries" of disciplines, and, like other scientific figures such as Galileo, he perceived the larger public as a means of disseminating his approach to scientific understanding (Holton 1987, 4). Like other scientist-writers, he was ideally situated to minimize such disciplinary boundaries.

Purpose

In view of the linguistic, textual, interpretive turn and of its embedding in culture, I propose to take a new look at a literary figure whose texts are often classified as popularization of science, but whose literary purposes and accomplishments mark a turn from the mythologized objective paradigm to a self-reflexive, interpretive paradigm in the study of science itself. In pursuing Loren Eiseley's role in the reexamination of science and its language, I hope to establish his role as an intellectual historian, as a student and interpreter of the history of science, and as a writer who anticipates the postmodern dilemma and postmodern techniques but who also emerges as a key figure in establishing a dialogue between the reflective scientist and the values traditionally associated with literature and philosophy.

By profession a physical anthropologist, university administrator (he was for a time Provost of the University of Pennsylvania, and later Benjamin Franklin Professor of the History of Science), and lecturer, Loren Eiseley was essentially a poet by temperament and both essayist and poet by choice. His works include fifteen published volumes: *Darwin's Century* (1958), a work of scholarship whose conclusion parallels his development as a popular essayist; six collections of popular essays—*The Immense Journey* (1957), *The Firmament of Time* (1960), *The Unexpected Universe* (1969), *The Invisible Pyramid* (1970), *The Night Country* (1971), and *The Star Thrower* (a 1978 collection bringing together several previously unpublished essays and some of Eiseley's most popular published work under the title of his best-known essay, which was originally included in *The Unexpected Universe*); a biography of Francis Bacon, originally published in 1962 as *Francis Bacon and the Modern Dilemma* but later republished as *The Man Who Saw through Time* (1973); a loosely structured series of reminiscences published as an autobiography, *All the Strange Hours: The Excavation of a Life* (1975); four volumes of poetry—*Notes of an Alchemist* (1972), *The Innocent Assassins* (1973), *Another Kind of Autumn* (1977), and *All the Night Wings* (1979); and a posthumous collection of essays, scholarly and popular, on which he had worked to establish the importance of Edward Blyth's contributions to the development of evolutionary theory, *Darwin and the Mysterious Mr. X* (1979). In 1987, a new collection of Eiseley's writings, including extracts from his notebooks, poems, a fragment of a novel, and photographs of the Eiseley family, was published under the editorship of Kenneth Heuer as *The Lost Notebooks of Loren Eiseley.*

Eiseley and the Postmodern Temper

Eiseley's texts have long been accepted as traditional humanistic essays that demonstrate sensitivity to nature and concern for the human role in nature. Studies of Eiseley's texts have centered on biography, on theme, on the development of *personas,* on his role as a natural-history writer and environmentalist, and on his development of narrative and autobiographical techniques. The texts, however, need to be situated in regard to the postwar and postmodern temper. A product of the Great Depression who spent his years as a college dropout riding the rails, living with the great horde of dispossessed Americans in hobo camps, Eiseley nevertheless returned to the University of Nebraska and went from there to graduate study at the University of Pennsylvania. When he finished his doctorate in 1937, he faced the uncertainties of the post-Depression academic market and the threatened disintegration of world order under Hitler's onslaught. Disqualified from military service because of his eyesight and a bad ear, Eiseley became increasingly more reflective against the backdrop of a world at war. His most characteristic essays, mostly written in the 1960s, fuse meditation on time and evolution with echoes of the anxieties of depression, world war, postwar world tensions, and the social turbulence of the 1960s. A constant theme for Eiseley is human potential—the ability of the human being to shape the "new evolution," cultural evolution—and the human choice involved. In the evolutionary understanding, he insists, humankind is not the final product of millions of years of change, but a stage in a self-fabricating evolutionary drama.

Through reflection and technique alike, Eiseley both writes the anxiety of his intellectual milieu and anticipates the postmodern sense of the loss of grounding—of the instability of self-created culture, of textual and scientific participation in this cultural construct. Eiseley wrote in the frame of time and cultural impact of Roland Barthes, who in *S/Z* made both the quintessential structural exploration of the system of codes that give meaning to a text and a move away from what had become "normal" structuralism.[5] In the words of Jonathan Culler, "Structuralists are convinced that systematic knowledge is possible; post-structuralists claim to know only the impossibility of this knowledge" (1983a, 22). As a physical anthropologist, Eiseley felt the impact of structuralism in linguistics and in human studies; he documented his readings in Claude Lévi-Strauss. Although he speaks for traditional humanistic values, his texts reveal, like those of Barthes, an increasing awareness of the instability of culture, of language, of knowledge, even of what we understand as nature itself.

In an essay first published in Eiseley's last collection, *The Star Thrower,* he describes a painting by Irwin Fleminger, "a modernist whose vast lawless Martianlike landscapes contain cryptic human artifacts" ("Science and the Sense," *ST,* 192). Eiseley's description of the painting, titled "Laws of Nature," demonstrates his awareness of the tenuous artifice of human culture:

> Here in a jumbled desert waste without visible life two thin laths had been erected a little distance apart. Strung across the top of the laths was an insubstantial string with even more insubstantial filaments depending from the connecting cord. The effect was terrifying. In the huge inhuman universe that constituted the background, man, who was even more diminished by his absence, had attempted to delineate and bring under natural law an area too big for his comprehension. His effort, his "law," whatever it was, denoted a tiny measure in the midst of an ominous landscape looming away to the horizon. The frail slats and dangling string would not have sufficed to fence a chicken run. (192)

The homely metaphor at the end does not diminish the sense of existential terror and isolation in Eiseley's description of the painting, which strongly suggests the flimsy instability of human culture. As in other instances, Eiseley turns to the larger landscape metaphor to delineate the fragility of human constructs in the midst of a constantly changing universe. Even Fleminger's title, "Laws of Nature," seems to suggest the decay of the Cartesian-Newtonian view, a decay that haunts Eiseley and impresses him with the tenuousness of human accomplishments—so much so that in a notebook entry dated June 7, 1953, he responds to Ernest Borek's *Man the Chemical Machine* with a comment that it is simply "a further extension of the Cartesian world view," which has moved from seeing humankind in terms of "pulleys and levers" to seeing humankind as a "*chemical* machine" (*LN,* 89). Reflecting on this extension of the mechanistic metaphor, Eiseley comments, "We must be near to reaching the end of this road" (89). Considering the effect of Fleminger's painting, he continues:

> The message grew as one looked. With all the great powers of the human intellect we were safe, we understood, in degree, a space between some slats and string, a little gate into the world of infinitude. The effect was crushing and it brought before one that sense of the "other" of which Rudolf Otto spoke, the sense beyond our senses, unspoken awe, or, as the reductionist would have it, nothing but waste. There the slats stood and the string dropped hopelessly. It was the natural law imposed by man, but outside its compass, again to use the words of Thoreau, was something terrific, not bound to be kind to man. Not man's at all really—a star's substance totally indifferent to life or what laws life might concoct. No man would greatly extend that trifling toy. The line sagged hopelessly. Man's attempt had failed, leaving but an artifact in the wilderness. Perhaps, I thought, this is man's own measure. Perhaps he has already gone. ("Science and the Sense," *ST,* 192)

As he realizes the flimsiness and arrogance of the human project of domination over nature, Eiseley sees the end of humankind as we know it, and he seeks comfort in a kinship with primitive cultures: "I felt the mood of the paleolithic artists, lost in the mysteries of birth and coming." (192).

There is in this image of the fragile grounding of human culture almost a parody of self-constructing, self-interpreting, yet self-destroying humankind. The image is as ungrounded as Borges's circular stairway to nowhere, which leads not to a predictable door, but to a lack of grounding (N. Katherine Hayles refers to Borges's stairway as an emblem that "speaks to the dangerous potential of metaphors to expose the ungrounded nature of discourse" [1990, 34]). This imagistic parody of the duality of humankind is in one sense as old as the ancient notion of the duality of good and evil in human nature, but it arises from Eiseley's sense of imminent doom for humankind—the destruction of humanity through the human cleverness that creates machines and explores space and the human lack of concern for our relation to the nature that produced us. Comparing Eiseley's sense of "lost orientation, lost control, and the premonition that human beings and their condition are part of a larger problem related to the changing paradigm and function of science" with Thomas Pynchon's, Ed Block notes that Eiseley, drawing from the traditional stage metaphor, "anticipates the direction that scientific and philosophical self-awareness had taken in the last thirty years" (4). He also notes Eiseley's warning, "'If the play has its magical aspect, however, there is an increasing malignancy about it'" (*FT*, cited in 4). Block argues that although *The Firmament of Time* "reads like a traditional humanistic text," it nevertheless "anticipates many postmodern claims: no nature, no self, no valid or defensible personal ethic" (1991, 11).

Essential to dealing with both self and nature in Eiseley's cosmogony is the recognition that both are constantly shifting and continually being made and remade through the individual's experience. And this experience, as translated into Eiseley's essays, suggests not closure but an ongoing dialectic of the self "then" and "now," of the self and nature "then" and "now." In "The Running Man" in *All the Strange Hours,* Eiseley writes, "To grow is a gain, an enlargement of life. . . . Yet it is also a departure. There is something lost that will not return" (1975, 31). Analyzing Eiseley's apparent motives, his manipulation of his audience through his dramatized recollection of his cruelty as a ten-year-old toward his deaf and gesticulating mother, John Clifford has noted that Eiseley writes of "departure" "almost deconstructively," for if writing brings growth and clearer understanding, it also brings distortion and "violence to the ineffable complexity of reality" (1986, 8). In the *Lost Notebooks* Eiseley writes, "If you see wire and rope and hinges and hairpins pinned on canvas, then perhaps this is all we are. . . . I have seen the declining art of too many cultures not to know—the crumbling ceramics, the design fumbling off without purpose" (Heuer 1987, 137). Yet in this frightening thought, the seed of the gain that, for Eiseley, always accompanies loss is the "marvelous quality of [the] universe, its latency" (137). "Perhaps this is all we are" is an ongoing theme that Clifford's reading would suggest is a rhetorical device to engage the reader in the same uncertainty and despair that Eiseley himself often projects.

Yet for all its undermining of meaning, Eiseley's "malign magic" or "almost deconstructive" insight nevertheless brings the text back to the human potential and the human responsibility to devise, before it is too late, a better way than contemporary culture has devised to meet that responsibility. The questions that he asks are those asked by traditional humanists, and the values he embraces center on the human potential. Yet Eiseley's persistent melancholia arises, in part, from an existential *angst* and from an understanding in which biological shape-shifting, the instability of living forms, parallels indeterminacy in physics and a growing, almost postmodern, sense of a similar instability of the text and of knowledge itself. And this concern with both textual and epistemological instability underlies both theme and method in Eiseley's texts, as will be evident when we consider the movement of the texts and of the *themata* that shape them.

Movement of Texts

The Early Texts: Scholarship, Popularization, and the Literary Turn

Eiseley worked on *Darwin's Century* and *The Immense Journey* concurrently, though several of the essays in *The Immense Journey* were published elsewhere beginning in 1946. *The Immense Journey* was first published in 1957; *Darwin's Century,* in 1958. Written after Doubleday requested a scholarly book to be published for the centennial of the publication of *The Origin of Species, Darwin's Century* represents painstaking scholarship and meticulous organization, with literary flashes emerging, particularly in the conclusion, in which Eiseley employs the landscape metaphor to which he will later return, extending it to portray Darwin's work as a rock surviving throughout the century. The essays in *The Immense Journey,* in contrast, are less complex than the carefully reasoned chapters of *Darwin's Century* and shorter than many of the later essays. In explaining and commenting on science, Eiseley relies on anecdotes that are less often extended for exploration of their literary possibilities than his later essays. He ranges from recollection of his own journey downward into a slit in the earth whose strata reveal the evolutionary history of life, to the "secret of life," which continues to escape human understanding and synthesis. In between, his excursions reflect the randomness of the journey that his synecdochic persona takes in the opening essay, "The Slit," including his decision to loll in a mountain stream in the attempt to feel the journey of water toward the sea and his musings on the human fascination with the machine, as revealed in a newspaper story about a mechanical mouse that may outperform the real thing in a maze (to which he concludes, "It's life I believe in, not machines" [181]). These two books, whose preparation overlapped, give insight into Eiseley's beginnings as a writer—in *Darwin's Century,* his interest in detailing evolution for the educated audience, but turning toward literary meditation in the conclusion; in *The Immense Journey,* his desire to explain science, but taking a more speculative turn, which was eloquently verbalized in his essay "The Enchanted Glass," published in the

American Scholar in 1957, the same year that *The Immense Journey* first appeared.

That Eiseley saw himself as moving from science to literature is documented in his notebook entries planning his autobiography as a "first volume" (*LN* 135). *The Firmament of Time* stands between the two early books and the highly personal, reflective, "literary" texts as a pivotal point of transition. Comprising six lectures that Eiseley delivered in 1959 at the University of Cincinnati and later prepared for publication, *The Firmament of Time,* which has been through more reprints than any of Eiseley's texts, addresses six apparently simple topics: "How the World Became Natural," "How Death Became Natural," "How Life Became Natural," "How Man Became Natural," "How Human Is Man?" and "How Natural Is Natural?" Yet the simplicity is deceptive, for in these lectures/essays Eiseley begins to explore more thoroughly the reflective, "literary" possibilities of his topic. Through these simple topics and questions, he begins to defamiliarize accepted and unquestioned terms such as *human* and *natural,* revealing them as the cultural constructs they are and reflecting on the implications of this constructedness of both the "human" and the "natural."

In *The Firmament of Time* Eiseley turns explicitly to language itself as the vehicle of understanding. He examines the role of metaphor in scientific understanding and explores the extended, apparently rambling, personal narrative that, in his term, "conceals" the philosophical reflection on science that underlies the texts. In his defamiliarization of terms and his explorations of extended metaphors, Eiseley presents an interactive universe in which the ancient dramatic metaphor is remotivated to suggest interaction, participation. Positivism, reductionism, and the particularization and fragmentation common in the accepted and dominant scientific epistemology gradually come under Eiseley's literary scrutiny, to be reexamined and undermined in favor of an *ethos* for science and for humankind that would view the human in relation to the world as system. Even the textual metaphor—anticipating understandings at least a decade away—is renewed and remotivated in an interactive, participatory sense.

The Critical Turn: Irony and Reflection

Eiseley's increasingly ironic and reflective texts, which often criticize elements of science, were written between 1969 and 1975 and include *The Unexpected Universe, The Invisible Pyramid, The Night Country,* and the autobiographical *All the Strange Hours.* In the essays in these collections, Eiseley pursues the model that he developed in *The Firmament of Time,* using often complex narratives, sometimes narratives embedded in other narratives, all based on personal experiences, to explore understandings that come from scientific experience or to question the scientific institution or the *ethos* that has emerged from science; always he examines the personal in relation to the universe and to the way we learn, repeatedly returning to the theme of the instability of knowledge itself. Beginning in *The Unexpected Universe,* having established the importance of metaphor in *Darwin's Century* and *The Firmament of Time,* Eiseley explores the uses of metaphor in science, sometimes deliberately undermining a conceptual metaphor, such as the machine, and attempting to remotivate it by undermining positivist, reductionist understandings and pointing out the probabilistic, systems-oriented meaning actually embedded in the metaphor.

Further, in *The Unexpected Universe* Eiseley begins to explore themes of strangeness—of the romantic, often frightening "country" at the margins of science and of mainstream human experience. Caught in the conformist academic community of the 1950s and the early 1960s, Eiseley felt himself very much a stranger— one who, in that time, might not even have been admitted to the universities where he was invited to lecture. Yet he was a stranger to the students of the late 1960s and the 1970s who demanded relevance and insisted on helping to design their own curricula. And he found himself a stranger, too, to science, meditating on the borders where science becomes inexplicable—in, for example, "The Innocent Fox" in *The Unexpected Universe,* where he tells of watching simulated lightning flashes from a neighboring attic.

The Invisible Pyramid openly questions the scientific institution and its connection to the military-industrial complex. Increasingly immersed in Emerson, Melville, and Thoreau, Eiseley feels, with Thoreau, that the railroad now rides *us.* Eiseley relentlessly questions the human impulse that leads to exploration of space when unsolved human problems threaten destruction even as technology enables us to pursue space. Eiseley's concern in *The Invisible Pyramid* is the twentieth century's construction of a monument, like the Egyptian or Mayan pyramids, to what it holds most valuable. In our case, Eiseley argues, the pyramid—associated with worship of the human, with the quest for immortality, and inevitably with death—that the twentieth century has constructed is "invisible"; it is distinguished by its absence, for it is a pyramid of method, of worship of the technological, and its chief symbol is a rusty earth-moving machine abandoned near a newly constructed shopping center and now covered with a prophetic snow that Eiseley sees as the inevitable end of our culture. Eiseley refers to his vision as "the terrible *deja vu* of the archaeologist" ("The Creature from the Marsh, *NC,* 156), the awareness that not only is life fragile and shifting, but so are culture and the very understandings that underlie that culture. "Beginning on some winter night," he writes in *The Invisible Pyramid,* "the snow will fall steadily for a thousand years and hush in its falling the spore cities whose seed has flown. . . . This has always been their end, whether in the snow or in the sand" ("The World Eaters," *IP,* 71).

The Night Country explicitly aligns Eiseley with a romantic gloom, with human beings marginalized by preference or by genetic chance, as is the man with a divided visage driving an overloaded hay wagon that bears down on the narrator and leaves him with a frightening vision of the dual nature of humankind in the essay "Strangeness in the Proportion." The narratives in this book are often personal reminiscences and tales from Eiseley's "bone-hunting" days, and he frankly engages in epistemological exploration. If the key terms in the first two of Eiseley's ironic or critical books were the "unexpected" and the "invisible," the key term in *The Night Country* is "strangeness," which engages the narrator and which he explores for its marginal and imaginative possibilities.

Eiseley's "autobiography," *All the Strange Hours,* is hardly an autobiography, but a fragmented collection of reflections on a life in science and at the boundaries of science, carefully structured in terms of chronology, but with much of the most personal censored. "Strangeness" remains a focal term for Eiseley's reflections— whether on his youthful experiences working in a hatchery or his observations in old age of the inexplicable behavior of sphex wasps. By the time of the autobiography Eiseley has largely abandoned popularization; now he focuses on episodes from a life in science and, especially, on the strange intermingling of chance, time, and randomness in science and in the individual life.

The Late Texts: Afterthoughts from a Life in Science

The Star Thrower consists of essays and a few poems that Eiseley selected, near the end of his life, to be included in a representative collection. Many of the essays had been published earlier, including the title essay, which originally appeared in *The Unexpected Universe.* Some, such as "Science and the Sense of the

Holy" and "Walden: Thoreau's Unfinished Business," were previously unpublished for whatever reason, though one suspects that Eiseley had never quite finished the former and had hesitated to publish it for fear of the scientific community's response. Often dubbed a "mystic" because of his speculations on the unexpected or the unknown, Eiseley disliked the label as a misnomer. He frequently finds a sense of wonder evoked by the natural world, but he does not write of the supernatural; he frankly refers to the "miraculous," which he defines in terms of a rift in the natural, or what cannot presently be explained. "As a scientist, I did not believe in miracles," he writes, "though I willingly granted the word broad latitudes of definition" ("The Innocent Fox," *UU,* 197). But what some take as reference to the supernatural is his ultimate skepticism, expressed in the autobiography in terms of his realization that "in the world there is nothing to explain the world" (*ASH,* 238). Others in this collection, including "Thoreau's Vision of the Natural World," had been published previously in another form.

Darwin and the Mysterious Mr. X is a series of essays published posthumously; the essays are clearly unfinished probings, most arising from a quixotic quest to show Darwin's debt to Edward Blyth. It was a quest that his colleagues refused to take seriously, and one doubts that Eiseley would have approved publication of these essays. Reviewing the book, Stephen Jay Gould called it "both an editorial error and a poor way to honor the memory and accomplishments of so fine a man as Loren Eiseley" (1987b, 61).

The Poems

Eiseley was hesitant to publish *Notes of an Alchemist,* which appeared in 1972. His poems were sometimes written independently, sometimes scribbled in the margins of his notes or on the drafts of essays. *The Innocent Assassins,* published in 1973, contains some of his best verse. *Another Kind of Autumn* appeared in 1977. A final collection of poems, including a number of unpublished early poems and some previously published in "little" magazines, appeared under the title *All the Night Wings;* hesitant as he was to publish poetry at all, one doubts that Eiseley would have approved of publishing the potpourri of poems that appear in this volume.

The early poems follow traditional forms, while the later poems, which are written in free verse, reflect Eiseley's sense of freedom from rules and clearly show his increasing sense of security as a poet. Several poems present interesting intertexts with the essays.

For example, "Arrowhead" in *Notes of an Alchemist* treats the endurance through multiple cultures of a flint chipped by a primitive carver who cared enough for his handiwork to embellish it; Eiseley treats the same subject in "The Illusion of the Two Cultures" (*ST*, 267–79), using the carver's "lingering" over his handiwork as an example of the absence of foolish distinctions between "art" and "technology" in the primitive mind. Eiseley's poetry deserves more study than it has received; the intertextuality of the poems and the essays could easily be the subject of another book. Because of the constraints of space, however, the present study focuses on Eiseley's prose.

Argument

With their focus on biography and the development of *personas,* on Eiseley's role as a natural history writer and environmentalist, and on his development of narrative techniques, scholars have not addressed what I consider to be central to Eiseley's thematic structure and his literary technique—his defamiliarizing of the Cartesian-Newtonian-Lockean view. Like scholars in all the sciences, Eiseley witnessed the collapse of the Cartesian worldview under the onslaught of twentieth-century physics. He was active as a physical anthropologist and as a popular writer when structuralism emerged to cut across a number of fields, from linguistics to anthropology to the literary world that was increasingly his metier. The collapse of the Cartesian view is, of course, of monumental importance, but what is more important for Eiseley the writer is the *trace* that this worldview has left behind—a trace that continues to have an impact, through human understanding of both natural and cultural worlds, on the physical world, on the *biota,* and on human culture and knowledge.

Themata in Science

In *Thematic Origins of Scientific Thought* and *The Advancement of Science, and Its Burdens,* Gerald Holton posits a dimension of scientific thought *other* than, or outside, the traditional view. Traditionally, he argues, "all philosophies of science agree" that two kinds of statements are meaningful in science: (1) "propositions concerning empirical matters," or "phenomenic propositions," and (2) "propositions concerning logic and mathematics," or "analytic propositions" (1987, 5). Such philosophies of science are rooted in empiricism or positivism and participate in a "silent but general agreement to keep the discourse consciously in the phenomenic-analytic plane," an agreement that is fostered by the way science is normally taught (6). For Holton, this view is based on a two-dimensional approach that fails to recognize thematic forces that affect the daily work of the scientist; this view, he contends, is not helpful in answering "questions every historian of science has to face consciously" (6).

Holton, then, argues for "a more sophisticated and appropriate" model, in which a metaphorical "third axis" arises that is "orthogonal to and not resolvable into the phenomenic or analytic axes" (1987, 13, 18). This third "axis" is the site of "those fundamental presuppositions, often stable, many widely shared, that show up in the motivation of the scientist's actual work, as well as in the end-product for which he strives" (18). When forced to decide among such presuppositions—what Holton calls "*themata* (singular *thema,* from the Greek θεμα, that which is laid down, proposition, primary word)"—the scientist must make judgments rather than decisions based on experiment or logic and hence decidable on algorithmic bases (18). Holton's scheme provides a graphic three-dimensional metaphor that allows the historian of science to chart the development of a given scientist's work and to distinguish different "volume-elements" of scientists who share basic similarities but work from different themata (18). Holton compares his themata to "old melodies to which each generation writes its new words" and classifies them into

> *thematic concepts* (such as evolution, devolution, or steady state); *methodological themata* (e.g., the practice of expressing regularities in terms of constancies or of extremes; or forming rules of impotency); and *thematic hypotheses* (such as the postulation of the discreteness of electric charge, or the wrong hypothesis of continuity for light energy, widely held for years after contrary evidence was at hand). (174)

Because such themata are basic presuppositions, they "are rarely verbalized," yet their presence helps to explain "the stability of the scientific enterprise" in spite of "profound changes during the past three centuries" (175).

Some themata, Holton finds, appear as dyads that mark oppositions or coupling (1987, 18). In constructing scientific theories, as the mind makes the leap from experience to axiom, it is free to make this leap, though not free "to make *any leap whatever.* The freedom is narrowly circumscribed by a scientist's particular set of themata that provide constraints shaping the style, direction, and rate of advance" (19).

Holton's argument is basic to the understanding that I pursue in analyzing Loren Eiseley's texts. Eiseley's project is to reexamine certain thematic assumptions of science rooted in the Cartesian view and to undermine them in favor of an *other* understanding. For Eiseley, all of science—and all of life itself—is dualistic. The Darwinian struggle for the fittest is offset by a tendency toward cooperation, both at the macro level and at the subcellular level. The human being possesses both the power of love and the power of hate. Life itself is both gain and loss. In undermining long-held assumptions, Eiseley enters into an ongoing dialogue of abilities and potentialities—a dialogue without closure, without an end to the experiments of evolving life.

In defamiliarizing basic elements of the Cartesian understanding that shaped Darwin's view and that have continued in the twentieth century to shape human understanding, Eiseley joins a group of important thinkers—including Paul Feyerabend, Thomas Kuhn, Michael Polanyi, Stephen Toulmin, and Michel Foucault—who have, throughout this century, addressed the myth of an objective science that works empirically through observation of a world whose existence is apart from humankind and can thus be examined and reported. That Eiseley perceives the impact of the Cartesian understanding and its offspring in the positivist program on the lay audience places him as an intermediary between the writer of traditional "natural history" essays and Kuhn, for example, whose work focuses on the history of science (though his followers have often been outside science proper). Eiseley's *genre* (which I consider specifically in chapter 2) is uniquely situated to carry out his program of reexamining, defamiliarizing, and suggesting alternatives to the understandings that have led twentieth-century humankind to accept aggressiveness as "natural," to build ever more powerful weapons, and to propel itself beyond the natural world into space—all without consideration for the natural "web" that holds all life in interdependency. Eiseley's undermining of these overarching *macro*themes can be seen in terms of *topoi* (literally "places," and Eiseley repeatedly turns to landscape metaphors) on which are based his exploration of the human role in an evolutionary world.

Eiseley's Themata

A powerful force in shaping Eiseley's scientific understanding is the set of themata that provide his constraints as scientist, observer, and thinker. I find four broad sets of themata: epistemological, methodological, evaluative, and ethical.

EPISTEMOLOGICAL THEMATA

Epistemological concerns are increasingly important in Eiseley's texts. Recurrently he questions what one must one know in an evolutionary world and how we come to know. Eiseley's epistemological themata emerge from the assumptions of the epistemologies that compete for the mind in the twentieth century. The empirical epistemology of science (observation, hypothesis, experimentation, confirmation of hypothesis through quantitative methods) is the privileged epistemology and the one that Eiseley, by profession, is compelled to espouse. Evident in most of his texts, however, particularly after *The Firmament of Time* and the overt rejection of his "scientific heritage" in "The Star Thrower" in *The Unexpected Universe,* is the pull of a competing epistemology—one based on human insight, on sympathy across the boundaries of form, on the intuitive understanding of poetry or the "connected" knowing or "connected epistemology, in which truth emerges through care" that Mary Belenky describes. In Belenky's scheme of a "connected" knowing, caring underlies the attempt to understand (1986, 102).

METHODOLOGICAL THEMATA

Eiseley's texts also reflect methodological themata, including the established methods of positivism, reductionism, and fragmentation that he finds embedded in the thinking of Western culture. Part of Eiseley's project is defamiliarizing the positivist view that characterized the "foundationalist program" of the Vienna Circle.[6] Where the positivist program perceives a positive, tangible, outside world "other" than, and separate from, the human being and subject to observation, calculation and manipulation by humankind, Eiseley undermines this understanding with a radical skepticism, which underlies his comment in the autobiography that, reflecting on a life in science, he writes because of "the loneliness of not knowing, not knowing at all" (*ASH,* 23). Later, as even his evolutionary understanding "wobbles" while he watches the sphex wasps' uncanny ability to select and immobilize their prey in a manner that he cannot explain by natural selection, Eiseley writes, "I have come to believe that in the world there is nothing to explain the world. Nothing in nature that can separate the existent from the potential. I start with that" (238). Between the "extremes" of the reductionist who would compress life "into the manageable compass of physics and chemistry" and the philosopher Henri Bergson, who sought to define life in terms of "a separate, indefinable principle, the *elan vital*," Eiseley writes in a passage that anticipates Holton's "thematic" insights, "we all flounder, choosing to close our eyes to ultimate questions and proceeding, instead, with classification and experiment," which are "apt to be colored by what we subconsciously believe or hope" (239).

Another alternative to positivism is, as early as *Darwin's Century,* a restrained constructivism. Humankind, "trapped . . . within the ominous and enigmatic present," without a conception of human and natural history, becomes "a fabricator of illusions" whose "restless and inquiring intellect will create its own universe and describe its forces" (*DC,* 28). Though the context specifies humankind *without* history, the awareness that we construct our own worlds, our own understandings, remains a major *antithema* to the positivist notion of reality in Eiseley's texts.

The antithema to reductionism and fragmentation is a holistic view that Eiseley often expresses in organic terms, but expresses exquisitely in the autobiography through the almost holographic image of the shattered mirror whose parts each reflect a whole that, though fragmented, continually seeks and expresses an identity. Eiseley also embodies the holistic, which is not fixed but flexible, in images of a society whose structure is "as taut and flexible as a spider's web" (*FT,* 147) or of the "vast web of life" that resembles "the legendary tree of Igdrasil, reaching endlessly up through the dead geological strata with living and related branches still glowing in the sun" (*UU,* 133).

Other methodological themata include exemplification through metaphors and reexamination of those metaphors that arose from the Cartesian-Newtonian worldview, particularly the mechanistic analogy that helped to shape the worldview of science, portraying the universe as a great machine somehow set in order by a causal force or being and operating in accordance with predetermined and fixed "natural laws." In a universe operating according to such immutable laws, the lawgiver stands figuratively behind all understanding. If the mechanism implies a grand maker and a grand design, the human being becomes the maker of human and cultural laws. The mechanism thus implies control, dominance, master-y (with all the sexually loaded connotations of the word *master* embedded in the notion of mastery).

Eiseley traces the mechanistic metaphor from its inception in

Greco-Roman culture to the twentieth century's quest to build a mechanical mouse that will outperform a "real" mouse in a maze ("The Bird and the Machine," *IJ*, 179–93). The mechanistic concept is implicated in the Malthusian notion of struggle and in the phrase that Herbert Spencer suggested and Darwin credited to him in the *Origin*, "survival of the fittest." Darwin himself participated in the mechanistic assumption, for his intellectual milieu was partly shaped by the machine metaphor: "the machine governed and sustained by an increasingly remote divinity, had been insensibly altered by the nineteenth century into a world of inflexible law, of Calvinistic gloom and necessity. It constituted the intellectual world into which Charles Darwin was born" (Eiseley, *DMX*, 208).

Against the mechanistic metaphor Eiseley explores metaphors of organism, of systems, of cooperation even in the midst of struggle. His exploration is ongoing, beginning in *Darwin's Century* with his discussion of the excesses of the Darwinians' stress on warfare (1977, 335) and his suggestion of the intertwining of aggressiveness and cooperation in the evolution of life (349). In attempting to subvert the mechanistic metaphor, Eiseley argues for the powerful potential in nature, in the human brain, in human societies. His antithema is the organismic view, which is also a systems view in which each part of a system—living, social, or ecological—is interdependent with the system's other parts. This systemic understanding forms the basic antithema in *The Invisible Pyramid*, Eiseley's position statement on the human relationship to the environment, which anticipates the current environmentalist movement.

The positivist program requires reduction of the world to its parts, a fragmentation of nature on the assumption that understanding the parts brings understanding of the whole. This fragmentation has been remarkably successful for modern science, especially in cell biology and in medical research,[7] but it tends to obscure the view of the whole. It is, as John P. Briggs and F. David Peat write, a "frag-mentation," a fractured way of thinking that they trace to Aristotle and a way of thinking that ignores the interrelationships in the natural world and the need to approach the world as a system rather than as a collection of parts (1984, 98–100). Eiseley's answer to such frag-mentation is a holism best exemplified in the cell's cooperation with other cells and in the humility of the primitives who paid obeisance to the spirit of the hunted before the hunt, recognizing, in spite of the need to kill, that they were part of the natural system rather than lords over it.

A corollary of this fragmented thinking is the separation of specialties and the separation of science itself from other endeavors and the tendency to establish a hierarchy of sciences, with the physical sciences at the top of the hierarchy. Such isolation of science makes "professionalized science" a deterrent to insight, to the creativity that underlies great science. The theme is one that Eiseley will develop in a variety of ways throughout his texts. In *Darwin's Century* he writes of a failure of professional science in its choice to ignore the work of Mendel. The story, he contends, "is worth perusal by all scholars, not alone because of what Mendel achieved, but also because the complete failure of communication in this particular instance was, to a major degree, the failure of professional science" (*DC,* 207). On the hierarchizing of sciences, Eiseley adds a wry but pointed comment as he suggests that, though he would not challenge the "superiority" of physics, we still might remember that, in the half-century argument that Lord Kelvin led against extended geological time and in favor of a solar system moving toward entropy, "the physicists made extended use of mathematical techniques and still were hopelessly and, it must be added, arrogantly wrong" (*DC,* 234).

Yet another outgrowth of reductionist fragmentation is the tendency toward cultural (including technological), racial, and gender hierarchization, according to which enlightened nineteenth-century northern European *man* saw himself as the head of a racial and gender hierarchy, science as the head of a hierarchy of knowledge, and the Western worldview as the head of a cultural hierarchy. Such hierarchizing led often to absurdity, as Eiseley exemplifies when he quotes Ernst Haeckel, who found the analysis of the cell the consummate achievement of science. "'With this single argument,'" wrote Haeckel, "'the mystery of the universe is explained, the Deity annulled and a new era of infinite knowledge ushered in'" (cited in *DC,* 346). In a similar outburst of foolish hierarchizing, Carl Vogt described a presumed racial hierarchy and concluded, "'We may be sure that wherever we perceive an approach to the animal type, the female is nearer to it than the male, hence we should discover a greater simious resemblance if we were to take the female as our standard'" (cited in *DC,* 263). For Eiseley, cultural hierarchizing is equally foolish and deserving of undermining, as he does in recounting his visit with a Mexican peon: "I had looked into his eyes and seen there that transcendence of self is not to be sought in the outer world or in mechanical extensions"(*DC,* 351).

EVALUATIVE THEMATA

Perhaps the single most important evaluative thema that emerged from the Cartesian framework was the determinism that permeated scientific thinking until twentieth-century discoveries made indeterminacy the genesis of a paradigm. Early science was permeated by Christian determinism, and nineteenth-century science, while it rejected supernatural intervention, nevertheless retained the deterministic perspective. So insistent is Sigmund Freud—who serves as a bridge between nineteenth- and twentieth-century thinking, between the Cartesian philosophy of consciousness, with its embedded mechanism, and the unexpected, often uncontrollable impulses of the *un*conscious—on the correctness of the deterministic view that he writes of the individual who accepts "small chance events," "If anyone makes a breach of this kind in the determinism of natural events at a single point, it means that he has thrown overboard the whole *Weltanschauung* of science" (1977, 28).

The deterministic and progress-oriented Cartesian-Newtonian view, particularly as it developed in the nineteenth century, unquestioningly embraces human manipulation of both nature and humankind. The mechanistic metaphor, with its implications of control and mastery, has combined with traditional Judaeo-Christian notions (the human as "master" of all creation) to create this perspective. One result is anthropocentric manipulation and destruction of the world that produced us. Another is more ironic: "Man the toolmaker," Eiseley contends, "grows increasingly convinced that he is only a tool" ("The Illusion," *ST,* 269). Although Leo Marx dismisses this attitude as only a *topos,* the *topos* marks a reality that is often stated because it is often observed. Acceptance of human beings as mere tools has led to unquestioning acceptance of a scientific institution that plays an increasingly political role, as Jurgen Habermas has argued (1970, 72–104).

In the Cartesian framework, positivism, reductionism, determinism, and mechanism all work together with the notion of progress, which Eiseley traces in *Darwin's Century* from its Judeo-Christian origins to its twentieth-century manifestation in faith in science, which he later calls "the twentieth century's substitute for magic" ("The Time Effacers," *IP,* 105). Such "progress" implies a teleology from which the Darwinian view should free the human mind. Yet Darwin himself, Eiseley contends, seemed to cling to a teleological view of evolution (*DC,* 349). The goal of the evolutionary struggle was clear: production of enlightened nineteenth-century northern European man—Tennyson's "roof and crown of things" (a phrase that appeared in "The Lotos-Eaters," published in 1832 and 1842, well before the *Origin*)—with other races and cultures representing mere steps on the ladder of evolution. Nineteenth-century northern European man was "industrial" man—confident in his ability to progress, to "master" nature. Yet "progress," Eiseley argues, is not a given, and with hindsight we can see that nineteenth-century northern European *man* was not the image of perfection.

Against this notion that evolution somehow works toward a goal, Eiseley interposes the dialogue of stability and instability in the life-world. Time and evolution, as Eiseley reexamines them, are both creative and destructive (*FT,* 180; *UU,* 78, 52). Against such notions of progress and teleology, Eiseley interposes an awareness of the role of chance in evolution, drawing constantly on metaphors that undermine the deterministic view—metaphors of randomness, of the random journey in which chance activates a feedback loop that is factored into chance itself, metaphors of the throw of the dice. And from the chance that is chaos or nothingness Eiseley perceives the creativity of evolution, the "formless and inchoate void of the possible" ("The Star Thrower," *UU,* 78). Against the teleological view, he sets notions of randomness, of the human ability to shape itself, of the infinite creativity of the evolutionary process (*DC,* 330), of the "failures" that, in a random world, become successes ("The Inner Galaxy," *UU,* 193).

The history of the Cartesian notion of human mastery and of its reversal so that the tool controls the toolmaker is lengthy indeed—recognized by Carlyle, Emerson, and Thoreau, as well as by modern historians of science. The notion of human control of an externalized world emerges from the mechanistic concept of the universe. As Charles Rosenberg notes, "attempts to use science to explain and to order social and individual behavior have been significant" (1966, 148). Mechanical metaphors of the nervous system in the nineteenth century were "clothed in the authority of science" and used "to shore up middle-class morale and to provide a rationale for moderation in every aspect of behavior" (139) in an age for which moderation was linked to morality and the nervous excitation of sensual pleasure was considered so dangerous that, as one writer suggested, "Every throb of pleasure costs something to the physical system; and two throbs cost twice as much as one" (Alexander Bain, cited in Rosenberg 1966, 139). So powerful and so tied to teleology was the mechanistic view that Ernst Haeckel proposed the notion of "dysteleology" in regard to the human body as mechanism: certain organs of the human body did not contribute to the functioning of the machine. Such organs, being "useless," sometimes became the focus of blame for illnesses and thus required removal (Rosenberg 1966, 182). Tonsils and appendixes were particularly vulnerable, even well into the twentieth century.

An apparently "useless" membrane beneath the tongue could be "clipped" if a youngster was declared to be "tongue-tied."

As a means of intellectual and social manipulation, the metaphors that emerge from the mechanistic view are a subtle, little-noticed force, especially in a society that, as Rosenberg indicates in an ironically fitting metaphor, looks at "science and the scientist as a kind of *deus ex machina*" (1966, 153). David Edge posits the centrality of ambiguous technological metaphors in contemporary thinking. Technological devices, he says, have become part of our everyday existence, "forming the literal basis of metaphors which give implicit, tacit structures to our thought and feeling" (1974, 136). Edge notes that the successful metaphor, "by radically restructuring our perception of the situation, . . . *creates new questions, and, in so doing, largely determines the nature of the answers*" (136; italics in original). Because successful metaphors are usually ambiguous, the result is an ambivalence in the public mind that leaves the public open to manipulation (137–39). Further, Edge argues, metaphor may "alter feelings and attitudes towards oneself and others, and the natural world," with Descartes's image of the body as machine a classic example (140). Most significantly, Edge suggests that a metaphor is likely to be accompanied by "attitudes appropriate to its literal referent" (141), so that in the popular mind the vehicle replaces the tenor. Society, then, responds with attitudes toward the vehicle instead of attitudes toward the tenor of the metaphor. In a kind of unwitting Derridean reversal, the vehicle becomes the tenor: assuredly the vehicle shapes the response to the tenor. Such response clearly occurred to the metaphor of the universe as a clock, with resulting shifts in attitudes toward nature from cooperation to exploitation—humankind as "master" of the machine (Edge 1974, 141)—and in the social application of the Darwinian metaphor of struggle in the tangled bank epitomized by Herbert Spencer's notion of struggle in the economic marketplace—survival of the "fittest" by any "useful" means. Thus metaphor itself becomes a means of "establishing and reinforcing moral and social control" (Edge 1974, 142).

Martin Heidegger's words reverberate in an equally ironic echo of one effect of technology: "So long as we represent technology as an instrument, we remain held fast in the will to master it" (1977b, 32). In the will to master technology, there is also the will to master and to manipulate the environment—a will to mastery that has turned back on the would-be masters so that we, as inheritors of the nineteenth century, its progress and problems and metaphors, may find with Thoreau that "the railroad rides us." A frequent reaction to social problems, says Edge, is to turn the blame toward controlling political institutions (mechanism) and then to try to strengthen our "controls" of these institutions (1974, 145). Perhaps, he suggests, in our notion of "control" we are "unwittingly perpetuating inappropriate metaphors" (146). The subject of metaphor and cognition is one that I discuss at greater length in chapter 2; undoubtedly we cannot expect to abolish the hold of metaphor on our thinking, but we can reexamine this hold and question the assumptions it causes us to bring to our understanding of humankind and the nature that produced it. Such an examination is, I believe, basic to Eiseley's project as a thinker and a writer.

In the attempt to undermine such metaphorical conceptions, Eiseley explores evaluative antithemata to determinism, progress, and teleology. The antithema to determinism is indeterminacy, which in the twentieth century is isomorphic in a vast array of fields. Eiseley treats indeterminacy metaphorically through the posturing trickster from assorted cultures, particularly among Native Americans, the trickster who stands behind the solemn priest and mimics his most serious moments or who, in the form of a tornado rolling across the great plains, undermines apparent order in nature. Eiseley's textual model of understanding is itself a model of indeterminacy, stressing as it does the uniqueness of individual insight and the way the individual makes, rather than discovers, meaning through either semantic or semiotic meaning (because of the complexity of this insight, I devote a subchapter, pp. 210–17, to the parallel of textuality and indeterminacy in Eiseley's usage). Indeterminacy is perhaps best expressed in probabilistic terms, which Eiseley embraces even in *Darwin's Century* as he explores the developing understanding of genetics and to which he returns in multiple explorations of the throw of the dice or the tornadic expressions of contingency on the high plains.

Eiseley's antithemata to progress and teleology are closely tied to his themata/antithemata of fixity/change. In his cosmogony, change is the one thing that we can depend upon. In "The Innocent Fox," he questions the human tendency to fear the unexpected, concluding, "There was no order," and later, "If there was order in us, it was the order of change" (*UU*, 202–3). The Eiseleyan metaphors of dice games, the trickster who mocks order, the shape-shifting of life forms, and textual indeterminacy all undermine notions of progress, fixity, and teleology; the constant shifting

of knowledge itself underscores Eiseley's concern with the unfixed
and the unexpected.

ETHICAL THEMATA

Eiseley's ethical themata are evident in his concern for the human relationship to the nature that produced it. His central concern is the development of humankind—biologically and
culturally—through evolution. Evolution is both subject of the
quest and source of the question, both subject and vehicle of understanding, for "evolution itself has become," he contends, ". . . a
figurative symbol" ("The Illusion," *ST*, 275). Cultural evolution has
brought humankind into a world of technology, of mechanization,
of "objectivity," of science viewed as magic. A frightening result
for humankind is the loss of humane concern, of pity, of the capacity for wonder that may accompany the Darwinian understanding.
"Like Kierkegaard, Eiseley demanded to know the inner, personal
meaning of the great intellectual schemes," write Gerber and
McFadden (1983, 113). He questions the individual's role in
evolving culture and evolving knowledge, turning to antithemata
such as the magical tales and rituals of primitive humankind, who
worshipped the bear and who prayed to the spirit of the hunted
before the hunt.

I argue, then, that in exploring, often in undermining and remotivating, the metaphors that underlie what he considers misunder-
standings in and of science—Eiseley explores the complex
implications of evolution to a public that he found particularly
skeptical toward the notion of ongoing evolution (Haney 1977, 201).
But in exploring metaphors and models he also explores the process of knowing and undermines or displaces certain popular assumptions that have largely arisen out of metaphor or codes, such
as the emphasis among Darwin's followers on the aspect of warfare
in evolution. Thus Eiseley explores other possible understandings,
even how we understand at all. If, as Gerber and McFadden argue,
the essay in its classical roots is "a vehicle for exploration rather
than demonstration" (1983, 21), then the heuristic function of Eiseley's metaphors only reinforces the larger heuristic function of the
essays. As metaphors and models of struggle are replaced by
models of cooperation, the possibility of exploring an *other* understanding is present.

For forty years, Holton contends, Albert Einstein saw his role
as not only that of a scientist but also that of "a popularizer,
teacher, and philosopher-scientist in the tradition of Henri Poin-

caré, Ernst Mach, and others of the generation before him" (1987, 29). Loren Eiseley's work, too—as teacher, popularizer, and philosopher-scientist—clearly anticipates what has become a great surge in productiveness of the scientist who writes.

Approach

In laying the foundation for a new understanding of Eiseley's work, I first focus on his *genre* and on the role of this *genre*. I argue that Eiseley, like a number of writers currently working, brings an ironic, self-*reflexive* and self-*reflective* perspective to science. In doing so, Eiseley and other practitioners of his *genre* attempt to bring before a public whose trust in scientists and in their technology, whether the scientists are engaged in genome-mapping or devising new methods of mapping subatomic particles or studying undersea life, leads to blind trust in those who determine public policy on weapons or disease control and to equally blind trust in what we know to be flawed technology—a much-needed perspective on science and on the individuals who do science.

In developing the themata that run throughout his oeuvre, Eiseley reexamines the language through which science is communicated, defamiliarizing some essential codes such as the notion of "natural" and reexamining a number of metaphors—from the omnipresent machine metaphors that continually reinforce the Cartesian understanding to more common but necessarily thought-shaping metaphors such as the "ladder" of evolutionary succession, which continues to reinforce an anthropocentric view. In examining Eiseley's attention to language and his attempts to undermine common *mis*understandings that have arisen since Darwin, I find throughout his oeuvre an awareness of epistemological instability, an increasing awareness of and attention to textual instability, and an insistence on the incessant dialogue between knowing and not-knowing; it is this epistemological dialogue that, for Eiseley, marks the human as a being of infinite potential for both knowledge and destruction. Important in this dialogue is Eiseley's treatment of the human body and the heuristic of knowing through attention to the interaction of body and mind as evidenced linguistically in the dominant root metaphors or philosophemes[8] associated with the body; hand, eye, and tongue serve as heuristic images and as root metaphors for his dialogue of knowing and not-knowing.

After establishing Eiseley's *genre* and its role, I trace the evolution of his critique, beginning with its important genesis in *Darwin's Century,* much of which was written concurrently with some of the essays in *The Immense Journey*—the one a reservoir of significant scholarship on the intellectual and historical context of the emergence of evolutionary thinking (in the introduction, Eiseley limits his subject to the nineteenth century and projects a second book on evolutionary thinking in the twentieth century) and the other one of his most widely read collections of popular essays. Treating the two as complementary, I find in *Darwin's Century* the genesis of the thought, themes, techniques, and style that distinguish Eiseley's career. The book is especially intriguing from a rhetorical perspective: written as a serious treatise by a scientist whose reputation was established enough that Doubleday asked him to do the work, *Darwin's Century* provides insights into the rhetoric of science and into what I call the "other rhetoric," which parallels Eiseley's description in "The Enchanted Glass" of "thought which is not science, nor intended to be, and yet without which science itself would be the poorer" (1957, 482).

In Eiseley's next volume, *The Firmament of Time* (1960), which has been described as "lyrical and meditative" (Angyal 1983, 57), I find an overt movement toward what was adumbrated in the earlier two volumes—what Hayden White calls the ironic view, the self-reflexive, self-reflective distancing of the author who perceives the instability of the subject itself, who turns to science itself as a subject for meditation that goes beyond history into irony. In *The Firmament of Time,* Eiseley develops some of his most elaborate interweaving of tales and reflection, of almost simultaneous text-making and textual analysis.

The next two volumes, *The Unexpected Universe* and *The Invisible Pyramid,* move toward affirmation of a different understanding—what might be called "poetic" understanding in "The Star Thrower"—and toward sometimes harsh criticism of the scientific institution in *The Invisible Pyramid,* which portrays humankind as analogous to exploding slime mold in the quest for space even at the expense of the ecosystem. From his extended use of the literary tutor text in "The Ghost Continent" in *The Unexpected Universe* to his yearning for a return to the life of a primitive on the high plains in *All the Strange Hours,* Eiseley increasingly affirms the sympathetic understanding that crosses the human-animal boundary, an understanding expressed by primitive peoples and in literature, and increasingly decries the "narrowness" of any profession, especially science, which allows itself to become so trapped in a single perspective that the imagination always associated with great science, from Bacon to Einstein to Watson and Crick, is stifled.

The writer's self is both subject and heuristic in *The Night Country* and *All the Strange Hours*. By now Eiseley is clearly comfortable with his identity as a writer, though he never abandons science. Sometimes thought to be his "autobiography" because of the personal reflections and simple joy in tale-telling, *The Night Country* marks Eiseley's full exploration of the self as subject. *All the Strange Hours: The Excavation of a Life* (1975), his appropriately discontinuous autobiography, continues the exploration of the self. Eiseley proclaims himself "every man and no man" as he reflects on a life in science. If in *Darwin's Century* he proposes "a long second look" at the implications of evolutionary thinking for the individual, in *All the Strange Hours* he takes "a long second look" at his own life in science, at the personal in relation to the professional. And this examination is the focus of the book, which is in no sense a standard autobiography, but a fragmented account in which the self is vehicle rather than focal point.

In examining the last book that Eiseley himself authorized (he selected these essays and poems from among other published texts and unpublished manuscripts and titled the collection after one of his best-known and most complex essays, *The Star Thrower*), I note that the collection speaks openly to what Eiseley has always treated—dramatically in the earlier "Star Thrower" and polemically in "The Illusion of the Two Cultures"—but has become increasingly important, the role of an "other" knowledge in exploring the human potential, the need for an ongoing dialogue between the epistemologies of science and of art. Finally, from textual analysis that ends with Eiseley's call for epistemological symbiosis, I conclude with a review of Eiseley's consideration of the limits of science and the limits of humanism, the boundary at which occurs the ongoing dialogue of knowing and not-knowing.

2

A Route for Critique: Eiseley's Genre

Eiseley is usually classified as a popularizer of scientific information whose purposes are not only scientific but literary as well. In his view, however, science and literature need not be separated; instead, science becomes a vehicle of literary understanding. His genre is a hybrid that participates in but also enriches the long-established tradition of "natural history." Eiseley's genre also partakes of a tendency toward hybridization in the modern discourse of knowledge, for whose roots Richard Rorty looks to the humanities in the nineteenth century:

> Beginning in the days of Goethe and Macaulay and Carlyle and Emerson . . . a kind of writing has developed which is neither the evaluation of the relative merits of literary productions, nor intellectual history, nor moral philosophy, nor epistemology, nor social prophecy, but all of these mingled together in a new genre. (1976, 763–64)

Jonathan Culler observes that the texts of the "heterogeneous" genre that Rorty describes have the "power to make strange the familiar and to make readers conceive of their own thinking, behavior, and institutions in new ways" (1983a, 9). It is this tradition of heterogeneity that Eiseley follows and expands.

Eiseley directs his essays toward an audience of literate readers, who may be specialists in their own fields but have an interest in the insights that an archaeologist and a scholar of early humankind can blend with literary insights in a merging of narrative and reflection that Andrew Angyal has called "a new genre—an imaginative synthesis of literature and science—one that enlarged the power and range of the personal essay" (1983, 39) and that E. Fred Carlisle has called "a new prose idiom for science and for literature" (1977, 112). Although we may not find in his texts a wholly "new" genre, certainly Eiseley can be credited with revitalizing a philosophically reflective and an ironically *reflexive* (turning to reexamine science itself) form of natural history.

With the personal approach as a starting point, Eiseley develops

> a highly elaborate form, with frequent literary references and allusions, numerous quotations, multiple themes, and an interwoven structure of contemplative concerns. This casual and informal, though sophisti-cated technique brings narrative and personal experience—essentially fictional and autobiographical tools—to bear on what is otherwise ex-pository material—scientific fact and hypothesis. (Angyal 1983, 39)

The form thus created is a synthesis of scientific and literary modes of expression that draws on the work of writers as disparate as the British parson-naturalists, American romantics such as Em-erson, Thoreau, Muir, Burroughs, and Melville; medieval mystics; and twentieth-century thinkers such as Whitehead and Santayana. As an interdisciplinary hybrid, Eiseley's genre engages not only autobiography, philosophical reflection, and popularization of sci-entific information, but literary intertexts and tutor texts, philoso-phy, and the politics of rocketry and education. "And," writes Angyal, "if there is a movement among scientists today to reassert the humane values that are intrinsic to science wisely practiced, the credit for that must also be due in part to Eiseley" (1983, 125).

Eiseley specifically calls for a hybrid that defies categorization in literature or science but is close to natural history. Such writing involves "a partial transformation of data" so that it "contains overtones of thought which is not science, nor intended to be, and yet without which science itself would be the poorer" (1957, 482). This genre interweaves "the realm of natural objects and the hu-man spirit" and records the "personal element, . . . the shifting colors in the enchanted glass of the mind which the extreme Baconians would reduce to pellucid sobriety" (480). Eiseley's liter-ary approach emerged during the 1950s, particularly during the research for *Darwin's Century,* which caused him, in Carlisle's terms, to "interiorize" Darwin's theory "so that it functions as a major structure for perceiving and comprehending experience," a structure through which he "makes contact with reality" (1974, 365). The work on *Darwin's Century,* Leslie Gerber and Margaret McFadden write, "served to crystallize Eiseley's entire intellectual outlook," and *The Immense Journey,* many of whose essays were written concurrently, emerges as "a companion volume" (1983, 50). Eiseley refers to his genre, in language that seems to echo Wordsworth's preface, as the "concealed essay, in which personal anecdote was allowed gently to bring under observation thoughts of a more purely scientific nature" (*ASH*, 177).

The Concealed Essay: Structure

In Eiseley's "concealed essay" the subject matter "is framed or 'concealed' by the personal approach, which serves as a rhetorical device to engage the reader's attention" (Angyal 1983, 39). But whenever science falters—fails to provide the answers it has seemed to promise, or gives insights that are nevertheless humanly incomplete—the text returns to the personal. (The model essay for this faltering is "The Star Thrower," which I shall examine in detail in chapter 7.) The faltering of science is crucial to Eiseley's strategy, for writing is generally equated in his texts with poetic or intuitive knowledge,[1] with what Mary Belenky et al. treat as "connected" versus "separate" or analytical knowledge (1986, 102). In addition to the personal and the "thoughts of a more purely scientific nature," there is a realm, Eiseley contends, "which can never be completely subject to prophetic analysis by science"; in this realm science may falter, for it is the realm of "indetermination" ("The Illusion," *ST*, 278), which evokes much of his reflection. The "concealed essay" thus becomes a complex form that combines "memory, landscape, and visual imagination" (Angyal 1983, 41). Like the natural history essay that influenced it, the form is interdisciplinary, providing the advantages of perceiving interrelationships of organisms and of cultures, of developing "holistic" attitudes toward nature that incorporate more concerns than the exact scientist may recognize (Haney 1977, 3–4). The form combines experiment and experience, so that "an extension of the sense has become an extension of science" (Carlisle 1974, 354–55). As it leads to meditation on subjects that science addresses, Eiseley's concealed essay depends on certain identifiable techniques.

The Two Rhetorics

Two traditions—Baconian observation and experiment and the "more gracious, humane tradition" associated with the English parson-naturalists (*DC*, 13)—are the sources of two rhetorics that Eiseley employs throughout his texts. The first of these rhetorics is the scientific, which is dominant in *Darwin's Century* and is called upon less frequently in the middle and later texts. The second—the emotionally charged rhetoric that looks at the human side of science, at the human being who does science, and at the responsibilities that human being brings to the task—is the rhetoric normally identified with Eiseley's texts: laden with adjectives, deliberately emotional, seeking the synthesizing view.

Eiseley's scientific rhetoric, which I shall examine more specifically in chapter 4, clearly calls upon the rhetorical conventions associated with the discourse of science—appeals to experimental competence, impartiality, and authority; impartial narrative and use of the third person; avoidance of emotionally charged insights, examples, and adjectives; reliance on enumerations rather than on description and personal narrative. Eiseley's "plain" or "scientific" style is marked by careful attention to detail, clear organization of the sequential development of evolutionary theory, and consideration of his audience in defining terms. His "plain" style thus echoes the rhetoric of the scientific article, in which, in Alan Gross's words, style is "a social imperative" rather than "a pattern of individual choices" ("Does Rhetoric?" 934). The scientific article, Gross contends, conveys not only information but "a view of the world as the causal interaction of physical objects and events" (1991, 935). Using the linguistic strategy of reserving the subject position for "physical objects and events," in effect treating them as "the causal center," the rhetoric of science "invests physical objects and events with the importance ordinarily bestowed on human beings" (936). Eiseley's "plain" style follows scientific convention, placing objects and events at the "causal center" and drawing on conventions such as enumeration for clarity. This style may be exemplified by a passage in *Darwin's Century* describing Johannsen's studies of 1903 and later, which showed

(1) that organisms with the same *genotype* (i.e., genetic composition) could differ *phenotypically,* that is, in their physical appearance; (2) that the selection of phenotypic characters without a genetic base would not yield hereditary change; (3) that selection of hereditary characters could induce some degree of physical alteration but the effect would attenuate and halt unless there were added mutations which are sometimes forthcoming and sometimes not. (*DC,* 228–29)

But even in the early texts, Eiseley turns frequently to the "other" rhetoric that will become his hallmark. This rhetoric relies heavily on his concern for the human side of science, for the role and responsibility of the scientist as a human being who is nevertheless a part of the evolutionary system that produced him, for what McMullin calls "an *interpretive* face" of science (1974, 670), for his creation of an ethos for science and for himself. Deriving from the Greek *ethos,* "custom, usage, trait," the concept of ethos comes to us primarily through Aristotle's emphasis on the orator as one who understands the Good in terms of established human values and considers his audience's understanding of this Good. Aristotle separated proofs into "artistic" (argumentative, rhetorical) and "inartistic" (evidential, but requiring interpretation for application to a particular case). He further contended that ethos should not be left for the audience to presume solely on the basis of the reputation of the speaker or writer, but should be established within the discourse itself; thus, for Aristotle, ethos works through logos (Hill 1983, 26–27). Today the Aristotelian concern with defining the self in terms of received notions of the Good has become mixed with the Ciceronian metamorphosis of *ethos* into pragmatic portrayal of the self as upright, as worthy of belief (Johnson 1984, 101–6).

If science, as we have come to understand, constructs its reality even as it constructs itself, this construction is closely related to the mixed Aristotelian-Ciceronian concept of the rhetorical ethos as construction of self for the audience. As Robert K. Merton defines it in his seminal work on science and ethos, the scientific ethos is an "affectively toned complex of values and norms which is held to be binding on the man of science"; its norms are expressed in the form of prescriptions, proscriptions, preferences, and permissions," and "they are legitimized in terms of institutional values" (1973, 268–69). Merton establishes four basic norms for the scientific *ethos:* universalism, the requirement that knowledge claims be understood in terms of impersonal criteria and established knowledge; communality, which requires that research be made public for the scientific community; disinterestedness, which minimizes self-interest in favor of the interest of the discourse community; and organized skepticism, which requires scientists to evaluate all knowledge claims in terms of logic and empirical data (Prelli 1989, 106).

This ethos of the scientist who writes to reexamine, to question, to explain, even to shape science itself underlies Eiseley's "other" rhetoric, which often turns to personal insights, anecdotes, and highly charged language expressing the kinship of human and other life forms. It is also, I believe, a reason for the expanding number of scientists who reexamine the role of science and the self. In a changing reality, the Good must be constantly redefined; thus the ethos of science is an ongoing construction. Hayden White's argument that the "ethical moment" in a historical work comes at the confluence of "an *aesthetic* perception" and "a cognitive operation," or argument, is, I believe, also relevant for the discourse of the reflective scientist (1987, 27). Rejecting the notion of "extraideological grounds," White perceives ethical choices in any narrative conception (26). White's argument is, I believe, especially applicable because Eiseley is concerned with intellectual history and with the *ethos* of the scientist. I turn, therefore, to a delineation of the techniques of Eiseley's "other" rhetoric—techniques that dominate the concealed essay and the quest for a synthesis of empirical understanding and empathy, a dialogue between the linear, cause-and-effect approach to knowing associated with science and the "other" ways of knowing associated with connecting, with feeling, and with literary understanding.

The Narrative of Science: Historicizing

Effective use of narrative—always emphasizing science in its historical context, but effectively expanded in the later essays to include, in addition, historical narrative, personal anecdote, or tales of the old West—is a hallmark of Eiseley's concealed essay. In his earliest essays and his research for *Darwin's Century,* Eiseley recognized the importance of historicizing science, of seeing science as a product of culture rather than an ahistorical vessel of "truth." Without history, he writes early in *Darwin's Century,* the human tendency is to *create* history, through myths of the supernatural or of heroic human deeds. Even so, the historian of science carries an almost overwhelming burden in supplying a narrative for scientific change and growth. Inevitably the historian of science *constructs,* through narrative, a reality for the past. Thus Eiseley's reflections on the history of science and on a life in science incorporate his growing awareness of the burden of historiography, of the errors and missteps regarded as truth in one age and as yesterday's error in another, with this tendency to think of the past as error and the present as truth or progress creating a powerful double bind—empowering the notion of scientific "progress" even as the historian undermines it.

Emerging as it did during a period when Eiseley was intensively studying the history of evolutionary thinking, his concealed essay is strongly informed with the history of science. Evolution itself cannot be separated from its historical matrix of Judaeo-Christian notions of progress, of Renaissance humanism and the "ladder" of nature, of the Malthusian struggle for survival, of the nineteenth-century notion of scientific "objectivity." Although the twentieth century has seen a growing reaction against attitudes associated with late nineteenth-century thinking—including positivism, reductionism, scientism, determinism, mechanism, and the myth of objectivity—attitudes that have already shifted in the scientific community may still be embraced by the general public. The public generally does not question the "given"—that which is evident, observable, available for scientific examination. In a significant parallel, the literary establishment was slow to question the positivist assumptions of the New Criticism (the poem as autonomous entity, the artistic unit as independent of historical circumstances, the text as focus of "objective" examination) and to recognize the importance of suspicious readings and questioning of the metaphysical assumptions embedded in the very language of concepts, including scientific ones. One of Eiseley's most important themes is the intellectual and historical context of evolutionary understanding of the human—particularly as humankind has come to assume dominance over, even destruction of, an externalized and particularized nature as its "right." Interrogating this *topos* of the human relationship to science is basic to Eiseley's ironic stance, whether through scholarship in *Darwin's Century,* through carefully structured lectures in *The Firmament of Time,* through themes of "strangeness" in *The Unexpected Universe* and *The Night Country,* or through literary meditations in his last collection, *The Star Thrower.*

Autobiography and the Observer-Participant

Basic to Eiseley's hybrid genre is his emphasis on the role of the observer-participant, especially through his use of autobiography. As an anthropologist and a student of nineteenth-century science, he notes the impossibility of "objectivity" and turns to individual experiences in his essays. Undermining the myth of "objectivity" is one of his tasks. For Eiseley, Carlisle contends, "science is a personal quest that recognizes the self as the origin of all knowledge—even scientific knowledge—but that also requires the systematic structure of science for its success—and that relies on investigation *and* imagination for its insights" (1974, 359). Eiseley recognizes what Heidegger calls "the necessary interplay between subjectivism and objectivism" (1977a, 128). And as Roger S. Jones notes, since Kuhn's concept of paradigm shifts and Holton's treatment of the thema-antithema by which science moves, almost no one would deny that "the observer has an uncontrollable and nonremovable effect on what is observed" (1982, 6). Jones cites Eiseley's work, in addition to that of Kuhn, Holton, Poincaré, and Polanyi, in reevaluating the role of subjectivism in science (207–8).

Eiseley uses personal narrative and privileges personal observation for literary purposes, but the autobiographical details do not appear for their own sake. Instead, the self provides a point of entry into complex material. The *I* of the statement is not necessarily the *I* of the author, for Eiseley's use of personal narratives involves creation of a persona who narrates and assumes other personas, such as wanderer, gambler, and fugitive.[2] In Lakoff and Johnson's view, "much of self-understanding involves consciously recognizing previous unconscious metaphors and how we live by them. It involves the constant construction of new coherences" (1980, 233). Eiseley refers to the "conscious 'I'" as insignificant to the phagocytes in the human body ("The Hidden Teacher," *UU*, 50)—insignificant except to provide access to the information presented.

Yet this *I* as a persona is a manipulator of the reader, constantly projecting the uncertainty of his own identity and thus drawing the reader into the quest for self or for knowledge. Eiseley's transcendence of the formal essay or the scientific tract depends on his ability to involve the self of the narrator and the self of the reader. Indeed, he parallels "the maturation of creative consciousness, the growth of the artist, to the evolution of the entire universe" (Schwartz 1977a, 11). That the "I" is often a fugitive may reflect the "elusiveness and final impossibility" of an identity in an "unexpected universe" (Gerber and McFadden 1983, 146).

This problematic of the conscious *I* is unavoidable in Eiseley's texts. The *I* is engaged in a journey toward understanding of self, but as Terry Eagleton writes, "not only can I never be fully present to you, but I can never be fully present to myself" (1983, 130). To seek an "identity" behind the *I* has often led to abandonment of the text in favor of history or biography. Eiseley attempted to evade those who would pursue a "critical biography" (see Carlisle 1983, ix–xiii). For Eiseley, the *I* is a literary or "shamanistic" device. In the primitive societies whose values he studied and respected, a shaman or mediator takes the responsibility for narrating events; the "author" is a creation of the modern world (Barthes 1977a, 142). Eiseley turns to the shaman and to primitive societies as models for the relationship of the human and the environment, and he prefers the function of the detached mediator or shaman rather than that of the "author." Shamanism, Ulmer suggests, serves as a model for "autobiography as a research tool applied to fields of knowledge beyond itself" (1985, 328, n. 22). Eiseley's method involves a shamanistic set of autobiographical "glimpses" through which to treat the elusiveness of being and becoming.

For his "autobiography," in fact, Eiseley chose an epigraph from Browning's *The Ring and the Book,* which deals with the impossibility of discerning "truth" because multiple perspectives make impossible a single identifiable truth. The epigraph reflects—in the sense of the shifting color spectrum—the changing perspective that comes with changing time:

> I' the color the tale takes, there's change perhaps;
> 'Tis natural, since the sky is different,
> Eclipse in the air now, still the outline stays.
>
> (Cited in ASH, iii)

Not only are the shifting colors present, as in the kaleidoscopic effect that Eiseley often employs metaphorically, but also the palimpsest effect in which the outline remains, to echo the subtitle, *The Excavation of a Life,* and the use of archaeology as a model of acquiring knowledge. Early in the "autobiography" Eiseley explains his use of the *I,* which he links to the epistemological uncertainty that underlies his texts:

> I am every man and no man, and will be so to the end. This is why I must tell the story as I may. Not for the nameless name upon the page, not for the trails behind me that faded or led nowhere, . . . not for the confusion of where I was to go, or if I had a destiny recognizable by any star. No, in retrospect it was the loneliness of not knowing, not knowing at all. (*ASH,* 23)

As "every man and no man," the *I* in *All the Strange Hours* is not the expected *I* of autobiography. Nowhere does Eiseley deal diachronically with his own life's story. The book treats recollections that lead to reflection on recurrent Eiseleyan themes, but it does not reconstruct a personal life. As the subtitle indicates, it is "an excavation of a life," with incidental artifacts uncovered from a life with what seems the same chance by which artifacts are uncovered in an excavation. As an autobiography, it is more an account of the journey of knowing than a reconstruction of a life. "Excavation" becomes the model of the epistemological quest, for human knowledge, like the individual life, is sedimented with meanings and quests for meaning. And "the fragmentary narrative" becomes "a way of depicting the tricks of memory," for the "continuously existing personality in time . . . is not accessible to memory" (Angyal 1983, 109).

The shattered mirror that recurs throughout the book is an apt metaphor for the autobiographical element in this text and others. The *I* is the *eye*—observer and homophone, the cracked fragment of a mirror reflecting the whole yet also participating in a reflection of the whole that the intact mirror would have reflected. In Eiseley's essays, the details from his life are as fleeting and as shifting as is the universe. These details shift as in a mirror or a kaleidoscope so that even the narrator who recalls emerging from a cave to see the modern world as "a little lost century, a toy" (*ASH*, 104) is uncertain who the *I* is. The *I* of Eiseley's texts, like the *I* in this example, is a dramatized observer—part self, part observer-participant, part fictional creation. The *I* allows a seemingly personal, though sometimes obviously fictional, narrator to draw the reader into the experience of a biologist who "dances" with frogs ("The Dance of the Frogs," *ST*, 106–15) or an explorer eye-to-eye with an owl in a cave ("Obituary of a Bone Hunter," *NC*, 181–91).

The *I*, then, is "every man and no man," the part that reflects the whole, the individual who recreates the experiences of humankind, even of life; it is the object of the search whose goal is impossible, for "the object is the search itself" (Pickering 1982, 21). The *I* is the synecdoche for humankind and its history, retold through tales whose "reality" or "fictionalism" is difficult to determine. But assuredly, the *I* is a strategy of gaining identification or sympathy and thus of drawing the reader into the epistemological quest that is basic to Eiseley's journey. What is important, he asserts as he attempts to explain why one person writes, is that "we all live in a moving stream, as surely as a catfish groping with its whiskers in the muddy dark" and that even an obscure writer of a book on aquarium-building may play a role in forming the inclinations of a writer: "That is the wonder of words. They drift on and on beyond imagining" (*ASH,* 170).

The *I* is able to involve the reader in recreating a journey—the long journey of life, the journey of a fish down the Platte River, the journey of knowledge. The journey is a wandering/wondering, and the *I* falters, as knowing falters. But faltering itself is a strategy to involve the reader.

Fictionalizing/Dramatizing

An important technique in the concealed essay is fictionalizing and/or dramatizing an event, creating a fictionalized setting for a historical occurrence or inviting the reader into the scene as a participant as the author is a participant. Even in the historically oriented *Darwin's Century,* Eiseley's fictionalizing of events emerges, and nowhere more effectively than in the vividly fictionalized account of Mendel's presentation of his findings in 1865:

> Stolidly the audience had listened. Just as stolidly it had risen and dispersed down the cold, moonlit streets of Brunn. No one had ventured a question, not a single heartbeat had quickened. In the little schoolroom one of the greatest scientific discoveries of the nineteenth century had just been enunciated by a professional teacher with an elaborate array of evidence. Not a solitary soul had understood him. (*DC,* 206)

Eiseley's recreation of the event, from the cold streets to the equally cold hearts of those "professional scientists" who failed to see the impact of Mendel's work, evokes sympathy for Mendel and prepares the reader for his observation that the scientific establishment's ignoring of Mendel the amateur "is worth perusal by all scholars, not alone because of what Mendel achieved, but also because the complete failure of communication in this particular instance was, to a major degree, the failure of professional science" (207). The theme is one to which Eiseley will return often and one that will form the basis of a chapter in his autobiography. His image of professionalized scientists is that of dancers in a fairy ring, a ring of mushrooms that has sprung up overnight (the metaphor implies that science is only a recent development in the history of humankind, which is itself only a moment in terms of geological history). The dancers in the ring provide Eiseley's metaphorical image of T. S. Kuhn's "normal science."

And Eiseley provides a sense of insight into Mendel's life as he writes of "the obscure priest who read the *Origin of Species* and carried on queer experiments with peas which he affectionately referred to as his children," working "by infinite patience alone in the solitude of a monastery garden" (*DC,* 206–7). Placing Mendel alone in the monastery garden with his "children" allows Eiseley to dramatize the expository point that while Lamarck and Darwin had focused on change, "Mendel was fascinated by stability" (208). Darwin's physical journey prepared him to look for change, but Mendel, the isolated priest working in a monastery garden, was prepared to look for the stability that works dialectically with change in evolution.

Eiseley further dramatizes Mendel's work as he recounts Fleeming Jenkin's challenge to Darwinism and invites the reader to participate in the physical and intellectual journey:

The answer to Fleeming Jenkin had been standing on library shelves in the Proceedings of the Brunn Society for the Study of Natural Science since 1866. Jenkin, the hardheaded engineer, and the gracious, dreaming naturalist who had been forced to retreat before him would both be gone before anyone blew the dust from those forgotten pages. . . . Yet if we are to understand him [Mendel] and the way in which he eventually rescued Darwinism itself from oblivion we must go the long way back to Brunn in Moravia and stand among the green peas in a quiet garden. (*DC*, 211)

In thus dramatizing moments from Mendel's life or Darwin's in *Darwin's Century,* Eiseley establishes a method evident in all of his texts. Later he will invite the reader to participate with him in feeling the flow of water from the muddy shallows of a high plains river ("The Flow of the River," *The Immense Journey*) or to enter imaginatively the silent undersea world of the porpoise, sacrificing hands for flippers and the ability to communicate with future generations through writing for awareness of only the present and communication through vibrations in a world of wavering green light ("The Long Loneliness," *The Star Thrower*).

Synthesizing/Dialogizing

Another characteristic of Eiseley's concealed essay is its condensing, synthesizing perspective. Assuming the role of meta-observer, he condenses great chunks of time into a fable-like narrative through which he can convey a perspective, an interpretation. In *Darwin's Century,* as he recounts James Hutton's realization of the continual decay and renewal of nature, he writes, "The man by the trickling brook had heard a roar like Niagara and seen a world go down into the torrent" (*DC*, 73). And in concluding his treatment of the quest for a "missing link," Eiseley condenses the account of the gradual realization that the gap between the human and the beast was larger than it seemed at first and the isolating effect of this realization:

It was then that [man's] isolation struck him more clearly. He stared thoughtfully at the tiny-brained among his kind. He dug in the earth and found bones beneath it. He began to sense that the wondrous chain was moving, climbing, perishing. He found his own lost, bestial skull in the drift by the river, and the flints that his hands had tried to shape. At first he sought to run away from the sight of these things or to tell the tale differently. In the end it could no longer be done. The tale will tell itself and man will listen. He is quite alone now. In spite of claims

that persisted into the beginning of this century, his brothers in the
forest do not speak. Unutterably alone, man senses the great division
between his mind and theirs. He has completed a fearful passage, but
of its nature and causation even the modern biologist is still profoundly
ignorant. (*DC*, 285)

This condensed account is a synthesis, an interpretation, an emotionally charged statement of Eiseley's recurrent theme of the intellectual loneliness of humankind. In interpreting, he draws
attention to what *he* is doing—telling the tale—as he notes the
human tendency to shape the tale according to how we *want* the
tale to read. (As he concludes that "the tale will tell itself," however, even though he has embraced the notion that the culture
writes the tale, Eiseley paradoxically embraces the dominant scientific notion of his own culture that "truth will out," that evidence
will require reexamination until the "correct" interpretation is
found.) When the tale has "told itself," what remains in Eiseley's
text is an interpretation—the "aloneness" of humankind. The position is more epistemological than eschatological, for the aloneness
is Eiseley's recurrent theme of "not knowing at all" (*ASH*, 23).
The epistemological position is one that remains constant throughout his texts. The "not knowing" is the nothing that the narrator
confronts in "The Star Thrower"; it is the not-knowing from which
springs interpretation, the rift that is also a joining, the vacuum
from which springs creativity. Time, Eiseley writes in the conclusion to *Darwin's Century,* which marks his turn to the synthesizing
"other" rhetoric, has become "a loneliness, an on-going" (*DC*, 334).

Tutor Texts

Given Eiseley's early desire "to be a nature writer," the increasingly literary style that he adopts is not surprising. Beginning with
the earliest texts, Eiseley uses the tutor text—a literary text that
he describes or quotes, that he may analyze in literary or linguistic
terms, and that he uses in drawing his own position. The "concealed essay" depends increasingly throughout Eiseley's writing
career on such tutor texts, often approaching science through literature or philosophy. *Francis Bacon and the Modern Dilemma* (a
source of much concern to him because of errors) also established,
as part of his study of Renaissance understanding, the notion of
the "text" of nature, of nature as "the one sure divine revelation"
(*FBMD,* 40). This code of nature as text or hieroglyph recurs frequently in all the texts. In *Darwin's Century* all the texts of the

pre-evolutionary figures resemble tutor texts, though their "li-
terariness" is marginal. Eiseley draws on tutor texts from literature
and, occasionally and even playfully, the text of nature, as a means
of seeking his own position. In *Darwin's Century*, for example,
he draws on Darwin's diary of the voyage of the *Beagle* and his
autobiography as important tutor texts. He divides Darwin's re-
marks into two classes—*"those bearing on the proof that evolution
has occurred, and those concerned with the actual search for the
mechanism by which organic change is produced"* (*DC*, 159; italics
in original)—and emphasizes the need to separate these two points
in order to avoid the kind of vacillation that Darwin himself engaged in as he grew older and encountered challenges to elements
of his theory.

Eiseley's pursuit of Darwin's tutor texts is necessary to fitting
together the pieces of what he calls the evolutionary "pirate chart,"
but it becomes also the foundation for a literary technique that
moves from Darwin's journals to Thoreau's; to Melville's *Moby
Dick;* to Keats; to Giovanni Pascoli's 1904 poetic treatment of the
Odyssean theme, "Ultimo viaggio" ("The Last Voyage"); to the
Biblical book of Job; and to C. P. Snow's essay "The Two Cultures." In some of the later essays ("Walden: Thoreau's Unfinished
Business," "Thoreau's Vision of the Natural World," and "Man
Against the Universe"), he adopts a tutor text and uses it to extract
a personal reading that differs from earlier readings of the text,
and he uses his new reading to illustrate his own point.

Tutor Structures

Another literary technique that appears in the concealed essay is
the use of what may be called "tutor structures" from literature—
narrative and other structures that provide an often unconscious
framework within the essay. From time to time Eiseley draws on
tutor structures such as the epic, Greek drama, the beast fable,
tales of the old West, travel lore, liturgical structures, and autobiography. This technique makes possible a dialogue between two
ways of knowing: in a single essay a tale of the old West may give
personal, human insight, while the alternative knowing is based on
archaeological findings. In *Darwin's Century* and in "The Ghost
Continent" in *The Unexpected Universe,* the tutor structure is the
epic: for the odyssey of life, of humankind, of knowledge itself,
Eiseley draws on structures associated with Homer's *Odyssey.* In
several essays in *The Firmament of Time* and in *The Night Country,* the tutor structure is the dramatic structure, with particular

emphasis on the relationship of the audience to the drama; Eiseley portrays the audience as creating through participating, in an image that parallels his point that we create the future by imagining it. In *The Night Country,* tales of the old West form tutor structures that Eiseley uses as a basis for telling of his own experiences as a "bone hunter," interweaving the tale and the human insight that it suggests.

Rhythms and the Model of Reading

In analyzing Eiseley's development of a consciously literary style, Carlisle suggests that he "began to develop a language of continuity or wholeness to counter the discontinuities and fragmentation that he found in science and scientific discourse, as well as in himself" (1983, 182). The key to this continuity is extension—of vision, of tactile experience, of participation in time. The result is that Eiseley's "new idiom" is characterized by "layers" of science, autobiography, figurative language, and speculation that create "rhythms" of mind and of prose (182–83). These rhythms—of identities of the self; of movement between inner and outer; of present, past, and future; of "intensive" analysis and "extensive" interpretation (as in giving details about the brain and then turning to man as a "dream animal"); of stability and change; of science and myth or science and ethics—identify Eiseley's idiom and enable him to "intertwine" the layers of discourse and to extend the senses (183–84).

Another rhythm, which is a useful device in explaining new material, is alternation—of the familiar with the unfamiliar, of the personal and subjective with the impersonal and objective, of the near with the remote, of the specific with the general. Thus Eiseley establishes the important rhythm of alternating explanation and persuasion with exploration and emotion, so that anthropology and autobiography, science and culture are made to intersect in the reader's mind.

Equally important in Eiseley's texts is the rhythm of reading, which is itself a journey. Wolfgang Iser describes the movement of reading as

> first, a repertoire of familiar literary patterns and recurrent literary themes, together with allusions to familiar social and historical contexts; second, techniques or strategies used to set the familiar against the unfamiliar. Elements of the repertoire are continually backgrounded or foregrounded with a resultant strategic overmagnification,

trivialization, or even annihilation of the allusion. This defamiliarization is bound to create a tension that will intensify the reader's expectations as well as his distrust of those expectations. . . . [W]e will find
ourselves subjected to this same interplay of illusion-forming and
illusion-breaking that makes reading essentially a recreative process.
(1980, 62–63)

The term *discourse* itself suggests movement, though not a pat-
terned, ordered rhythm; deriving from the Latin *discursus,* it indicates "a running back and forth." Serres writes that "to read and
to journey are one and the same act" (*Jouvences,* 14, cited in Harari and Bell 1983, xxi). In portraying life as process and using the
journey as a root metaphor, Eiseley turns to a literary or "textual"
model. Iser suggests that through Ingarden's "*Konkretisation*" we
construct the "virtual" reality of literature, which is parallel to the
"virtual" reality of living. The *becoming* of the literary "reality"
thus involves the rhythm of interaction between reader and text.
Ultimately, Eiseley's texts describe the human being in the process
of reading (creating) her own literary work of art. Unlike the
eighteenth-century Book of Nature, it is an incomplete, shifting,
changing, *interactive* reality. The individual shapes it. Just as man,
having escaped physical determinism, shapes his *becoming,* the
becoming of the literary text is the model for the human journey
even as the journey is the model for the text. The model involves
the individual in a creative process, an engaging of nature and life,
a participation in the process rather than a passive observation or
a relegation of participation to a scientific or military-industrial-
political elite.

The Concealed Essay: Metaphorical Techniques

In addition to the structural characteristics of Eiseley's concealed essay, specific metaphorical techniques are evident. Eiseley's textual turn and his metaphorical techniques are directly
related to the cognitive view of metaphor. Before turning to the
specific metaphorical characteristics of the concealed essay, I shall
examine some recent views of the relationship of metaphor to the
human cognitive process.

Metaphor and Cognition

Traditionally, metaphor has been relegated to the role of orna-
ment, considered a means of attracting attention and influencing

the audience, "a matter of words rather than thought or action" (Lakoff and Johnson 1980, 3). Plato saw rhetoric and poetry as suspicious, allied to public oratory and to illusion and thus a threat to "truth." The cognitive view of metaphor was actually suggested by Aristotle when he argued that "it is from metaphor that we can best get hold of something fresh" (1941, 1410b). The most cogent recent explanation is given by George Lakoff and Mark Johnson in *Metaphors We Live By*. Reflecting contemporary linguistic, textual, and cultural studies, this text echoes Eiseley's explicit statements on the essential metaphoricity of human thinking and on the effects of metaphor on human actions. As opposed to the "ornamental" view of metaphor, Lakoff and Johnson hold that our basic conceptual system is "fundamentally metaphorical in nature," shaping both thought and action (1980, 3). Metaphor underlies our basic concepts, which in turn create our realities. Because our metaphorical concepts are systematic, we are able to understand part of a concept in terms of another. Yet, as metaphor allows us to focus on one aspect of a concept, it also "hides," or suppresses, other aspects (10). In this view there is a sense of both gain and loss that echoes postmodern ambivalence and Eiseley's emphasis on the dialogical nature of understanding. Lakoff and Johnson argue, for example, that the metaphorical concept ARGUMENT IS WAR suppresses the elements of accommodation that some cultures include in arguments. Not only the concept itself, but our actions in arguing, are grounded in metaphor. The "conduit metaphor" similarly governs both thought and action; it is complexly structured as follows:

IDEAS (OR MEANINGS) ARE OBJECTS.
LINGUISTIC EXPRESSIONS ARE CONTAINERS.
COMMUNICATION IS SENDING. (10)

Thus "the speaker puts ideas (objects) into words (containers) and sends them (along a conduit) to a hearer who takes the idea/objects out of the word/containers" (10).

Lakoff and Johnson contend that metaphors may be (1) "structural," determining the structure of one concept in terms of another; (2) "orientational," suggesting culturally based spatial orientation such as up-down, front-back, in-out; or (3) "ontological," providing ways of seeing "events, activities, emotions, ideas, etc., as entities and substances," as in thinking of rising prices as "inflation" or the human mind as a machine (1980, 14, 27–28). Through such ontological metaphors we actually emphasize different as-

pects of the mind's experience; thus the visual field becomes a
container, and what we see is "inside" that field (28–30). In a statement that Eiseley specifically anticipates, Lakoff and Johnson argue that "purely intellectual concepts," as in scientific theory, "are
often—perhaps always—based on metaphors that have a physical
and/or cultural basis" (18–19). Metaphors, then, "may create realities for us, especially social realities" (156).

Further, metaphor may help to justify inferences, to give sanc-
tions for actions, and to supply goals for future action (Lakoff
and Johnson 1980, 142). Thus the "chemical" metaphor that an
international student perceived in Americans' "solutions" to problems could be explored as an approach to problems that would
suggest finding methods that would "dissolve" present problems
for a period of time without "precipitating out" greater ones (144).
For Lakoff and Johnson, the conventional cultural metaphors that
structure our thought and actions are rooted in the "myth of objectivity," which has flourished in the rationalist and empirical traditions (195) and, I contend, underlies the privileging of the
epistemology of science as it is commonly viewed today—science
as unmediated observation of facts recorded in a transparently
referential language. Although "objectivity" in science was intended to eliminate what Lakoff and Johnson call "the effects of
individual illusion and error" (194), the problem that dominates
any consideration of science today is that in the West we take both
the metaphors and the myths of our culture as "true." Language
and culture create a dangerous double bind, for mythical objectivity not only "purport[s] not to be a myth, but it makes both myths
and metaphors objects of belittlement and scorn: according to the
objectivist myth, myths and metaphors cannot be taken seriously
because they are not objectively true" (186).

A basic element in the myth of objectivism (and in the corres-
ponding "myth of subjectivism") is again one of Eiseley's primary
concerns, the separation of humankind from nature. In a culture
dominated by Western science, "reality" is seen as external, objective reality; to study reality is to study the physical world. Thus
the separation of humankind from the environment is linked to the
quest for mastery over nature: "successful functioning is conceived
of as *mastery over* the environment" (Lakoff and Johnson 1980,
229). In our culture of objectivism those who "impose their metaphors on the culture" are privileged to create the definitions of
"truth" (160).

In their quest for an alternative view, Lakoff and Johnson suggest an "experiential myth," which sees humankind as part of, not

separate from, the environment. In a move that Eiseley anticipates in his whole project, especially in his later involvement with process philosophy, Lakoff and Johnson suggest that the experientialist myth "focuses on constant interaction with the physical environment and with other people" (1980, 230). Such a myth does not abandon science or the quest for a degree of objective knowledge, but it does suggest an awareness of the role of metaphor in thought and a shift in deeply ingrained attitudes. And abandoning the myth of absolute truth in science may lead to a "more responsible" practice of science and to a "more reasonable" sense of scientific knowledge and its limitations as a result of the "general awareness that a scientific theory may hide as much as it highlights" (227). Such an "experientialist" view of the world requires "developing an awareness of the metaphors we live by," acquiring "experiential flexibility," and attempting to see life "through new alternative metaphors" (233).

Similar awareness of the role of metaphor in shaping concepts, goals, and actions has been voiced, though not always with the powerful insights of linguistics that Lakoff and Johnson bring to the subject, over a number of years. Although the expected use of metaphor in the scientific text is consciously exegetical, as reflected in Niels Bohr's notion that sometimes "we can fully understand a connection though we can only speak of it in images and parables" (cited in Wechsler 1978, 6), M. H. Abrams wrote in 1953 that "even the traditional language of the natural sciences cannot claim to be totally literal," for "its key terms often are not recognized to be metaphors until, in the course of time, the general adoption of a new analogy yields perspective into the nature of the old" (1971, 31). The "facts" of a scientific understanding may be chosen and shaped by the figures that convey the understanding. "For facts are *facta,* things made as much as things found, and made in part by the analogies through which we look at the world as through a lens" (Abrams 1971, 31). In other words, "images generate ideas and ideas clarify images" (Gruber 1978, 133).

Richard Rorty argues that the image of "the mind as a great mirror," with "knowledge as accuracy of representation" of an external reality, "underlies much of our thinking" (1979, 12). If we accept Rorty's notion of the controlling role of metaphor in our thinking, then we recognize that as the "picture" or the metaphor of the world changes, so does the image of the human relationship to the world. Biologist Owsei Temkin notes that metaphors from science and technology are "not always mere figures of speech," but "integrating concepts" (1949, 185) used by biologists or psychiatrists or physicists and reflecting a larger general view of the world. Addressing the integration of metaphor and subject matter that underlies our basic concepts, physicist Roger S. Jones treats space, time, matter, and number—the foundational concepts of physics—as metaphors by which we structure the world. Some metaphors, argues Richard Boyd, actually "constitute, at least for a time, an irreplaceable part of the linguistic machinery of a scientific theory" (Boyd 1979, 359–60). Such "constitutive" metaphors include the notion of the brain as a kind of computer and the notion of consciousness as a kind of "'feeedback' phenomenon" (360).

Drawing on the role of metaphor as model, Jones contends that metaphor may go beyond its exegetical use "to extend theories and even to make new ones," continuing that "scientists (and indeed all who possess creative consciousness) conjure like the poet and the shaman," and that even scientific theories are metaphorical; intertwined with what we see as reality, such metaphors function conceptually to create for us a structured world, "the novelty and unlikelihood of order in the midst of chaos" (1982, 45, 49). Using the example of Darwin's tangled bank, Howard Gruber argues that metaphor may reveal "an intimate connection between visual and poetic imagery and productive scientific thought" (1978, 124). In such uses, the organizing power of metaphor is vital. Complex imagery can provide organization of information "in complex packages, schemas, or frames," which help to control behavior (136). Indeed, metaphors functioning as models in effect write the story of human thinking, according to Fredric Jameson:

> The history of thought is the history of its models. Classical mechanics, the organism, natural selection, the atomic nucleus or electronic field, the computer: such are some of the objects or systems which, first used to organize our understanding of the natural world, have then been called upon to illuminate human reality. (1974, v)

And Jacques Derrida argues that "metaphor is never innocent. It orients research and fixes results" (1978b, 17). The "impression" of the image may remain even if we forget the image itself (Rapaport 1983, 60–61). Remaining unknown to the conscious mind of the perceiver or the creator, image, metaphor, analogy can, then, provide a framework for understanding, shape results, and serve as a means of understanding for even the scientist who cannot physically see the waves or particles that he comes to view as complementary theories.

In view of such powerful insights into the cognitive functioning of metaphors and models, we see clearly the need to examine their uses in scientific texts. "When the understanding of scientific models and archetypes comes to be regarded as a reputable part of scientific culture," Max Black contends, "the gap between the sciences and the humanities will have been partly filled. For exercise of the imagination, with all its promise and its dangers, provides a common ground" (1962, 243). I shall explore the significance of these arguments in Eiseley's project as I turn to Eiseley's notion of metaphor and its use in his texts.

Eiseley and the Role of Metaphor

Eiseley is concerned with the role of language, especially of metaphor, in communicating scientific knowledge. Linguistic tropes are essential to science, he contends: "It is only by the hook of the analogy, the root metaphor, . . . that science succeeds in extending its domain" ("How the World Became Natural," *FT,* 20). Such tropes also serve as bridges of understanding from one field to another, Eiseley argues, for "it is the successful analogy or symbol which frequently allows the scientist to leap from a generalization in one field of thought to a triumphant achievement in another" ("The Illusion," *ST,* 274). And creating scientific analogies, he contends, is a creative act not separate from but identical to the creative act of the literary imagination: "Such images drawn from the world of science are every bit as powerful as great literary symbolism and equally demanding upon the individual imagination of the scientist who would fully grasp the extension of meaning which is involved. It is, in fact, one and the same creative act in both domains" (274).

Accepting the essential metaphoricity of scientific thought, Eiseley nevertheless expresses the scientist's ambivalence as he writes of the "escape" of such metaphorical concepts from the privileged, circumscribed scientific discourse community. For example, both evolution and the expanding universe have become "figurative symbols," which some may accept as if they could be proved or disproved experimentally. Yet, like Freud's unconscious, "such ideas frequently escape from the professional scientist into the public domain. There they may undergo further individual transformation and embellishment. Whether the scholar approves or not, such hypotheses are now as free to evolve in the mind of the individual as are the creations of art" ("The Illusion," *ST,* 277). As in the life world and in the world of knowledge, with such metaphors there is both gain and loss. Once outside the discourse of science, "as figurative insights into the nature of things, such embracing conceptions" are subject to a dialogue, and they "may become grotesquely distorted or glow with added philosophical wisdom" (277). Whatever turn they take, they may now evolve freely, and "all the resulting enrichment and confusion will bear about it something suggestive of the world of artistic endeavor" (275). Creativity and enrichment, distortion and confusion—metaphorical insights themselves echo the duality of human achievement and self-undermining.

At this point the term *metaphor* itself deserves further exploration in the Eiseleyan context. Eiseley is concerned with root metaphors; his encompassing metaphor is the journey—of evolution, of life, of man, of the individual, of knowledge, of the scientific institution, and of reading. *Metaphor* itself is vehicular, as Jacques Derrida underscores in "The *Retrait* of Metaphor" with a half-page of architectural symbols of vehicles from limousine to double-decker bus. Metaphor, he writes, "occupies the West. . . . *Metaphora* circulates in the city, it conveys us like its inhabitants, along all sorts of passages. We are . . . the content and the tenor of this vehicle: passengers, comprehended and displaced by metaphor" (1978a, 6).

Derrida's circulating vehicular philosopheme reappears in *The Post Card* (the card is from Socrates to Freud), in which the post card becomes a model of logocentrism. The post card, as Gregory Ulmer notes, exemplifies logocentrism because it represents teleology; the post card is "destined," and "the entire history of the postal *techne* rivets 'destination' to identity" (1985, 126–27). Derrida, however, uses the postal model to explore "the possibility that a letter might *not* be delivered," viewing the model "from the side of dysfunction" (143). In this view, the post card is a model for the age, with its teleological expectations, which Derrida's program is to deconstruct. The link between Derrida's interpretation of the "destining" that characterizes the postal era—the postal *techne*—and Eiseley's treatment of metaphor is the emphasis on discontinuity. Metaphor is a transfer, but, as we discover in Eiseley's texts, the transfer is not always exact: the root metaphor brings both gain and loss; both creativity and distortion emerge as metaphor "escapes" into the public mind. Thus root metaphors require reexamination and undermining.

Metaphor, from the Greek *metapherein,* "to transfer," has its roots in *meta-,* suggesting change, and *pherein,* from Indo-European *bher,* "to bear." *Metaphor* itself, then, is a method of moving from one point in space to another, or from one *eidos* to another. The method of metaphor is the metaphor of method.

Metaphor carries, bears (also "to bear children"), endures (for-bears), becomes or bears a burden, brings, shakes, confers, defers, differs, suffers, and transfers. Metaphor itself makes a journey. The metaphor of the journey is the journey of metaphor, which is the journey of striving to know. The journey of metaphor and of knowing is filled with chance, change, undecidability, indeterminacy, unexpectedness—by now familiar terms from Eiseley's oeuvre. Metaphor is not a journey involving ends or closure, whether of struggle or machine or knowledge. If we expect the journey to have a destination, we encounter the possibility that discontinuity will interrupt, that the post card will not reach its appointed destination.

Through the journey of metaphor and the metaphor of the journey, the means of knowledge become indistinguishable from the knowledge itself. The journey enters the epistemological dimension. Through the journey Eiseley—aware of sidetracks and detours—attempts to undermine (mis)understandings of science in the public mind as well as to effect the transfer of information from the discourse of science to the general reader. The journey is a transfer, but also a shuttle, weaving the texture of the text. The journey of wandering—as opposed to the postal journey (and how often does a letter wander, Odysseus-like, before reaching its destination?)—is a model for the way we read[3] or the way we learn—stopping, starting, sometimes moving without direction, occasionally lost, wandering like Odysseus and Darwin among the islands of knowing.

In Eiseley's view, the path of assumed objectivism, metaphorical at best, has been expanded into scientific myth. Although an analogy may be false, Eiseley writes in a passage that adumbrates Thomas S. Kuhn's notion of the "paradigm shift,"[4]

> yet so potent is its effect upon a whole generation of scientific thinking that it may lie buried in the lowest stratum of accepted thought, or color unconsciously the thinking of entire generations. While proceeding with what is called "empirical research" and "experiment," the scientist will almost inevitably fit such experiments into an existing comprehensive framework, an integrative formula, until such time as that principle gives way to another. ("How the World," *FT,* 20)

Having studied thoroughly the scientists whom he later calls the "dancers in the ring" (*ASH,* chap. 18)—the thinkers of Bacon's time and their responses to empiricism, the forerunners of Darwin and their difficulty in seeing outside their own framework of analogies, and the Darwinians and their tendency to emphasize struggle to the detriment of the larger view—Eiseley is, both in his explicit comments and in his textual methods, clearly aware of the scientific, social, and cultural importance of metaphor. For him, a primary task is to defamiliarize the old metaphors and to remotivate them, or to provide new insights through alternative metaphors. At times, as in the metaphor of the stage in *The Firmament of Time,* he defamiliarizes a metaphor, but then he deliberately resituates the renewed metaphor in a new context. The process of reexamining and defamiliarizing metaphorical understandings, I submit, is an important key to his literary method.

Defamiliarizing and Reshaping Metaphor

At issue in Eiseley's texts, then, is not the elementary explication of a theory so well known that it has become a root metaphor (an interesting parallel is Stephen Jay Gould's interest in the iconography of cartoons and advertising, which, he contends, give evidence of the accepted notions of the ladder or the inverted cone of evolution, both suggesting the nineteenth-century notion of progress[5]), but an attempt to explore and *engage* the theory, to internalize it in terms of the individual's response to other individuals and to the natural world that produced us all, to undermine and demystify misconceptions that often have arisen through the evolution of metaphor. If in Eiseley's view evolution itself has become a root metaphor, bringing with it important cultural baggage that both clarifies and distorts, his technique of defamiliarization focuses on culturally embedded metaphors that have arisen not only from Darwinian but also from the Cartesian-Newtonian views embedded in nineteenth-century evolutionary thinking, and on the method or epistemology of science, which has itself assumed metaphorical, even mythical, force.

The essay in its classical roots (in which the "weighing" or "assaying" of a thesis is important, as well as the "attempting" of such a task) is "a vehicle for exploration rather than demonstration" (Gerber and McFadden 1983, 21). The heuristic function of Eiseley's reexploration of metaphors reinforces the larger heuristic function of the essays. As models of struggle are replaced by models of cooperation, the possibility of exploring an *other* understanding is always present. Eiseley attempts to evoke the sympathy with and concern for the human relationship to the natural world that will lead to cooperation rather than struggle, to understanding of the "living screen" rather than dominance over an externalized inanimate nature.

The metaphors that Eiseley defamiliarizes are those that reinforce notions of struggle, decidability, determinism, fixity, reductionism, and teleology long dominant in Western thinking. Eiseley
sees the machine metaphor as retaining a powerful impact long
after its time, suggesting control of nature as well as a teleological
view. As alternatives to the machine metaphor, Eiseley explores
metaphors of magic, games of chance, the "magic" theater, and the
random journey—all of which suggest continuous change, indeterminacy, undecidability. He displaces metaphors of struggle and
hierarchy (both the power hierarchy, whether in nature or in human
affairs, and the hierarchy of fields of knowledge) with metaphors of
coexistence and symbiosis (especially, in cultural and intellectual
domains, of the symbiosis of science and art). Metaphors of reductionism and positivism are subverted by metaphors of magic, alchemy, and primitive shamanism. Thus the planned, ordered,
machine-like universe gives way to the "hazard" of a dice game.
The mingling of genetic traits that sexual attraction brings is the
shuffling of a deck of cards. The reductionism of Ernst Haeckel,
who proclaimed in the nineteenth century, before modern cell biology, that all mystery is annulled by knowledge of the components
of the cell, is displaced by the inexplicability of alchemy.

Eiseley reexamines the origins of "natural" as an essential root
metaphor in Judaeo-Christian thought, as well as its relation to
Newtonian mechanistic thought and the nineteenth-century merging of the two. To understand modern thinking requires a reexamination of the human fragmentation of knowledge, and of the
division into internal/external and man/nature. In this reexamination, Eiseley explores the role of the machine analogy and the
Western notion of progress, which incorporates both positive and
negative qualities, both "enrichment" and "confusion." In the
West, he notes, Christianity led to the notion of unreturning time,
which has in turn led "to that belief in progress and the uniqueness
of the historical process which has led on to the achievements of
modern science" (*LN*, 191), but has also led to a mythologized
notion of inevitable "progress" and to blind faith in a technology
that may destroy the planet. Thus humankind has become separated from the nature that produced it and increasingly dependent
upon machines.

An alternative view that Eiseley explores emerges from a tension
between the supposedly unmediated vision of science and the necessarily mediated vision of textuality. Instead of seeing humankind
as separate from and needing to "control" the environment, Eiseley
explores an understanding based, not on parts of the whole, but

on the whole as an interactive system. For this "experiential" or "interactive" understanding, the textual metaphor is essential— most effectively explored in the autobiography in terms of the artist and her materials. The textual metaphor, which I shall develop at greater length in chapter 7, signifies the human relationship to nature as one of reader and text, a relationship not of consumption, exploitation, or control, but of interaction, of interpretation, questioning, indeterminacy, and openness to the unexpected.

In displacing metaphors of struggle with by those of symbiosis, metaphors of reductionism with those of inexplicability, metaphors of teleology with those of the unexpected, metaphors of hierarchy of systems of knowledge with those of equality or symbiosis of science and art, metaphors of fixity with those of change, and metaphors of decidability with those of undecidability, Eiseley attempts to displace the caste system of cognoscenti and "others" and to displace the animal willingness to be manipulated in favor of a willingness to participate in a dialogue ideas. Gerber and McFadden compare Eiseley's questioning of intellectual schemes with Kierkegaard's: "What Hegelian thought was to Kierkegaard, Darwinian evolution was to Eiseley. Both men ask the question, Where shall *I* live in this embracing system?" (1983, 113) Eiseley, I would add, expands the question into epistemology: How shall I *know* where I shall live? The answer comes in an epistemological dialogue rather than in the "way" of science alone.

In reexamining the metaphorical embedded notions of fragmentation of knowledge, the belief in "progress," and the separation of the human from the natural, Eiseley reexamines "meaning" itself. His concern with "meaning" echoes that of Roland Barthes, his contemporary and also a student of society, nature, literature, and the system of signs. Responding to Hegel's account of the ancient Greek fascination with "the *natural* in nature," Barthes writes that ancient humankind "constantly listened to [nature], questioned the meaning of mountains, springs, forests, storms" and "perceived in the vegetal or cosmic order a tremendous *shudder* of meaning, to which he gave the name of a god: Pan" (1972a, 153). Since then, Barthes continues, "nature has changed, has become social"; everything natural "is *already* human, down to the forest and the river which we cross when we travel" (153). For Barthes, then, the meanings are less important than their creation. Positioned similarly between structuralism and the oncoming postmodern world, both Barthes and Eiseley, as students of society and signs, of humankind and the "nature" of meaning, return to the ancient "shudder" by which humankind creates meaning. In his reexamination of the basic assumptions of science, Eiseley, through the basic journey metaphor, leads the reader through a complex quest for a dialogue between the Baconian epistemology of observation and induction which serves as the foundation for modern science, and the intuitive, often nonconceptual, noninductive knowledge of poetry, between the "assertive" element of science (as the discourse of knowledge) and the "evocative" element of poetry (Ulmer 1985, 213), between the analytic, "separate" knowing of science and a synthetic, "connected" knowing. And the epistemological dimension of Eiseley's project is reinforced by his exploration of the metaphorical dimension of the physical means by which we know.

Reexamination of the Dominant Philosophemes

Essential to Eiseley's project of undermining and displacing metaphors that distort the understanding of the individual in relation to an evolutionary universe is his exploration of the organs of the human body that serve as dominant root metaphors or philosophemes—hand, eye, and tongue. In the *Lost Notebooks,* he outlines "How Man Came," a plan for a book emphasizing the human relationship to inanimate and animate matter; the second part is described as a "tale of man's organs, but emphasizing the real unity, the real taproot," which is "life as represented in the genetic chain" (*LN,* 108). All of life, Eiseley contends, is caught in "little intangible prisons compounded partly of the nature of the senses, partly of the mind that uses the senses" (113). The struggle of all life is to reach beyond these prisons, and the struggle of human life is to understand its prison (language itself appears in Eiseley's texts as a Nietzschean "prison house"). Precisely because of the neurophysiological link of the human brain and the organs of knowing, "for a moment all this endless light-year universe hangs in a little cup between the eyes and ears, so dreadfully is man composed" (112).

Anthropologically, hand, eye, and tongue are the organs that have evolved specialized uses that make further physical specialization unnecessary. Not only have these organs brought us to where we are culturally, but they are also the primary organs with which knowledge-gathering is associated. The human eye provides a depth perception which, with brain and hand, makes possible a conscious, directed coordination leading to hunting, tool-making, building, map-making, journeying in its multiple senses. The tongue and associated "speech" organs, through what linguists call their "overlaid" linguistic powers, have enabled humankind to displace the present for the future, to create an "other world" of human culture. And through writing, these organs have enabled humankind to break the time barrier, to read and write both past and future. Because of the biological and evolutionary importance of these organs, they dominate Eiseley's texts as they shape Western philosophemes. As shapers of the voice-ear, inside-outside, eidetic concepts, these organs constitute the epistemological circuits by which the human journey and the journey of knowledge are possible.

As tropes, these organs begin metonymically, but each also takes on metaphoric overtones in a kind of oscillation between Jakobson's "metaphoric and metonymic poles" ascribed to romantic and realistic texts, respectively. As Umberto Eco has argued, "metaphor can be traced back to a subjacent chain of metonymic connections which constitute the framework of the code and upon which is based the constitution of any semantic field" (1984, 68). Anthropologically, humankind is near the beginning of the journey, despite popular perceptions to the contrary. In Eiseley's epistemological journey, anthropology itself is the intersection of biological and cultural knowledge. And Eiseley explores where humankind is, how we came to be there, and how we can consciously become more than we are. The epistemological circuits provide a vehicle for both explanation and exploration, as well as for defamiliarizing misconceptions and displacing misleading metaphors.

"The body-self," writes Thass-Thienemann, "has set the primary patterns for all subsequent understanding of the world" (1968, 211). Eye, hand, and tongue—as the organs of internalizing (visualizing), grasping or probing (taking to oneself or reaching beyond oneself), and containing (language as a "container" for "ideas")—are basic physical codes of knowing; as codes, they evoke unconscious responses. The intelligent general reader may not think of the hand as a root metaphor for concept, but the reader knows what it is to reach, probe, grope, grasp. The reader is unlikely to be aware, as are Derrida and Rorty, for example, of the dominance of the visual in Western philosophy, but she finds light/dark, day/night, inner/outer oppositions comfortably comprehensible.

Further, with hand, eye, and tongue a dual image of humankind as self-creating and self-undermining emerges in Eiseley's texts; for these organs have dual capabilities in their cultural roles, and the essential metaphoricity of language provides dual meanings for their functions. The hand is capable of cooperation and destruction; there are "two eyes"—of summer and winter, of harshness and gentleness. The tongue is a "little member" with great potential for both positive and negative effects. The positive cultural possibilities of these dualities are undermined by the seeds of their own negation. Eiseley's emphasis on these dualities or tensions is part of his project of remotivation; he defamiliarizes the standard anthropological qualities that separate humans from other animals, displacing the standard theme (embedded in the philosophemes themselves and emerging out of the mechanistic notion of the universe as absorbed in nineteenth-century organicism and even into evolutionary thinking) of mastery over the environment, over the "natural," in favor of the ancient ability of life to "grasp" or to "reach" for something, to be more than it presently is.

HAND

The hand is traditionally the metaphor of concept because it grasps, holds, gathers together, speaks to the notion of thinking as having.[6] *Concept* itself derives from the Latin root *concipere,* "to take to oneself," "to take into the mind." Ultimately it derives from *capere,* "to take or seize," which derives from the Indo-European root *kap,* "to grasp." The hand is a model for reaching, probing, grasping, touching (tactile internalizing of information), and for gathering together or collecting. James Bunn suggests that the asymmetrical form of the hand gives a freedom of movement and a precision of grip that enable it to function as a model of grasping but also lead semiotically to evocation (1979, 48–49). Eiseley's suggestion is just this evocation, this reaching and drawing out.

For Eiseley, the hand further suggests the means of reaching, of probing, into the universe. Its ability to probe and its link with the eye make the hand an intermediary between the "chemical senses" of taste and smell, through which we experience stimuli without idealizing or theorizing them, and the "theoretical" or "idealizing" senses of sight and hearing, which transform external stimuli into mental images.[7] The hand, then, is the organ of a spatializing philosopheme. The sense of touch is both abstract and concrete, both spatial and idealizing. It is a means of making tangible the abstract. The word *tangible* itself derives from the Latin *tangere,* "to touch," from the Indo-European root *tag,* meaning "to touch" and also "to set in order," thus linking the hand and the mental ability of classification.

Bunn notes the importance of the evolutionary aspects of neurophysiology in semiology, suggesting that "the sensory-motor qualities of 'handling' are crucial for the modeling of discoveries" (1979, 47). He distinguishes between the abilities of the crafting hand to devise "cutting tools" and "combining tools" and links handling of objects with classifying and thinking about objects (47, 49). Bunn notes that in classical mechanics space and time are separated, as are subject and object in classical philosophy, with classical language revealing "both kinds of separation" (54). In primitive art, for example, a "spatialized form of abstraction—in Latin *abstrahere,* 'to drag away,' 'to divert'—forms the basis of substitution magic," yet the earliest humans seemed "more oriented toward the temporal process of abstraction: tools and signs were implements of planning, of anticipation magic, where making the implement deferred immediate gratification in order to concentrate on a future goal" (52).

The relevance of Bunn's comments in the present context is the parallel between what the hand *does* and the mental activities that follow these manual operations. In his metaphorizing the hand in terms of reaching out or of life's quest to be more than it presently is, Eiseley attempts to displace the modern metaphorical emphasis on tool-making as a means of control in favor of the primitive sense of displacing immediate gratification in order to achieve a goal. Not only is the hand associated with conceptualizing, but it is also the organ by which cutting and combining take place, as well as classification, measurement, and ordering of mass, surface, and weight. "Hands, not eyes," writes Bunn, "define the phenomena of mass, weight, surface, and volume" so that they are directly related to the process of planning, ordering, and classifying (1979, 49).

Further, the hand in Eiseley's texts is the heuristic philosopheme of groping, probing, reaching to be more than it presently is— hence of the whole journey. On the notion of groping, Bunn cites Henri Focillon's "In Praise of Hands," in *Life Forms in Art,* in which the Centaur tells of discerning objects through feeling and of lifting his hands to feel breezes (1979, 48). The Centaur's "groping," Bunn suggests,

> is the most common metaphor for the prelude to discovery. In the dark, a trial and error of the hand feels its way to understanding: touching, spanning, grasping. However, as a metaphorical state of incipience, "groping" is a condition that one longs to replace with some larger law in logic, geometry, and semiotics, so that everyone need not repeat the particular instance of an apple falling on one's head. Groping is that wordless state, fraught with expectation, prior to the successful articulation of word, phrase, or formula. (48)

In the multiple senses of the journey, the hand is the essential metaphor for this "incipience" and "expectation." The reason the hand is so apt "as a latent metaphor for the discovery of signs" is that it "signifies what it does: it gropes for and grasps other probes, and it reaches, in the fashion of Tantalus, for intangible connections with the whole" (48).

The hand is often referred to in terms of its anatomical capabilities: it is "prehensile" or "grasping"; it is capable of making tools, of chipping a flint or wielding an ax or designing and constructing shelter. By catachresis, the hand is an emblem of humankind, and for Eiseley it is a symbol of the whole human journey: "It is not a bad symbol of that long wandering . . .—the human hand that has been fin and scaly reptile foot and furry paw" ("The Slit," *IJ*, 6). But it is also a figure for the human reaching out to other beings, out of the compassion poignantly expressed in the mad "star thrower's" attempts to rescue living creatures and return them to the sea or in some unknown builder's labor to prepare the abandoned shelter that the troubled Thoreau described in his journey. And it is a figure for the human expression of self and of culture through art, history, and science.

EYE

The eye provides a model for internalizing or "filling," as Lacan notes in regard to the physical response as light reaches the eye (1981a, 94). Like the other organs of knowledge that take on metaphoric functions, the eye begins metonymically. It represents not only the simple visual function, but the whole problem of observation and cognition. Not only is the eye the organ of the philosopheme of "idea" (from the Greek *eidos*, "form"), but it is a means of exploring the problem of knowing. As the means of human observation and participation in the world, the eye is the center of the light/dark, inner/outer oppositions that are basic to Western thinking. The link between sight and hearing, the "idealizing" philosophemes, is evidenced by the similarity in derivation of Germanic words for *eye* and *ear*. Words for *eye*, such as the Old English *erage* and the German *Auge*, begin with a diphthong likely derived from words for *ear*, as in the Gothic *auso* (Thass-Thienemann 1968, 210). And in his stress on the neurophysiological link of eye and hand in the process of evolution, Eiseley frequently returns to the link between eye and hand in the emergence of writing (hieroglyphic or alphabetic).

From the "terrible crystal" of genius to the "eyes on stalks" of poets, the eye is both physical reality and metaphor, both a means of observing and the focal point for a series of metaphors that suggest the eye's role in shaping observation. Strongly linked to the Western metaphysical tradition that unites light (the sun) and the sojourn, light is essential to the journey. Indeed, the Irish language uses the same word, *suil,* for both *sun* and *eye* (Thass-Thienemann 1968, 262). Through visual images or the "inward eye" there emerges an "inner" world. Indeed, writes Eiseley, "the world cannot be said to exist save by the interposition of that inward eye" ("The Star Thrower," *UU,* 88), a continuation of the Baconian division into "inner and "outer." The Eiseleyan eye, however, is not a source of transparent seeing; its vision is influenced by a mixture of genetic proclivities and cultural influences. Eiseley remains within the visual imagery that leads Derrida to suggest the impossibility of "wean[ing] language from exteriority and interiority," which are actually "embedded . . . at the very heart of conceptuality," with a resulting "original and irreducible" equivocality in the Western languages (1978b, 112–13). And he remains within the vocabulary that Richard Rorty links to seventeenth-century visual imagery (1979, 6, 369). Yet he displays an awareness of the tensions implicit in this vocabulary, and he attempts to undermine the accepted, "stable" referentiality of the vocabulary as he demonstrates that this vocabulary, like the world and the languages that describe it, participates in the continuous shifting of meanings that parallels the continuous shape-shifting of the natural world.

Further emphasizing equivocality, the eye is also a homophone for the "I." Eiseley's use of autobiography, his scientist's awareness of the observer-participant, and his dramatizations of events from his own past as well as from the human past lead to an omnipresent "I" who is also an eye. For the eye and textuality are inextricably bound together, inescapably equivocal. As Mary Ann Caws suggests, "the point is not the text in the eye, but the I too, in the text itself. From this perspective, this dual-faced image should illuminate the double textual interest . . ." (1981, 35). What Caws calls an "obsessive desire of the onlooker to be included in the scene" brings about "the constant addition of the 'je'" (35). The "I" of the statement, as in Jakobson's concept of the shifter, becomes commingled with the "eye" and the "I" of the observer. The "I" as the "I/eye" of all humankind brings an extension of perception into the reader's domain, as the reader both creates and participates in the text.

The eye itself, Jacques Lacan suggests, is both a receptacle and a labyrinth (1981a, 93). (In the association of sight and hearing, the "idealizing" philosophemes, the ear, too, is —at least contains—a "labyrinth.") The light that the eye receives "may travel in a straight line, but it is refracted, diffused, it floods, it fills—the eye is a sort of bowl—it flows over, too, it necessitates, around the ocular bowl, a whole series of organs, mechanisms, defences" (94). Thus in Lacanian terms the ambiguity of the visual philosopheme is inescapable: "the relation of the subject with that which is strictly concerned with light seems . . . to be already somewhat ambiguous" (94). Both the Lacanian ambiguity of "light" and the eye as a receptacle are relevant to Eiseley's exploration of the relationship between sight and meaning, for the visual philosopheme itself reflects the thematic indeterminacy and ambiguity explored in the texts. Further, the eye as a receptacle parallels the functions of the other two organs, which suggest an epistemological taking to oneself, or internalizing—a grasping or groping for meaning.

Like the other dominant philosophemes, in Eiseley's texts the eye also carries the thematic ambivalence of the self-undermining of human knowledge. There are the "two eyes" of microscope and telescope, the "double vision" of a fish and a poet, and the two ways of seeing—two "eyes"—at Walden Pond. As overt metaphor and as philosopheme, the eye continues Eiseley's displacement of notions of fixity, truth, and reality for the shifting, unfixed, uncertain forms that are only illusions of the time dimension and that further displace notions of science as unmediated and fixed or unerring observation. Like the hand that reaches but often only probes or explores what is ungraspable, the eye may see only to find the vision ambiguous as the mirror or the kaleidoscope shifts. Thus Eiseley's epistemological exploration leads to a realization that knowing is not-knowing, that the world is process rather than fixity, that the world as text, like the written text, undermines itself. Through the dialectic of "two eyes," a tension appears that parallels the similar tension of the hand's dual capabilities; the two eyes suggest alternating perceptions and occasionally suggest the synthesis that Eiseley would propose.

On the most basic level, the figurative eye is a metonymic extension of seeing. Both telescope and microscope appear repeatedly as "eyes." Quoting Thoreau's question in his *Journal,* "Who placed us with eyes between a microscopic and telescopic world?" ("The Mind as Nature," *NC,* 216), Eiseley explores mechanical extensions of the eye as means of further extending human "vision," though he always returns to poetry as the source of an alternative extension of "vision." The metaphorical eye of humankind on Mount Palomar began metonymically, he suggests, in the eyes of myriad creatures: "A billion years have gone into the making of that eye; the water and the salt and the vapors of the sun have built it; things that squirmed in the tide silts have devised it" ("The Great Deeps," *IJ*, 45). This mechanical/metaphorical eye is a palimpsest of evolving sight apparatuses. It haunts the observer who remembers the first seeing apparatus and is reminded by a frog's eye "warily ogling the shoreward landscape" of "those twiddling mechanical eyes that mankind manipulates nightly from a thousand observatories" (45).

This awareness of the palimpsest of eyes is, for Eiseley, more important than mechanical extensions of vision, for "the most enormous extension of vision of which life is capable [is] the projection of itself into other lives. This is the lone, magnificent power of humanity" ("The Great Deeps," *IJ*, 46). This metaphoric extension of vision he later calls "widening the eye of the world" (*ASH*, 235) and suggests that it can be accomplished by unwinding a snake from a pheasant or by the communication of Odysseus and the dog Argos through the recognition of life across the boundary of forms. In a metatextual reference, he suggests that this extension of vision can be accomplished by art, by the "timeless eye" of the essayist, whose vision can "reduce us to minuscule proportions" (*ASH*, 155). Such a metonymic extension of vision, which becomes metaphoric, is clearly a poetic, mediated vision that, ideally, serves as a balance to scientific vision and displaces notions of perfect, unclouded, or unmediated vision.

TONGUE

In his outline for "How Man Came" in *The Lost Notebooks*, Eiseley's proposed third chapter, "How He Found Language (His Tongue)," links human identity to language. The tongue in its catachrestic sense of language suggests containing. For Eiseley, this internalization occurs neurophysiologically because of the "remarkable" human brain "linked by neural pathways to . . . [the] tongue" ("Science and the Sense," *ST*, 196). Like other twentieth-century thinkers deeply impressed by the epistemic shift from being to language, Eiseley perceives the tongue as a metonym for the most significant human ability. Thinking, even awareness of existence, is inseparable from language. As thinkers such as Ernst Cassirer and Benjamin Whorf have suggested, language shapes our perception of the world and our understanding of it. Cassirer links language and science as "the two main processes by which we ascertain and determine our concepts of the external world" (1971, 907). Eiseley's texts, as is especially evident in the *Notebooks,* are increasingly preoccupied with what Thomas S. Kuhn calls the "mutual accommodation between experience [Kuhn's term for 'the world'] and language" (1981, 418). Eiseley links language to the creation of value or meaning out of nothingness (explicitly developed in "The Star Thrower," which I address in chapter 7) as he finds in language the source of an "invisibly expanding universe which man had unconsciously created out of nothing" ("The Invisible Island," *UU,* 166). In this approach he parallels thinkers such as Heidegger, Barthes, and Bachelard—all of whom are similarly concerned with the relationship of science and language, with experience and understanding.

Anatomically, the tongue is the philosopheme of inside/outside; it can be seen as a model for internalizing through its paradoxically internal-external action. "The tongue," Thass-Thienemann suggests, "is an external part of the body as well as an internal one," thus linking internal and external perceptions of the body-self and invoking both anatomical and psychological realities (1968, 247). Associating the tongue with licking (the Latin *lingua* should have been *dingua* except for a possible fusion with *lingere,* "to lick"), Thass-Thienemann finds "a blending which fused the objective anatomical reality of the tongue with the subjective experience of 'licking'" and continues to associate the tongue with "licking" flames and their association with the sexual, as in the German *Brunst,* "fire, sexual drive" (247). Exploring the tongue as the philosopheme that opens a new world for humankind, Eiseley treats language as the first mediator of the individual's relationship to the world.

Among the organs of the philosophemes, the tongue has a unique position. Not even the eye, says Thass-Thienemann, is linked to meanings as widely as the tongue. The use of *tongue* for languages, he suggests, is not just metonymic, "not simply the anatomical reality in question. It is rather the subjective kinesthetic feeling of the motion of the tongue while speaking" (1968, 247). The kinesthetic is similarly at the base of Barthes's "aesthetic of textual pleasure," which would involve *"writing aloud,"* a "vocal writing"—emphatically not speech—that stresses emotions in an "articulation of the body, of the tongue, not that of meaning, of language" (1975, 66–67).

The tongue is traditionally linked to creativity, especially in Western cultures based on the Word of Creation, the hierarchy of voice over writing that Derrida attempts to undermine. Thass-Thienemann suggests that the complex of meanings surrounding the tongue "absorbed the attributes which belonged originally to the creative and generative spirit" (1968, 249). Thus "the creative word appears as a primary act of the generative spirit," which in early thinking was assumed to precede existence (Thass-Thienemann 1968, 249). Therefore, the word—the unmediated presence of truth or reality—is related to the Presence of the Word. In primitive societies cutting out the tongue, like castration, was a means of "depriving man of his generative power" (Thass-Thienemann 1968, 250). The tongue, then, is an emblem of the traditional creative power, but it also carries the suggestion of the human ability to destroy. Like the other philosophemes, it is a two-edged thematic sword.

The tongue as catachresis for language participates in Eiseley's continuing exploration of the relationship between the senses and "sense" as meaning. But *sense* is also related to the meaning "direction" (from the Latin *sentire*, "to go mentally") that is important in Eiseley's epistemology of journeys. The mental journey is dependent on, mediated by, language—the "inner speech" of psychoanalysis. Speech itself is, in linguistic and anthropological parlance, an "overlaid function"—that is, the organs used for speech, with the exception of the vocal cords and the brain centers that make speech possible, have other functions actually necessary to the survival of the organism. Speech is a step removed from necessity, hence an "overlaid" function. Language, Eiseley writes, is the source of "a unique superstructure," a "superorganic world," constructed upon a biological base (*LN*, 152). Since language functions as a metonym for culture, it is significant that culture, too, is an "overlaid" function—an addition that has altered humankind and the planet we inhabit. Any consideration of language necessarily involves an interrogation of the relation between the two.

Eiseley's interrogation of speech as an "overlaid" function parallels his exploration of the other two philosophemes. Speech, he writes, is the first great tool for "cutting up and delineating environment and time" (*LN*, 50), thus paralleling the functions of the tongue and the hand. In response to Garet Garret's statement that "man reached his hand into emptiness and grasped the machine," Eiseley asks, ". . . but did he not equally grasp the first tool in ancient days, and to do so did he not have to conceptualize them with another tool?" (*LN*, 189). The word that best expresses the Eiseleyan "reaching" for knowledge by means of this conceptualizing, parallel to the hand as the organ of reaching, probing, and grasping, is Bunn's synesthetic use of *groping* in relation to language: groping as "that wordless state, fraught with expectation, prior to the successful articulation of word, phrase, or formula" (48). In Eiseley's terms, articulation is seldom successful, often tentative, but what is important is the groping itself. Eiseley's explorations of the voice-ear circuit are interlinked with his explorations of hand and eye—just as the organs themselves are interlinked with the brain in the neurophysiological development of man, just as hand and eye work together with tongue and brain in writing that preserves cultures.

The effects of language are linked to Eiseley's exploration of a linguistic concept, displacement, which he explores through multiple senses. With the beginning of language, Eiseley writes, the human being gained "the uncanny power of symbolically reworking his surrounding environment in his head," the power to "displace or transform the existent world for a prospective emergent reality reoriented in the mind"; significantly, though, this reality requires, "through speech or writing, the aid of other individuals" (1971a, 6–7). Eiseley defines *displacement* in terms of what it makes possible: it allows humankind "to make use of the imaginary in order to control reality," to "talk about what is absent" (145). (As Bunn suggests of the hand's abilities, linguistic displacement makes possible postponement of the immediate in consideration of the future; see Bunn 1972, 48–53.) The positive thematic side of linguistic displacement, then, is its contribution to the cooperation without which human culture could not have evolved. And the human capacity for wonder—both awe and curiosity—is tied to language. The tongue, like the eye, both shapes and limits understanding. It makes possible the exploration of both the "inner" world of consciousness and the outer world. Like Barthes, who maintains that "the exploration of language, conducted by linguistics, psychoanalysis, and literature, corresponds to the exploration of the cosmos" (1972b, 167), Eiseley correlates language as heuristic and the human relation to the cosmos (this point is particularly evident in Eiseley's critique of the quest for space in *The Invisible Pyramid,* which I treat in chapter 8). Language not only links present, past, and future, but it also offers an alternative to the violence of Darwin's tangled bank, through the image of the gentleness of early human beings who poured gifts of flint into the grave of a loved one and talked of that loved one's destination ("How Natural Is Natural?" *FT,* 180–81). Language remains the vehicle of wonder, understanding, and cooperation.

Yet the converse side of linguistic displacement, like the physical displacement of the human from nature, is that of loss. For Eiseley, the tongue not only speaks the distinguishing ability of the human being, but it is also the source of increasing separation of humankind from nature. If language enabled humankind to survive in and to manipulate nature, to create a "second world" of culture, it also made possible the technology that would allow humankind to devise a system for leaving the earth itself and, in focusing on that quest for space, to ignore the interdependence of all life forms ("The Last Magician," *IP,* 144). If, through symbolic communication, the human being can exist "at least partially within a secret universe of his own creation" (*DC,* 120), the price is a doubling of displacement—a separation from other forms of life and loss of sensitivity to the environment, a separation that has redefined the "natural" in terms of what can be manipulated, understood, and used rather than in terms of a whole to which humankind belongs. Yet another tension that linguistic displacement brings is the ability of language—itself unfixed and changing—to fix a concept, to freeze a definition of the "human" at a given stage, and thus to interfere with cultural evolution. Language is thus a means of confinement, metaphorically both an island and a prison. Language nevertheless provides a means of escape through writing, but writing, too, has a dual capacity: it can help create the future for good or for ill, and as a record of the human quest for power it is "man's devastating power to wreak his thought upon the body of the world" ("The Long Loneliness," *ST,* 43).

In treating the tongue, Eiseley struggles between the traditional privileging of speech (truth, science, logos, logic) and privileging writing, which is associated with ambiguity, with the uncanny and the unexpected, with both poison and cure, as Derrida has noted.[8] Eiseley privileges "reading" as a metaphor for the way we make sense of the world. Thus he speculates that language may have originated in the human "reading" of nature, in the quest for meaning, in what Gombrich would call "making" meaning before "matching" meaning to nature. Without denying the traditional notion that writing emerged as a device of "practical economics," Eiseley suggests "the possibility of an earlier phase involved with oracles and the interpretation of material markings increasingly 'read' by man"—that is, primitive readings of the scapulae of animals or of markings on shells—he speculates that perhaps "this unconscious search for symbols . . . played a 'natural' role in the creation of writing." Thus humans began in other words to "read" nature, "and reading nature involves other things than economics"

(*LN*, 176). In this speculation, "natural reading" would lead to writing, and writing would lead to language.

In many of his texts—"The Star Thrower" (*UU*), "The Golden Alphabet" (*UU*), "The Running Man" (*ASH*), for example—Eiseley clearly privileges writing. He notes the power of humankind to create a future through the written word, as Shakespeare's witches in *Macbeth* exteriorize a self-fulfilled prophecy ("Instruments of Darkness," *NC*, 51). Life itself, Eiseley contends, cannot be comprehended from its elements alone—by scientific observation alone—and to express this mystery, he turns to the time-transcending mystery and ambiguity of writing/reading: "Finally, as the greatest mystery of all, I who write these words on paper, cannot establish my own reality" (51). The self-referential text and the self-referential author are equally insubstantial; the substantiality/insubstantiality of the text itself exemplifies the greater mystery. Like life, which bears the seeds of its own destruction, or the text, which contains ambiguities that undermine it, the thread of speech/writing/reading is subject to a tension that runs throughout Eiseley's texts. Writing becomes for Eiseley "a more fruitful heuristic" than speech, as John Clifford has noted (1986, 6), because writing echoes the ambiguity that Eiseley finds in all of life and all of knowledge. As Eiseley writes in the autobiography, the wonder of written words is that "they drift on and on beyond imagining" (Eiseley, *ASH*, 170). And in privileging the attempt to "read" nature (in the sense of creating through reading), Eiseley privileges a semiotic dimension. Thus in the heuristic of the tongue the texts explore the relationship of language, reading in its multiple senses, and the semiotics of this inside-outside organ.

Epistemologically, the tongue shares another problem with the other organs of the dominant philosophemes: the elusiveness of knowledge itself. The principle of linguistic displacement underscores the shape-shifting of verbal knowledge, which is parallel to the shape-shifting of visual images or of the tangible world that the hand grasps. As the "unexpected," to which we as humans owe our very existence, is to the universe, so is ambiguity or equivocality to language: "Anything achieved by man has been created first by words, and words . . . partake of human ambivalence" (1971a, 9). Language, in Eiseley's project, functions—almost in a postmodern sense—as the means of both understanding and misunderstanding. Like the "meaning" that it is assumed to carry, language is capable of deconstructing itself.

In Eiseley's own strategic displacement—of misleading metaphors and misconceptions—linguistic displacement is part of the "indetermination" that enables humans to escape from evolutionary struggle and determinism. Eiseley also uses the tongue in displacing the notion of direct or unmediated observation of nature or unmediated creation of ideas. Because of language, humankind is no longer able to accept the universe as given, for "it has to be perceived and consciously thought about, abstracted, and considered" ("The Unexpected Universe," *UU*, 32). And the tongue as metaphor also displaces metaphors and notions of fixity in meaning. For if the tongue produces both "the blessing and the curse," Eiseley considers this observation "just as germane to the field of science, for science is also subject to the frailties and fallibilities of human endeavor" (1971a, 9).

Eiseley contends, however, that ambiguity can be as useful as the "unexpected" to the sensitive individual. If all human beings are "castaways" on a "world island," as he writes in the beginning of *The Immense Journey* (*IJ*, "The Slit," 14), still, as Blanchot suggests, language is what drives the human into exile (Josipovici 1982, 15). If the model for the Eiseleyan journey is the wandering of Odysseus, then the unexpected and the equivocal are to be accepted, even celebrated. The tongue itself helps displace notions of the fixity of knowledge. Tied to the use of language, which is essentially ambiguous, knowledge cannot be fixed. In his exploration of the relationship of knowledge and the senses, Eiseley finds that the shift of knowledge is a shift of the kaleidoscope, a shift or a gap in what are often assumed to be fixed understandings.

More than the other organs, the tongue is problematic in the metaphorics of a visually oriented artist, yet it contributes a constant tension between the visually ambiguous and the verbally ambiguous. In his own early experiences, Eiseley admits the pull of visual impressions, left by his deaf artist-mother; but he recalls, too, that his father, who had once been an itinerant Shakespearean actor, "in that silenced household of the stone age—a house of gestures, of daylong facial contortion—produced for me the miracle of words when he came home" (*ASH*, 22). His problem is metaphorizing language or the tongue without visual associations. As he oscillates between visual and verbal philosophemes, Eiseley uses the tongue to center on the temporal dimension that language brings. Hand and eye lend themselves to visual, spatial metaphors. The tongue introduces the temporal dimension—hence the metaphors of "invisibility," with language portrayed as "spores of thought," as an "invisible island," as essential to creating an "invisible pyramid" of technology. Eiseley's attention to the tongue develops as he explores the theme of the human escape from evolutionary struggle, which is linked to communication. He separates language from conscious achievements such as the discovery of fire and emphasizes the neurophysiological link of tongue and brain. Although the manifestation of language is cultural, its potential, he argues, is genetically coded,

> written into the motor centers of the brain, into high auditory discrimination and equally rapid neuromuscular response in tongue, lips, and palate. We are biologically adapted for the symbols of speech. We have determined its forms, but its potential is not of our conscious creation. . . . Speech has made us, but it is a human endowment not entirely of our conscious devising. ("The Angry Winter," *UU,* 115)

This passage suggests a doubling of the code of "writing" that echoes the notion that language derived from human attempts to read the codes found in nature itself, and if the code of language is "written" into the brain, then this "writing" underlies language.

In explaining the human ability to rework the world by transforming exterior reality, as well as the ambiguous dual consequence of this transformation, Eiseley turns to a biblical allusion for the ability of the tongue, recalling that the writer of the epistle of James "observed long ago that although the tongue is a little member it sets the course of nature on fire" (1971a, 7). The biblical allusion further develops the tongue's duality and introduces its link with indeterminacy as he continues:

> James fully recognized the ambivalent and frightening shapes that can be summoned up by language when he remarked that the tongue "can no man tame" and that it was capable of loosing deadly poison into the world. Thus, though man's full life is acquired and his culture and institutions transmitted by the word, speech has not been an unmitigated blessing. Men can distort or manipulate the meaning of words. (7)

The tongue thus is an unpredictable, uncontrolled element of human culture that parallels the brain's insertion of indeterminacy into evolution as it inserts indeterminacy into culture.

In Eiseley's exploration of metaphors related to the tongue, he suggests the definition of language, the source of language, and the creations of language. The temporal rather than spatial quality of language leads to metaphors that suggest movement, process, continual change—all qualities more easily associated with the temporal than the spatial domain. Language is defined as a "world of streaming shadows" ("The Dream Animal," *IJ,* 121) in a metaphor that describes man as a "dream animal" and his "other world" of culture and evokes the flickering shadows of what we know as reality as seen on the walls of Plato's cave. The Platonic intertext, however, in this context does not evoke an ideal world, but the unreliability of language, which may seem real or fixed but is subject to constant change. Writing, too, which links language to the visual/spatial realm, is unfixed and subject to individual interpretation. Thus the tongue is always linked to movement and process. Similar "shadows" appear also in a complex metaphor based on the dialectic of negentropy-entropy in which the human being is compared to a water strider dancing over the surface tension of water. Eiseley suggests that, as the water strider manages to avoid breaking the tenuous film of the water's surface, the human being also precariously "dances upon shadows," which are linguistic shadows ("The Invisible Island," *UU,* 154). The added precariousness of the insect's dance on the water's surface further reinforces the metaphor's suggestion of language as unfixed and constantly changing. Entropy is more likely than negentropy, and the surface tension of the water is easily broken, just as the surface order of language easily dissolves into ambiguity.

Eiseley's most elaborate exploration of language is "The Invisible Island" in *The Unexpected Universe.* In this essay, which I shall address in greater detail in chapter 7, he links language, itself indeterminate, to the quintessential indeterminate journey, that of Odysseus. The human being is a "dream animal" in *The Immense Journey,* and Eiseley focuses extensively on language in *The Unexpected Universe,* on language and its relationship to the future that we create in *The Invisible Pyramid,* and on language and its relationship to knowledge in *The Star Thrower.* The language-created world is also linked to the journey, for it has made the human being a "wanderer" who exists "between an instinctive mental domain he has largely abandoned and a realm of thought through which still drift ghostly shadows of his primordial past" ("The Lethal Factor," *ST,* 256–57). Eiseley equates culture with order, with the creation of structures—dependent on verbalizing—that replace instinctive guidance, though at the expense of suspicion toward other tribes who speak other tongues. In the inner world the human being is especially subject to uncertainty; she is both within and without time and history. The metaphor of the "other world" continues the suggestion of a spatial and visual domain created by the organs of the dominant philosophemes—a domain that, like language, can be changed or enlarged only through the mediation of art and history. The tongue, though a source of duality, is also the means to the understanding that may preserve humanity.

The tongue, then, as the organ of the voice-ear philosopheme and articulator of the "word," is the source of both metonymic and metaphorical explorations in Eiseley's texts. As a metonym for language, which in turn is a metonym for culture, the tongue is associated with metaphorical explorations of intersections in the human journey and the journey of knowledge. Language mediates and shapes what we learn, though Eiseley finds writing a more profitable heuristic than language. But like humankind, language contains the serpent and the bird, the force of destruction and the force of freedom. As the philosopheme that partakes of both inside and outside, the tongue is the source of unfixed temporal and spatial metaphors that reinforce the undermining of fixity, decidability, and determinacy.

Conclusion

Eiseley's view of knowledge as self-deconstructing and his recurrent theme of "growth" and "departure" reflect the scientist's philosophical struggle to know. The struggle between the mediated vision of poetic or connected or dialogic knowledge and the "truth," or supposedly unmediated knowledge, of science, is central to the Eiseleyan view. What the Russian statistician and philosopher V. V. Nalimov calls the scientific "stereotyped vision of the world," emerging from the language of determinism, causes us to see the world in terms of cause and effect, of logic, of human and natural "mechanisms," of hierarchies, of human separation from the environment. "Within this vision," Nalimov contends, "a human being is nothing more than a block of matter which became so sophisticated that it managed to master the logic built into the foundation of existence" (1982, 3–4). Eiseley clearly recognizes the problematic of understanding that we have inherited from the nineteenth-century view of an "objective" world that could be experienced without mediation. Such positivist views of "truth," "reality," and "representation," as historians of science have demonstrated (see chapter 1, p. 21), are often dependent on what Eiseley calls "styles" of thinking, which emerge in science as in any human endeavor ("Strangeness," *NC*, 140).

Basic to Eiseley's project is what Stephen Jay Gould calls "debunking"[9] of myths of scientific understanding. Science, as both Gould and Eiseley claim, should not belong exclusively to an elite cadre of specialists, but should be accessible to thinkers of any specialty; indeed, major scientific insights have often come from nonspecialists and amateurs. Darwin himself was strongly influenced by the literary naturalists, and Eiseley argues that the role of the amateur in the emergence of science is not to be discounted (*DC*, 13). Eiseley's essays cross borderlines and engage the question of interdisciplinarity in subject and in metaphor. They are a blend of literature with social, historical, and scientific history and criticism, philosophical reflection, and increasingly frequent use of Eiseley's own literary readings. But the vital element that makes Eiseley's whole project worthy of reexamination is his questioning and demythologizing of the Cartesian-Newtonian-Lockean paradigm and its impact on scientific thinking, particularly through his reexamination of dominant metaphorical concepts.

3

The Need for the Hybrid Genre

Darwin did not make the world "scientific" in his own terms
even if he thought he did. He made it poetic, saved it from
"induction." . . . He had made it a new symbol, freer, open at
the ends—a symbol of indefinite departure.
—Eiseley, *The Lost Notebooks*

As an interdisciplinary hybrid, Eiseley's genre—which has established a precedent for the work of others such as Lewis Thomas,
Stephen Jay Gould, Fritjof Capra, Ilya Prigogine, Oliver Sacks,
John P. Briggs and F. David Peat, Rupert Sheldrake, Richard
Selzer, and V. V. Nalimov (who specifically credits Eiseley with
the kind of holistic quest that leads Nalimov to write)—merits
recognition in an age when "the solidarity of the old disciplines"
breaks down (Barthes 1977b, 155) and even in the humanities hy-
brids such as what Rosalind Krauss calls the "paraliterary"[1] challenge assumed notions of "representation" and "reality." In literary
theory Gregory Ulmer finds collage/montage the model of hybridization adopted by "post-criticism" (postmodernist, poststructural,
Derrida's model of the age as a post card), "replacing the 'realist'
criticism based on the notions of 'truth' as correspondence to or
correct reproduction of a referent object of study" (1983, 86). As
a technique in the discourse of knowledge, metaphor, like collage,
displaces the piece to a new context, where it nevertheless "retains
associations with its former context" (1985, 59). Collage, indeed,
may serve as a model for interdisciplinary discourse, which relies
on the techniques of selection and combination (59) and trails associations from other contexts. Eiseley not only embraces interdisciplinarity, but he writes of the human being as "a mosaic of odd
parts drawn together as one might rifle a cosmic junkyard to make
a more than usually complicated tin woodman or a scarecrow,"
some of whose parts are "obsolescent," while others are "bent" to
new purposes (*LN,* 106). My point is that Eiseley's technique of

drawing from multiple contexts to make an interdisciplinary discourse draws on the processes of selection and combination—both of which are at once purposeful and haphazard—that natural selection has followed in the evolution of the human. The discourse parallels its subject; this point itself speaks to the essential appeal of the genre.

If thinkers from diverse backgrounds such as Einstein, Holton, Barthes, Rorty, Gould, *and* Eiseley have all called for a kind of popularization—whether through an interdisciplinary hybrid of philosophy and social criticism, or of science that can be absorbed by an educated general audience, or of science and art "for the uses of life"—then the hybrid genre is not to be overlooked. And if only poetic, "connected," "evocative" knowledge is adequate to convey a full range of understanding, a hybrid genre may provide a forum for a dialogue of epistemologies and a means of transferring from the discourse of knowledge information that is not just vital, but emotionally moving as well. For Eiseley, "knowledge without sympathetic perception is barren" ("The Ghost Continent," *UU*, 18), and the "poet's mind attempting to see all" ("Science and the Sense," *ST*, 200) is the way of viewing the universe that may protect humankind from itself.

My concern in this chapter is the need for the hybrid genre—specifically, the isolation of specialists, the scientist's response/responsibility, and the role of a reflective union of science and art in creating an epistemological dialogue. Underlying Eiseley's purpose as a writer is a concern for how and why any scientific theory gains currency, especially the idea that "any scientific theory which becomes widely popular . . . offers the possibility of support to some popular ideology"—a notion that he calls "ideological science" (*LN*, 97–98). Elsewhere, as noted in chapter 1, Eiseley comments on the evolution of a scientific idea or metaphor once it has "escaped" from the domain of the professional ("The Illusion," *ST*, 275). Underlying his genre is a concern for what he calls the need for a "re vision" of science (*LN*, 137). For Eiseley, the role of the writer who is also a scientist is "to open man's eyes to the human meaning of science, to find his path in the open society, to prevent his relapse into 'aloneness in the universe'" (133).

Institutionalization of Science and Estrangement from the Public

Eiseley contrasts institutionalized science with science "as a dream and an ideal of the individual," regarding institutionalized science as a cultural construct that is "subject, like other social structures, to human pressures and inescapable distortions" ("The Illusion," *ST*, 272). Historically, Western science has grown out of the medieval assumption of the rationality of God and out of the human separation from other life ("How the World Became Natural," *FT*, 9–30). Thus science looks upon the natural world "as might a curious stranger" and then turns to the human being with the same sense of estrangement ("The Last Magician," *IP*, 144). As a cultural expression that can create tools, science still cannot control these tools ("The Star Thrower," *UU*, 81).

Especially strong in the public mind, and not uncommon among scientists, is the assumption that science is the repository of truth and of mysterious methods. Increasingly, Eiseley writes, science has become the twentieth century's "substitute for magic" ("The Time Effacers," *IP*, 105), and scientists themselves form a kind of priestly elite. "The power of science consists . . . in its conflation of knowledge and truth," writes Stanley Aronowitz. Not only knowledge, but the scientific *method* of acquiring knowledge, becomes part of this conflation. The assumption that everything can be quantified and objectified is part of an often erroneous linking of science and rationality, as well as a rejection of intuition as foreign to rational thinking, writes Geoffrey Vickers (1978, 145). Eiseley argues that modern science assumes "that the accretions of fact are cumulative and lead to progress, whereas the insights of art are, at best, singular and lead nowhere, or when introduced into the realm of science, produce obscurity and confusion" ("The Illusion," *ST*, 272). Unquestioning faith in institutionalized science and its epistemology leads, in Eiseley's view, to loss of vision, of humanity, of wonder, for "when one has destroyed human wonder and compassion, one has killed man, even if the man in question continues to go about his laboratory tasks" (Eiseley, "Science and the Sense," *ST*, 298).

An isolated and authoritarian scientific institution, in turn, takes on mythological overtones. For Eiseley, both scientists and the general public are guilty of what Roger S. Jones calls "scientific idolatry," characterized by an "implicit assumption that an external physical world exists as an objective reality independent of the human mind and that the business of science is the discussion and description, not the creation, of that world" (1982, 206–7). Rather than looking at the kind of reality science creates, people look to science as earlier humans looked to magic. Science, in Michel Serres's terms, "is grasped as myth, it becomes myth" (Serres, *Feux*, 18, cited in Harari and Bell 1983, xix). Institutionalized and isolated science thus easily becomes authoritarian and narrow, with what Eiseley calls a "deliberate blunting of wonder, and [an] equally deliberate suppression of our humanity" ("The Illusion," *ST,* 271). Such an attitude Eiseley calls a "puritanism"—a rigidity and inflexibility characterized by an "authoritarian desire to shackle the human imagination" (269).

With the emergence of a scientific elite, appearing in the public mind as possessors of truth who move unfailingly toward further truth and, sometimes unwittingly, assuming political power, the populace is inevitably affected. The masses, Jurgen Habermas argues, "accept their own depoliticization" (1970, 104), though, in a free society, decision-making must involve an informed populace that is aware of its relationship to other life forms and to the environment that continues to shape life itself. David Edge's argument that scientific metaphor can restructure the way we perceive a situation is significant here. If, as he contends, attitudes associated with the vehicle of the metaphor are transferred to the tenor, then metaphor itself may be used to control the populace. Such control is evident in the nineteenth-century "electrical" model of the nervous system, which was imaged as a closed system through which a fixed amount of electrical "nervous energy," decreed by heredity, flowed (Rosenberg 1966, 138). In a popular metaphor, the brain was imaged as the central headquarters of a telegraph system; blockage of a "channel" could cause the nerve fluid to rush through the remaining "channels of sanity . . . with heightened velocity" (George M. Beard, quoted in Rosenberg 1966, 138). Such metaphorical understandings allowed the power structure to emphasize "balance" and moderation—both good Victorian values—and thus "to shore up middle-class morale" and control human behavior (Rosenberg 1966, 139).

Gerald Holton cites Norbert Wiener's work on cybernetics to emphasize the need to bring the individual "into a cybernetic relationship with the social organizations on which he depends," and thus to undermine "the Promethean drive to *omnipotence through technology* and to *omniscience through science*" (1987, 182, 183; italics in original). Citing Sir Peter Medawar (one of Eiseley's mainstays), Holton suggests that one way of accomplishing this task is to promote the realization that political decisions rather than scientific concerns frequently determine the direction that science will take (183). An important step toward this realization is "bringing science and history together . . . in scholarly research and in the classroom, for scientists and for nonscientists" (208). Holton supports this suggestion with his recollection of studying, in Vienna in the 1930s, history in terms of human conquests, with a map that was changed for each new era, while the "other map," that of the Periodic Table of the elements seemed to embody order and sanity; and just as the class was caught up in Hegel's concept of freedom as the motivating force of history, the map of Austria became brown, and the history teacher appeared in a Nazi uniform (203–4).

The realization that institutionalized science is also tied to other institutions underlies Holton's powerful argument for bringing history and science together. Habermas links military secrecy to secrecy concerning research findings with practical applications and argues that a formerly free and expected contact between the scientist and the public is no longer possible (1970, 76). The client of research is no longer an interested public, but merely an agency concerned with technological application of research findings. Memoranda and research reports replace scientific reflection; and mediation through public opinion, which could lead to confrontation of technical capability with human understanding, is replaced by communication between political authorities and scientific consultants (Habermas 1970, 72–76). Thus instead of abolishing war and cruelty and corruption, science has enabled them to thrive.

The public, then, willingly participates in its own depoliticization, accepts the scientific institution as "right," and remains isolated from the scientific elite. Eiseley's concern is not only for the human and political effects of such depoliticization, but for the effects on human knowing. "Man, the tool user," he writes, "grows convinced that he is himself only useful as a tool, that fertility except in the use of the scientific imagination is wasteful and without purpose" ("The Illusion," *ST*, 269). Such a populace willingly abandons imagination, accepts the scientific path to understanding as "right" and "true," and cooperates in destroying the environment. The far-reaching effects of new technologies on the physical environment have been treated from Hawthorne and Thoreau to the present; Leo Marx documents how the machine, particularly through the railroad metaphor, is asserted in literature as manipulative, even destructive, technology.[2] The point is the need to examine the human relationship to a world that human beings help to create.

Isolation of Specialists from One Another

Even the specialist is affected by the increasing isolation of science. C. P. Snow's glib pronouncement in 1956 that the modern world is divided into two cultures, with literary intellectuals and scientists, especially physical scientists, representing separate poles, is of special concern for the scientist who writes. Snow's concern with the "literary" culture's "total incomprehension" of science (inability to describe the Second Law of Thermodynamics is the equivalent of never having read anything by Shakespeare) leads him to argue that nonscientific intellectuals in the West "have never tried, wanted, or been able to understand the industrial revolution, much less accept it" (1964, 22). Eiseley's undermining of Snow, especially in "The Illusion of the Two Cultures," which I address in detail in chapter 11, is part of his quest for a dialogue of understandings, for an accommodation between the analytical epistemology of science and the synthesizing epistemology of art.

The problem of isolation and lack of understanding continues even as communication between specialists becomes more limited to the initiated. Many specialists fear that the individual who hopes for communication between different sciences may be attempting "to put scientific discussion on a mass basis and thus to misuse it ideologically" (Habermas 1970, 69). Indeed, Margaret Mead notes that we may be in danger of doing what other civilizations have done—developing "special esoteric groups who can communicate only with each other and who can accept as neophytes and apprentices only those individuals whose intellectual abilities, temperamental bents, and motivations are like their own" (1959, 142–43). Ironically (since he is often accused of obfuscation), Jacques Derrida similarly decries what happens when "a writing made to manifest, serve, and preserve knowledge . . . encrypts itself, becoming secret and reserved, diverted from common usage, esoteric" and eventually "becomes the instrument of an abusive power, of a caste of 'intellectuals' that is thus ensuring hegemony, whether its own or that of special interests" (1979, 118). Eiseley expresses the same concern anecdotally, recalling the top-heavy Mayan system headed by astronomer priests who calculated time as no civilization had done, but who seemingly became too burdensome and were toppled by revolution; their descendants worshipped their upended mathematical tablets ("Man in the Autumn Light," *IP,* 131).

The Scientist's Response/Responsibility

A major concern for Eiseley and for other scientist-writers is the responsibility of the scientist in a technological society. "The understanding of science—one need hardly repeat the litany—becomes ever more crucial in a world of biotechnology, computers, and bombs," writes Stephen Jay Gould (1987a, 7). Gould has argued that "science can only be harmed in the long run by its self-proclaimed separation as a priesthood guarding a sacred rite called *the* scientific method" public (7). Indeed, he contends that science should be "accessible to all thinking people because it applies universal tools of intellect to its distinctive material" (7). Benjamin S. P. Shen writes of the need for "science literacy," which he defines in terms of "an acquaintance with science, technology, and medicine, popularized to various degrees, on the part of the general public" (1975, 45). One advantage of scientific literacy is its enabling the general public "to take advantage of science's many benefits while avoiding its many pitfalls" (46). Shen discusses "practical, civic, and cultural" (46) science literacy—the practical leading to problem-solving and subject to development even in the absence of alphabetic literacy, the civic suggesting citizens' awareness of issues related to science so that the individual will be willing to participate in democratic processes, and the cultural suggesting, for example, artists who read *Scientific American* for information on DNA or individuals who enroll in physics courses for nonscientists (46–50). Though limited to a rather small number of individuals, cultural science literacy can affect human affairs by influencing leaders of public opinion and by reducing the impact of the "pseudosciences," such as diet fadism and astrology, which many Americans seem eager to embrace (Shen 1975, 50). Shen calls for an "ordinary-language science" (parallel to ordinary-language philosophy) that will draw scientists and the general public closer together in the realization that similar logic prevails in science and in everyday thinking and decision-making (51–52).

What Shen and Gould propose is a popularization of *concepts* that will lead to more effective practical influences of science, to more knowledgeable civic participation, and to an interdisciplinary cultural awareness of science. Others have suggested that wider scientific literacy requires careful illustration of the role of science in intellectual history (Arons 1983, 93, 105–7; Holton 1987, 182–204). In their arguments, these writers posit the need for what Eiseley and a number of scientists writing today actually do and what, even in the early years of this century, scientists such as Einstein and his contemporaries were able to do. Einstein and his contemporaries, influenced not only by other scientists but by wide reading in philosophy as well, saw themselves as carriers of culture in a sense beyond the scientific (Holton 1987, 164–65); Einstein himself, for forty years, was not only a scientist but also "a popularizer, teacher, and philosopher-scientist in the tradition of Henri Poincaré, Ernst Mach, and others of the generation before him" (29).

Eiseley and the Scientist's Responsibility

As a physical anthropologist with a background including extensive field work in paleontology and archaeology, Eiseley could easily have retreated into investigations of interest primarily to other specialists. However, his concern with the history of science, with the evolution of thinking, and with the impact of evolution on world history led him into writing for a larger public. As part of the scientific institution, Eiseley is nevertheless the scientific "heretic"[3] who questions that establishment. Yet because he stands as part of the scientific establishment, but with a deep interest in the humanities, Eiseley is directly concerned with analyzing the dangers of institutionalized science: its single-minded pursuit of one kind of knowledge at the expense of the consequences, its Ahab-like pursuit of space at the expense of other avenues of knowledge, its tendency to remain enclosed within its own specialties like dancers in "fairy rings," its acceptance of idolatry, its dependence on the military-industrial complex.

But Eiseley is also concerned with fostering a sense of perspective on time and creative change, on the human responsibility in a world that, in a sense, evolutionary thinking has created. Because human "volition has taken its place in the world of nature" (DC, 350), reexamination of well-known concepts in the life sciences and exploration of an alternative view to that of "frag-mented" science is just as important as—indeed, is foundational to—understanding about nuclear science. If we understand the interrelationships of life, we should be better able to make decisions that will affect it. Even Richard Dawkins, who contends that the basic evolutionary unit is the gene and that the gene's survival is based on its essential selfishness, suggests a separation between "the level at which altruism is desirable" and the genetic model (1978, 11). In other words, if the gene is above all selfish, human culture need not perpetuate this trait at the cultural level. But we need to know the difference. Essentially, this is what Eiseley is saying when he suggests that the human brain is able to insert further "indetermination," to affect evolution for better or for worse.

To comprehend the principles of evolution requires one level of understanding[4]; but to imagine, to visualize, the implications of evolution for human life today and in the future in terms of process and change requires additional understanding. Psychological, ethical, social, and individual implications are inextricably interwoven in any complex understanding. To comprehend the theory in Eiseley's sense of "the uses of life," in a synthesis of science and art, requires of the general reader far more than naive understanding of the process. Thus what we find in Eiseley's texts, as opposed to specific details of phylogenetic descent, is what has been called a "phylogeny of knowledge," a reversed phylogeny in which the concepts of (1) the age of the earth, (2) extinction, and (3) natural selection are seen as the specific branches leading to modern evolutionary thinking, with Mendelian genetics and molecular inheritance as contributors along the way (Olson and Robinson 1975, 227). Three of the branches of this phylogenetic tree—the age of the earth, the concept of extinction, and the variety of life—are three of the four points that Eiseley develops in scholarly detail in *Darwin's Century* and in popular form in *The Firmament of Time* and that underlie many of his later essays. Eiseley presents a phylogeny of biological thought rather than phylogenies of vertebrates or of hominids, though much scientific information is gathered from biology, anthropology, archaeology, and paleontology.[5] Eiseley's treatment of evolution includes defamiliarization of determinism, mechanism, teleology, struggle, and hierarchy in order to stress a holistic view or a dialogue of epistemologies and to deemphasize hierarchical thinking.

Eiseley also explores the notion of the *movement* of science, its struggle for evolution even though "styles" of seeing may sometimes block change. In fact, this movement is often tied to metaphor. Eiseley is intensely concerned with the difficulty that specialists themselves have in accepting views that challenge their philosophical framework. Like the peripheral figures whom Richard Rorty regards as keeping alive the notion "that this century's 'superstition' was the last century's triumph of reason" and that the latest terms borrowed from science "may not express privileged representations of essences" but may be just another vocabulary for describing the world (1979, 367), Eiseley repeatedly responds to scientists as a conservative body dependent on narrow professionalism for the institutional perpetuation that may become their greatest detriment. He notes that the literary figure may anticipate a concept even when scientists are restrained by their intellectual frame. Coleridge, he notes, recognized "'the way in which the intellectual climate of a given period may unconsciously retard or limit the theoretical ventures of an exploring scientist'" ("How Life

Became Natural," *FT,* 61), for Coleridge suggested that there is "a sort of secret and tacit compact among the learned, not to pass beyond a certain limit in speculative science" (cited in p. 61). In *Darwin's Century,* Eiseley comments on the way that Western philosophy, strongly influenced by theology, "caused men to look upon the world around them in a way, or in a frame, that would prepare the Western mind for the final acceptance of evolution" (6). This notion of a "frame" that influences thought or enables one to accept a new thought more easily than he would otherwise is basic to Eiseley's accounts of the emergence of evolutionary thinking. Eiseley develops the point further in *The Firmament of Time* as he comments on the importance of distancing oneself from the intellectual frame of his age: "The individual who learns how difficult it is to step outside the intellectual climate of his or any age has taken the first step on the road to emancipation, to world citizenship of a higher order" (*FT,* 6–7). Eiseley notes, too, the way that "even the scientific atmosphere evolves and changes with the society of which it is a part" (7). But Eiseley's most significant early statement of the frame (and the one that most closely resembles Kuhn's) comes later in the same text: "While proceeding with what is called 'empirical research' and 'experiment,' the scientist will almost inevitably fit such experiments into an existing comprehensive framework, an integrative formula, until such time as that principle gives way to another" (20).

The notion that science is a conservative institution that retards the emergence of new insights is one that Eiseley pursues throughout his texts. In *The Invisible Pyramid* he comments that "scientific training is apt to produce a restraint, laudable enough in itself, that can readily degenerate into a kind of institutional conservatism" ("The Spore Bearers," *IP,* 78). The context of this comment is his argument that modern humankind, enchanted with technology, seems determined to abuse the environment and to pursue space exploration even while ignoring what Eiseley considers more pressing environmental concerns. The scientific establishment emerges as "conservative," blindly pursuing its own ends and its own methodology regardless of the consequences. In *The Night Country,* Eiseley continues to explore the limitations of the age in which the scientist works, writing that "even the great visionary thinkers" have difficulty escaping the "limitations" of their own age ("Strangeness in the Proportion," *NC,* 131). He suggests that today we have substituted "authoritarian science for authoritarian religion," and that authoritarian scientists have a "totally erroneous impression that science is an unalterable and absolute system"

(139). He argues that authoritarian scientists "adhere to a dogma as rigidly as men of fanatical religiosity. They reject the world of the personal, the happy world of open, playful, or aspiring thought" (139). Eiseley concludes, then, that the modern world has produced "new class of highly skilled barbarians" who manifest a kind of "puritanism" and who, lacking "grace or humor, have found their salvation in 'facts'" (140, 142). And in his autobiography Eiseley returns to a metaphor that was only hinted in *Darwin's Century,* the metaphor of the "fairy ring" in which scientists seem to dance. In the chapter titled "The Dancers in the Ring," he comments, "Professional academic science tends to strengthen the mutual pull of the dancers already circling in the ring, not, on the whole, those trying to dance out," and he expresses the hope that our understanding of this creation of a "frame" can help us "to combat instilled scientific conservatism more successfully" (189).

Further, Eiseley contends that scientists are not exempt from the pull of emotions as well as the pull of tradition, philosophy, style, and professionalism. Against the popular notion that science is "a sustained, undeviating march toward some final truth," Eiseley notes the "ambiguities, fears, and trends which may play upon and influence severely disciplined minds" (*ASH,* 186). There is always the likelihood, he comments later, that experiments may "be colored by what we subconsciously believe or hope" (239). The scientist cannot escape the nonobjective world, for "in the supposed objective world of science, emotion and temperament may play a role in our selection of the mental tools with which we choose to investigate nature" ("Science and the Sense of the Holy," *ST,* 186–87). Again, Eiseley's text anticipates a well-known explanation of the movement of science—Gerald Holton's suggestion of the impact of "personal struggle" (1987, 19) and of the elements of science that Holton treats as "themata" or "(usually unacknowledged) presuppositions pervading the work of scientists" (29). Like Eiseley, Holton argues that there is often "quite flagrant neglect of 'experimental evidence' when such evidence is contrary to a given thematic commitment" (14).

Technological application of science is often perceived as the greatest threat to modern culture, both because of its impact on culture and its ability to destroy both culture and life. But at least as dangerous as political manipulation or the physical threat of destruction is the role of humankind as "master" of technology, the determination to fulfill the possibility suggested by the vehicle of the essential technological metaphor, the machine. For Martin Heidegger, the danger is not in technology itself, but in its "Enframing," which "threatens man with the possibility that it could be denied to him to enter into a more original revealing and hence to experience the call of a more primal truth" (1977b, 28). Richard Rorty echoes Heidegger in asserting that what is most frightening about science is the elimination of "the possibility of something new under the sun, of human life as poetic rather than merely contemplative" (1979, 389). Eiseley contends that when "the human realm is denied in favor of the world of pure technics," the appearance of the "unexpected" and the human need to feel wonder in an "unexpected universe" are stifled ("The Illusion," *ST*, 269). In his view, the function of the poet and the function of the scientist overlap in these two areas; if they are worthy of their appellations, they function as does Rorty's philosopher. Darwin, Einstein, and Newton, as well as Leonardo, "show a deep humility and an emotional hunger which is the prerogative of the artist" (276).

Reflective Union of Science and Art

What Habermas calls "the self-reflection of the sciences themselves" is a means by which philosophizing can serve as "interpreter" between specialists (1970, 8). Heidegger writes that "every researcher and teacher of the sciences, every man pursuing a way through a science, can move, as a thinking being, on various levels of reflection and can keep reflection vigilant" (1977c, 181–82). Eiseley's answer to the danger of technology, to what he calls the "frame" or the "fairy ring" and Heidegger calls the "Enframing" that limits insight, is a reflective (and self-reflexive) union of science and of art, and a dialogic, interactive epistemology that does not abandon the "way" of science, but incorporates into it the human (humane) insights of a synthesizing knowledge. In the reflection that asks the questions—in the sensitivity to both physical and animate surroundings, in the awareness of the web that connects all life—we at least refuse "to close our eyes to ultimate questions," and we reject mere "classification and experiment" as escape ("The Coming of the Giant Wasps," *ASH*, 239).

If "Enframing" is the danger of technology, for Heidegger there is still, within the essence of technology, a "saving power" against Enframing. Thus, he writes,

because the essence of technology is nothing technological, essential reflection upon technology and decisive confrontation with it must hap-

pen in a realm that is, on the one hand, akin to the essence of technology and, on the other, fundamentally different from it.
  Such a realm is art. (1977b, 35)

Reflection through art, then, is one answer. Like Heidegger, Eiseley turns to poetry—his poetry includes informal essays that blend science and the images and reflection of poetry—and to an epistemology that blends the scientific method and poetic understanding. Science is the vehicle for understanding: in learning about time, evolution, nature itself, one learns about humankind. The reflective scientist can play a crucial role. Eiseley's task is in part to question the role of the scientist in disseminating scientific information to the public and to doubt science itself even as he explicates it—in a sense to "renounce" science as an isolated domain (though this renunciation is never complete) in favor of an understanding that also incorporates synthesizing, connected "other" knowing. As scientist and as poet, Eiseley uses the self as a means for questioning science and for seeking values that science itself cannot give or that have been overlooked or distorted in the transmission of information from one generation to another.
  Eiseley concludes that art and science are essentially not separate; the rift that Snow perceived is not a rift at all. In a stone carved by an anthropoid who might frighten us today, Eiseley finds a meeting of the forces of art and technology: "There had not been room in his short and desperate life for the delicate and supercilious separation of the arts from the sciences" ("The Illusion," *ST,* 271). What is important is that the carver had "the kind of mind which, once having shaped an object of any sort, leaves an individual trace behind it which speaks to others across the barriers of time and language" (271). In utilitarian terms, the carver had "wasted time"; in a world filled with dangers he had shaped a practical tool and then "with a virtuoso's elegance, proceeded to embellish his product" (271). Art and science, brought together in this object, are not separable. Indeed, Heidegger reminds us:

  There was a time when it was not technology alone that bore the name *techne.* . . .
  Once there was a time when the bringing-forth of the true into the beautiful was called *techne.* And the *poiesis* of the fine arts also was called *techne.* (1977b, 34)

In Heidegger's sense, *poiesis* is truly "making," whether the product is statue or poem or tool; the etymological connection, through *poiesis* and *techne,* speaks to the time when art and technology were not thought to be separate, when vision was not enframed. For Eiseley, the carver's stone "tells the story"; both science and art demand "a high level of imaginative insight and intuitive perception" and result from "the same creative act in both domains" ("The Illusion," *ST,* 279, 273). Both are, as the primitive model suggests, essential in creating humankind as "human" rather than as animal.

Although both artist and scientist often ask the same questions, they may be questions that "the rigors of the scientific method do not enable us to pursue directly" (Eiseley, "The Illusion," *ST,* 274). Thus the dialogue of "ways" of knowing can lead to a new kind of interactive knowledge. In the artist's brain there can exist "the momentary illumination in which a whole human countryside may be transmuted in an instant" in an "absolute unexpectedness" ("Strangeness," *NC,* 137). Eiseley uses Melville's Ishmael and Ahab to symbolize opposite uses of science and art. The whole of *Moby-Dick* is a vehicle for contrasting two ways of looking at the universe—not "two cultures," but the sensitive view as opposed to an obsessed view. Ishmael, who says, "'I only am escaped to tell thee,'" represents the sensitive view of the universe, whereas Ahab symbolizes science in monomaniacal form. Ahab symbolizes science without the insight of the poet. He is the monomaniac who has lost all sense of awe, who has given himself over to extreme reductionism and aims only for "facts" that will "explain" the inexplicable. Although Melville's

tale is not of science, . . . it symbolizes on a gigantic canvas the struggle between two ways of looking at the universe: the magnification by the poet's mind attempting to see all, while disturbing as little as possible, as opposed to the plunging fury of Ahab with his cry, "Strike, strike through the mask, whatever it may cost in lives and suffering." ("Science and the Sense," *ST,* 200)

Ishmael and Ahab thus symbolize, respectively, the poet, who can be both scientist and artist, but who is open to the oncoming, unexpected process of the universe, and the reductionist, who is mindful only of his own pursuit, whatever the consequences. For Eiseley the "real business" of both artist and reflective scientist is to contribute to the human's "understanding [of] his ingredients" ("Thoreau's Unfinished," *ST,* 237).

As one who attempts "to see all, while disturbing as little as possible," the artist functions as a catalyst to reunite humankind with the nature from which it emerged. In this process a kind of catharsis emerges. Only the human being is aware of individual and species death, and only the human "knows the way he came" ("Thoreau's Unfinished," *ST,* 243). The role of the reflective scientist-artist is to participate in this catharsis and thus to render the other catharsis—the technological "utter cleansing" (243)—untenable.

Eiseley, then, is deeply concerned with the interconnectedness of the roles of scientist and artist. His notion of his genre, as indicated in his foreword to *The Firmament of Time,* includes a sense of sharing; these lectures were designed "to promote among both students and the general public a better understanding of the role of science as its own evolution permeates and controls the thought of man through the culture" (*FT,* v). This statement suggests much of Eiseley's overall purpose. He treats more than scientific facts; his interest is in the evolution of science itself, in the role of science as it affects human thought, in the individual's relation to science as an institution, and in discovering the underlying assumptions and limitations of science in order to engage in a dialogue that involves both the scientific method and connected, synthesizing, poetic understanding.

4

The Genesis of Method I: *Darwin's Century*

They lured him to a place he wanted to avoid and there, hidden between the pages of *The Odyssey* which had become their grave, they forced him—and many another too—to undertake the successful, unsuccessful journey which is that of narration—that song no longer directly perceived but repeated and thus apparently harmless: an ode made episode.

Maurice Blanchot, "The Sirens' Song"

Darwin's Century holds a pivotal position in Eiseley's work. A carefully wrought work of scholarship, it delves into the sources and effects of the Darwinian theory and treats Darwin himself in relation to scholarship among his contemporaries. When Doubleday requested a scholarly book to be published for the centennial of *The Origin of Species,* Eiseley began the research and writing that would help to shape the rest of his career. Indeed, Gerber and McFadden write that "*Darwin's Century* served to crystallize Eiseley's entire intellectual outlook," helping to create a perspective that emerges from Eiseley's study of evolution, but nevertheless retaining a perspective that is "distinctly and uniquely Eiseleyan, a set synthesis of ideas and feelings" that will reappear throughout the texts (1983, 50).[1] The book represents painstaking scholarship and meticulous organization, with literary flashes emerging, particularly in the conclusion, in which Eiseley employs the landscape metaphor to which he will later return, extending it to portray Darwin's work as a rock surviving throughout the century.

The book marks the genesis not only of Eiseley's dominant thought, but of his metaphorical and narrative techniques, and of his style. In the book he demonstrates his concern with the impact of a deterministic view on human thinking, with nineteenth-century assumptions of teleology and of the stability of science, living forms, and knowledge. The basic *themata* emerge, as does his emphasis on the latent creativity of life; on the relationship of time and life; on the dialogue of aggression and cooperation in nature; on "nature's hieroglyphs" as models for the ambiguity of human encoding, decoding, and understanding; on the confining and separating effects of "professionalism" on science and scientists; and on the confusion of progress with technology.

The Journey: Framework Metaphor

The dominant metaphor in *Darwin's Century* is the journey; it is *the* conceptual metaphor that chooses Eiseley (as the truism in studies of metaphor tells us, "We don't choose the metaphors; the metaphors choose us"). As a traditional code, the journey has a goal and a purpose. Yet in *Darwin's Century* Eiseley begins his ongoing dialogue with this metaphor: he draws on the journey as a foundational concept, a code, a root metaphor, but as a scientist he cannot demonstrate a purpose for the evolutionary journey. Evolutionary thinking breaks with the traditional code because it requires that the journey be a random journey, and this nonteleological journey also becomes Eiseley's means of exploring our method of understanding. The epistemological impact of the metaphor derives from its identification with the way we learn—stopping, starting, moving without always having a direction. The metaphor of the journey is thus the journey of metaphor as well. At issue is the epistemological quest for understanding both the complexity of the random process and the individual's relationship to that process.

Eiseley clearly recognizes, as we have seen, that evolution itself has become a contemporary root metaphor—not only scientific fact but a tool for expression of the implications that come from the fact (see "The Illusion," *ST,* 275). Because the movement of science depends, in his view, on root metaphor, contemporary thinking is necessarily colored by evolution. Just as the concept of evolution assumes a structure, gradually changing, with old physical forms embedded in humankind's present stage, the metaphor of evolution also assumes a gradually changing structure, with old experiences embedded in the present human consciousness. Both the physical forms and the referent of the evolutionary metaphor—psychological, social, cultural, and intellectual forms—are constantly changing. Eiseley links the metaphor of evolution to the metaphor of the journey. That evolution itself is a journey is unremarkable. The significance of Eiseley's metaphorical journey is his remotivation of the metaphor to displace the embedded Western concept of teleology, of movement toward a goal, of movement toward a destiny or a destination.

As a metaphor for method or a point of organization, the journey is as old as literature. Homer, Dante, Chaucer, Cervantes, and Bunyan all used the journey. Eiseley, however, parallels the journey to Odysseus's wandering rather than to Dante's or Chaucer's or Bunyan's movement toward a goal (though Chaucer's Wife of Bath knew much of "wandering by the way," the "path" of Chaucer's characters is clear). Outside the Judeo-Christian tradition, Odysseus, sidetracked and wandering, is the literary model for a journey filled with unexpected adventures and loss of a sense of direction or purpose. Odysseus, who at one point refers to himself as "Nobody" and who is both "Nobody" and "Everybody," symbolizes both "man's homelessness and his power" ("The Ghost Continent," *UU*, 24). Power, for Eiseley, is undermined, and gain is always accompanied by loss. In an echo of Odysseus, Eiseley in his autobiography refers to himself as "every man and no man" (*ASH* 23).

A model of the Western concept of the journey, specifically the choice of paths, is evident even as post-Darwinian thinker Sigmund Freud describes his plan to lead the reader on "an imaginary walk"—through the "dark wood of the authorities (who cannot see the trees)" and finally onto "the high ground and the open prospect," which leads to the question: "'Which way do you want to go?'" (1965, 55, n. 1). The path, then, leads through confusion and darkness into the clearing or light of day, which is itself linked to the word *journey*. Ultimately, the word *journey* derives from the Latin *diurnum*, "daily portion," which derives from *dies*, "day," which in turn derives from the Indo-European *deiw*, "to shine." Thus *journey* is etymologically related to the notion of the day's traveling, the space covered in a day, the "daily portion," as well as to the notion of luminosity, the traditional root metaphor for understanding, as employed in Freud's model of the journey.

As Derrida has contended, metaphors of presence always return to "the circle of the heliotrope," with "natural light" as "a kind of ether of thought and of the discourse proper to it" (1974, 68–69). Thus, Derrida continues that the sun is "interiorized . . . in the eye and the heart of Western man," for he "sums up, assumes, and fulfills the essence of man 'illuminated by the true light'" (71). In Western thought, then, the human being is the one who finds "light," "truth," and "reality" by following the dominant epistemology, which is based on "observation." Playfully, Derrida pinpoints the quest for closure and the summation of illumination in Western thinking. Paul Ricoeur adds that "by being images for idealization and appropriation, light and sojourn are a figure for the very process of metaphorizing and thereby ground the return of metaphor upon itself" (1979, 289). Here is the paradox for Eiseley's journey—the return of metaphor upon itself. Yet the "shining" of *dies* is linked not only to the journey or to metaphor as method, but to all metaphors of light or looking; and it is this "looking" that Eiseley reexamines, directly and metaphorically, in all his texts.

The word *odyssey* itself interpenetrates the texts as a synonym for *journey*. As W. H. Auden has noted, humankind is the quest hero (Introduction, *ST,* 18); the common "man" parallels the common noun. For Eiseley, the mental and cultural journey, like Odysseus's journey, is solitary and mysterious, undirected but subject to the "magical self-delineating and mind-freezing" word *man,* which, by freezing human self-definition, could have fixed us at any point in evolution—could "have held us hanging to the bough from which we actually dropped" ("Instruments of Darkness," *NC,* 54). But what makes Odysseus the synecdoche, and archetype, for humankind is his capacity for overcoming the "mind-freezing word," for wandering on the epistemological path. Because metaphor makes a journey and is a journey, the means of knowledge become indistinguishable from the knowledge itself; vehicle and tenor become interwoven in the texture of the text as the journey enters the epistemological dimension. Through the journey Eiseley—aware of sidetracks and detours—attempts to undermine (mis)understandings of science in the public mind as well as to effect a transfer of information from the discourse of science to the general reader. The Western, Christian assumption of progress and "the uniqueness of the historical process" is, Eiseley argues, directly linked to modern science (*LN,* 191).

In the Western tradition, this notion of progress is embedded in metaphors of the journey. Yet "the student of nature is confronted with innumerable novelties distributed through time but proceeding constantly toward an unseen and unique future" (*LN,* 192). It is this uniqueness and unpredictability that Eiseley attempts to inject into his journey as he defamiliarizes *telos*. Odysseus's journey is a model for a journey without reliance on technology—a legendary journey more akin to the primitive than to a technology that links destination and identity.

With the evolution of the human brain comes a cultural journey from what Eiseley calls the security of instinct to the insecurity of questioning. Life is both a quest and a questioning, "for man [has] fallen out of the secure world of instinct into a place of wonder" (Eiseley, "Man against the Universe," *ST,* 221). If we have lost instinct, we have "replaced it with cultural tradition and the hard-won increments of contemplative thought" ("The Ghost Continent," *UU,* 7). Cultural evolution has brought a world of technology, of machines, of "objectivity" or supposedly unmediated knowledge, of scientism. A frightening result for humankind is the loss of humane concern, of pity, of the capacity for wonder. The fall from instinct, which Eiseley analogizes in terms of the Judeo-Christian fall, is a fall into the journey of knowledge (in a sense, a fall into metaphor). But Eiseley reminds us that "knowledge, or what the twentieth century acclaims as knowledge, has not led to happiness" (5). If humankind has been sidetracked into the dead end (the dead-letter office of Derrida's *La carte postale*) of scientism, the random journey provides an alternative heuristic.

Eiseley's journey is multifaceted: it begins as a personal journey, with the individual using his own experiences as the basis for knowledge, and it thus becomes an epistemological journey. But it is also the journey of life itself, the journey of the human as an evolved and evolving creature, the human inner journey that leads to culture, a quest (both a wandering and a wondering), a journey of exploration, and the journey of science. Finally, in a blending of rhetoric and metaphor, the journey becomes a rhetorical technique as the narrator asks the reader to accompany him on a journey into a strange, wavering realm of underwater sight and insight ("The Long Loneliness"), into a movement backward into time ("The Slit"), or into the physical experience of water flowing to the sea ("The Flow of the River"). Eiseley thus plays on the journey of reading, not infrequently calling the reader's attention to the process in which she is engaged.[2] Although the multiple uses of the journey suggest its vast significance as a motif and a metaphor in Eiseley's texts, my concern is chiefly with Eiseley's undermining the teleological baggage that the metaphorical concept carries.

The Odyssean voyage, like the Eiseleyan "journey" of a writer, is also the symbol of the creative journey. As a metaphor, it links the human, cultural, epistemological, and scientific journeys to the journey of textuality. Blanchot finds Odysseus in every writer—Odysseus as creative, erring, wandering man. Tied to the mast, Odysseus hears the sirens' "imperfect" song, which draws him "towards the space where singing really begins" (1982b, 59). The practical Odysseus, who listens but devises his means of avoiding the sirens' temptation, is nevertheless lured "to undertake the successful, unsuccessful journey which is that of narration," to make a song that is "ode made episode" (61). The artist, Blanchot argues, must be transformed "into nobody, the empty, animated space where art's summons is heard" (1982c, 197). Blanchot's Odysseus is like Eiseley's "castaway" whose exile is his essential condition. He has but one rule: "to avoid the slightest allusion to a purpose or a destination," and he experiences, not the "presence" of the event, "but the beginning of that endless movement which is the encounter itself" (1982b, 61, 65). Without destination, wandering, Odysseus moves toward the infinite space of creativity itself.

Because *metaphor* itself suggests a vehicular philosopheme (*metaphor* as bus, "taxi" in Greek), the vehicle of thought is a method of transportation of concepts from one discourse to another—from the discourse of knowledge to the general audience. Eiseley links the journey and reading, and reading is the dominant means of knowledge in Eiseley's texts—not only reading of the printed text, but also interactive reading of nature as hieroglyph, which suggests that the reader interprets largely in terms of the meanings she brings to the text. In Serres's terms, "Science is the totality of the world's legends. . . . To read and to journey are one and the same act" (*Jouvences,* 14, cited in Harari and Bell 1983, xxi). Barthes writes of reading as one journeys: "We read a text (of pleasure) the way a fly buzzes around a room: with sudden, deceptively decisive turns, fervent and futile" (Barthes 1975, 31). In Barthes's terms, the reader's response to the text is one of stopping and starting—a kind of push-pull, stop-start, insight-aporia that occurs as responses occur. As the reader becomes aware of an insight, there is a contrary pull to stop and play with that insight, a tendency to buzz off into new directions of response. "The double movement of revelation and recoil will always be inherent in the nature of a genuine critical discourse," writes Paul de Man (1985, 289).

The journey is not only reading, in both textual and semiotic senses, but literature itself, as Maurice Blanchot suggests in linking "the reality of literature" with "the illusion of infinity," for in the literary journey one is "on the road without the possibility of ever stopping" (1982a, 222). The traveler on this road is like the wanderers in the Biblical wilderness, and the reader finds his space "truly infinite even when he knows—and the more so when he knows—that it is not" (222). Blanchot's journey has no beginning: "even before starting it restarts, before finishing it is already harking back" (222). Thus the world is a book and the book is the world. The journey never comes to closure, for it is linked to textuality.

Discourse is ultimately about discourse itself; woman reading is woman engaging in the most human of activities.

In the Eastern tradition, the journey is the *Tao,* the way to an enlightenment that consists of intuitive knowledge and spontaneous action. Reflecting, as Fritjof Capra notes, the qualities increasingly evident in the "new" physics, it is "the cosmic process in which all things are involved," and our world is "a continuous flow and change" [1977, 92, 95]). In the context of his "rocket book," *The Invisible Pyramid,* in which he argues for human awareness of and participation in nature and against mechanized control over nature, Eiseley quotes the words spoken in ancient Buddhist cities—"'Thou canst not travel on the Path before thou hast become the Path itself'"—and explores the notion that "written deep in ourselves is a simulacrum of the Way and the mind's deep spaces to travel" ("The Spore Bearers," *IP,* 77). As a journey of the mind, Eiseley's journey does not move toward closure or final illumination or *telos,* as do most Western journeys. The illumination of the Tao, in Barthes's words, is a linguistic blank, "a panic suspension of language," so that, as in haiku, "what is abolished is not meaning but any notion of finality" (1982, 75, 82). "Mercurial and shifting," humankind has "at heart no image, but only images," writes Eiseley ("The Lethal Factor," *ST,* 263). The illumination of the journey is recurrent, varying, and textual; it is a creation of meaning out of the linguistic blank.

In Eiseley's demystification of evolutionary concepts, the underlying question is one of a choice of conceptual paths: whether to choose the image of determinism or of flow, the image of the beast, which is the image of humankind as fixed and determined by its origins, or the image of humankind as continuously changing and able to insert "indetermination" into the evolutionary process. This choice is directly related to a choice of methods: the empiricist, reductionist "way" or epistemological method of many scientists leaves little room for humane speculation, and Eiseley explores the notion of a "way" that does not reject the method of science, but incorporates intuitions, feelings, sympathy. As in Lakoff and Johnson's view of the "experientialist myth," in which scientific knowledge remains possible, though without claiming absolute truth (a view that should "make scientific practice more responsible" [1980, 227]), in Eiseley's view, "knowledge without sympathetic perception is barren" ("The Ghost Continent," *UU,* 18). The means to knowledge must keep sight of need for human sympathy. For Eiseley, the path to awareness is essential to the understanding;

without traveling the path, there can be no understanding. Style and understanding are interwoven.

Eiseley's journey, then, is a wandering, an *errance* (from *errare*, "to wander"). In an essay such as "The Invisible Island," the path seems tangled as the narrator shifts from the earth as a "creature" to Herman Melville's comments on the placement of a whale's eyes, to the survival of failures rather than the fittest, to the role of the human brain in helping failures survive to the "web" or "screen" by which nature holds the status quo in place, and finally to the linguistic "island" created by speech, with all the implications of isolation that go with the concept of an island (*UU,* 147–71). In such wandering Eiseley's journey of knowledge follows a tangled path, a kind of objective correlative of the way knowledge emerges. In Eiseley's view, the movement of knowledge, like that of evolution, is not always orderly. The wanderings of evolution, through minute mistakes in DNA, have brought us to where we are. Without "the capacity to blunder slightly," writes Lewis Thomas, "we would still be anaerobic bacteria and there would be no music" (1980, 23). Whether in evolution of forms or in evolution of knowledge, the factor of *errance* is basic to the journey; it is the objective correlative by which Eiseley's text creates the experience.

What is suggested in this sojourn, in addition to the sense of movement, is that the journey is a *spatializing* of understanding. Lakoff and Johnson argue that we think of the "visual field as a container and conceptualize what we see as being inside it" (1980, 30). Conceptualizing the path spatially, we also spatialize temporality, creating a means of dealing with surfaces that can be touched; even the word *concept* suggests touching, gathering, grouping, holding, grasping, or "taking to oneself" in the mind. The journey, then, is a means of giving spatial borders to something that cannot be "grasped"—time. (Conceptualizing time as a fourth dimension is more difficult than visualizing it—hence the popular metaphors of time as an arrow or a cycle.)

The random journey, then, is an open-ended spatial model for the way we learn—stopping and starting again, moving sometimes without direction, wandering and wondering and open to discovery as long as we are aware of the frame of a worldview or an epistemological method and not totally encircled by it, never leading to closure. Eiseley's journey is not movement toward a fixed goal; instead, it subverts the common notion of the teleological movement of evolution—a notion that the nineteenth century upheld in its image of northern European *man* as the epitome of evolutionary development and that twentieth-century thinking often continues.

The tenor of this journey is evolutionary change, which is in turn a vehicle—a movement through "unexpected" rather than predetermined stages, a quest whose object is unknown.

The evolutionary journey is also a cultural journey, in what George Gaylord Simpson calls the "new" evolution.[3] A recurrent Eiseleyan motif is the human potential, whether realized or not. "Latency" is perhaps the most fascinating part of the journey, and "latency" is another way of expressing the indeterminacy, the lack of destiny, in the journey. Again, the journey as Eiseley construes it is a displacing metaphor. The cultural implications of this view of the human part in the process are significant because awareness of our role in determining the future gives us the chance to consider the kind of future we want to create.

Eiseley also focuses on the journey of science. In *Darwin's Century* he writes of early students of time and life as "time voyagers" who constructed a "pirate chart" and, like Darwin, journeyed both literally and in libraries. E. Fred Carlisle notes the parallel between explorers piecing together maps and nineteenth-century scientists piecing together the aspects of evolutionary understanding; even scientific discoveries, he notes, are "pieced together from the results of uncertain thrusts into the unknown" (1974, 362). But theories also serve as tentative "maps," and the journey suggests a combination of "the adventure, risk, and mystery" of science and the individual's "passion, commitment, and risk" (363). The metaphor, then, is not only the method, but also the message.

With the random voyage, then, Eiseley draws together the individual, evolutionary, human, epistemological, scientific, and creative journeys. The journey, then, is the metaphor of metaphor; it is a transfer, a displacement, a method of exploring the wandering of evolution and the wandering/wondering of knowledge. Eiseley's is an epistemology of journeys, a heuristic of intersections between disparate metaphors from diverse fields of knowledge, a displacement of the journey from its traditional sense of fixed destinations to a random voyage. Finally, the reader is returned from the mechanism to the text—from the mechanical model of fixity and determinacy to the textual model of interaction and indeterminacy.

Beginning the Journey: Darwin's Century

At the outset of *Darwin's Century,* Eiseley establishes the journey as the basic metaphor as he links Darwin's discovery to the Renaissance voyages of discovery. But this text also links the journey of discovery to the classical journey, the *Odyssey,* which will be a recurrent motif in Eiseley's essays. *Darwin's Century* is Eiseley's quintessential odyssey—as a journey of wandering, of wondering, of discovery and disappointment, of the individual's confrontation with the unknown and the unknowable. Chapter titles in *Darwin's Century* keep the motif of the odyssey before the reader, particularly in the first three chapters, titled, respectively, "The Age of Discovery," "The Time Voyagers," and "The Pirate Chart." The first chapter links the voyage of discovery with the voyage of knowledge, and the second chapter places the early preevolutionists in the role of voyaging, but through the extended time of geological discovery. The third chapter evokes notions of adventure, romance, and lawlessness that are associated with the great Renaissance voyagers and that Eiseley subtly transfers to the intellectual voyagers who, like treasure hunters who have found fragments of a pirate chart, must attempt to piece together a meaning and must, eventually, cooperate in solving the puzzle.

The "pirate chart" also carries associations of a map secretly, even ambiguously, inscribed, whose meaning requires deciphering. Implicit in these associations of the early evolutionists with "pirate charts" and maps and piecing together meaning is the metaphor of textual meaning that becomes a *topos* and an insistent intertext in all of Eiseley's writings. Once again he works to defamiliarize the Renaissance notion of the Book of Nature as the single source of God's revealed meaning, and even the nineteenth-century myth of textual meaning. Instead, Eiseley's map, chart, or text (whether a printed text or a Thoreauvian "text" of nature) must be understood cooperatively and interactively; the meaning depends greatly on the observer's construction, not on a fixed or teleologically intended meaning somehow emanent in the text.

Linking together Darwin's literal journey and the physical and intellectual journeys of humankind, Eiseley focuses—in the structural center of the book, chapter 6—on Darwin and "The Voyage of the *Beagle.*" Darwin, as Eiseley is careful to note, combined both the literal voyage and the intellectual one. The epigraph to this central chapter is from Darwin himself and links Darwin's "observations" on the voyage to the intellectual context that made them possible: "The force of impressions generally depends on preconceived ideas" (quoted in p. 141). On boarding the *Beagle,* Eiseley writes, Darwin carried with him "the three fragments of the lost chart of Smith, Cuvier, and Hutton"—that is, of geological and biological uniformitarianism, and of awareness that the worlds of the past were all the result of the same forces operating over long periods of time (*DC,* 151). In 1832 Darwin received in Montevideo the second volume of Lyell's *Principles of Geology* (160), and his *Diary* indicates "a mind fresh from the European geological and biological controversies" (158). In contextualizing the emergence of Darwin's understanding, Eiseley both employs and undermines a common myth of Darwin as a Romantic, a man of genius operating in isolation; although he portrays Darwin as one who lived the Romantic adventure, Eiseley nevertheless demonstrates the cultural embedding of Darwin's work.

Other Metaphors

Not only is the voyage established, but other metaphors that "choose the writer" emerge in *Darwin's Century.* The ladder has long been associated with evolution, arising out of the medieval and Renaissance conception of the *scala naturae,* or the scale of life, which portrayed a fixed, immutable hierarchy of special creation, from "uncreated atoms" to *man* (woman was merely a vehicle for producing man). As Eiseley demonstrates, the *scala naturae* gave a framework, already in place, for a scale or ladder of evolution, though the conceptions have in common only a sense of hierarchy. Eiseley approaches the notion of "two ladders"—one "a ladder backward into time which involves the careful anatomical comparison of existing forms of life at various levels of complexity," and the other "that of paleontology itself, the analysis . . . of the organic remains of all those once living orders which have left bones or impressions of their bodies encased in the substance of ancient land surfaces or sea bottom" (*DC,* 5). But the ladder as metaphor threatens basic evolutionary understanding, and Eiseley, while using the code in introducing his subject, begins to undermine it. In describing Cuvier's accomplishments, he writes of using the comparative anatomy of the living "'as a ladder to descend into the past'" (cited in p. 85). One of Cuvier's major achievements was his break with the Scale of Being to conceive of four major groups—Vertebrates, Mollusca, Articulata, and Radiata. In departing from the unilinear hierarchy, Cuvier demonstrated "that there were many stairways of life rather than one" (87). Once Cuvier had opened the path to multiple orders of life, "each was unique and ramifying along its own evolutionary corridor. Man was not the creature toward which the worm was striving.[4] Life was a bush, not a ladder" (88). One factor that led to failure of early evolutionary speculations was "the arrangement of life in terms of a single scale with man at its head," Eiseley argues (117). Cuvier's rejection of such unilinear progress was, for Eiseley, "a necessary preliminary to the kind of branching evolutionary phylogeny which is now everywhere accepted" (118). Noting the linguistic role in determining the survival of ideas, Eiseley points out, in fact, the similarity of certain phrases in the Scale of Nature and in pre-evolutionary and evolutionary thought, a similarity that occurs because "we, with our modern evolutionary ideas, have unwittingly inherited so much from this preceding era of thought" (119).

The metaphor of the ladder carries with it a particularly devastating transfer of the vehicle to the tenor in Victorian notions of racial hierarchies, which Eiseley attacks repeatedly, arguing that "*long before the clear recognition of fossil forms of man there existed in the minds of western Europeans a notion of racial gradation, and a conception of that gradation as leading downward toward the ape*" (*DC*, 264; italics in original). One writer argued for the Caucasian as the highest on this scale, for the Negro as possessing the low brow, projecting jaw, and "'slender bent limbs'" of the Caucasian child well before birth, the Native American reproducing the Caucasian child nearer birth, and the Oriental reproducing the Caucasian newborn (263–64). Darwin, however, though he saw the gap between savages and civilized human beings, was nevertheless aware that the savage is far closer to modern human beings than to a "missing link." Only years after his voyage and under the influence of others did Darwin make hierarchizing observations. When, later, Darwin himself fell victim to using "the living taxonomic ladder" to interpolate "marked differences in the inherited mental faculties between the members of the different existing races," he turned to an assumption that "unconsciously reflects the old Scale of Nature" (288).

Another metaphor that Eiseley begins in *Darwin's Century* is that of the drama. Linking pre-Darwinian Western notions of time to the Christian tradition, he writes that "Christian thought had long contemplated eternity but it had been the shadowless, changeless eternity of God. Earthly time had been seen by comparison as the brief drama of the Fall and Redemption, the lowly world of Nature merely the stage setting for a morality play" (2). It is a metaphor to which he will return for artistic elaboration, as we shall see particularly in *The Firmament of Time*. Later he compares the Christian sense of time with that of primitives and with "the endless cycles over which Greco-Roman thought had brooded," concluding again that the Christian world portrayed time in terms of "the human Fall and Redemption . . . played out upon the stage of the world" in a drama that "was unique and not repetitious"

(80). The uniqueness of the Christian drama is its introduction of time as unreturning, the concept of time absorbed by and fundamental to Western science.

Techniques

Although *Darwin's Century* is primarily a work of scholarship intended for an audience already familiar with at least some of Darwin's texts and theories, it nevertheless is a watershed not only of *topoi* or *themata* but also of Eiseley's literary method. When Eiseley makes the "rhetorical turn," as he begins to do in this book, he blends insights of literature and literary analysis with those of intellectual history, archaeology, and paleontology. Because *Darwin's Century* is the point of emergence of several techniques, I shall examine the techniques of the concealed essay more specifically in this chapter than in future ones.

Early in *Darwin's Century,* Eiseley writes of two powerful influences on the study of nature; significantly, as he specifies these two influences on English thought, he also calls attention to two styles of writing that have shaped his own texts.

> In English thought since the time of Bacon two influences have been paramount in the study of living nature. One stems directly from the purely scientific and experimental approach of Bacon. . . . The other more gracious, humane tradition descends through John Ray and Gilbert White, two parson-naturalists, to the literary observers of later centuries, men such as Thoreau and Hudson. The two streams have at times mingled, influenced and affected each other but they have remained in some degree apart in method and in outlook. Though Darwin is generally claimed by the scientists, it is worthy of note that he did not remain uninfluenced by the literary tradition in natural history which is so strong in England. He was a devoted reader of Gilbert White. (13)

The Baconian spirit of observation and experiment and the "more gracious, humane tradition of the parson-naturalists" are clearly the traditions that Eiseley values, and they are the sources of the "two rhetorics" that he employs throughout his texts. Although the scientific rhetoric is basic in *Darwin's Century,* Eiseley nevertheless oscillates between this reasonably neutral, "scientific" rhetoric and an "other," more literary rhetoric that draws on emotion, on the human ability to feel empathy across the species barrier. In the emotionally charged concluding page of *Darwin's Century,* he writes of the twenty-eight-year-old Darwin's speculation that animals and humans "may be all melted [later emended to *netted*] together" and concludes:

> If he had never conceived of natural selection, if he had never written the *Origin,* it would still stand as a statement of almost clairvoyant perception. . . . It is for this, as much as for the difficult, concise reasoning of the *Origin,* that Darwin's shadow will run a long way forward into the future. It is his heritage from the parson-naturalists of England. (352)

This emotionally charged rhetoric that looks at the human side of science, at the human being who does science, and at the responsibilities that human being brings to the task is the rhetoric normally identified with Eiseley's texts: laden with adjectives, deliberately emotional, seeking the synthesizing view. These two rhetorics deserve further exploration.

Because *rhetoric* has historically been relegated to the status of flourishes, of nonlogical persuasive use of language, the rhetoric of science has only recently received scholarly attention. Lawrence J. Prelli, in his study of "the rhetorical dimensions of *creating* and *evaluating* scientific communication" (1989, 1), of the inevitable use of persuasion in science, argues that "scientific discourse is accepted or rejected on grounds of its *reasonableness*—given the issue at stake, the knowledge conditions of the scientific community, and the perceived expertise of the makers of the claims" (7). Avoiding the totalizing claim that everything is rhetorical, Prelli nevertheless identifies "topics of discussion that characterize doing science regardless of specialty" (185), which he calls the "*topoi* of scientific reasonableness" (186). The first set of *topoi* (literally meaning "place," from the "commonplace" or often-used situation or structure in classical rhetoric [Holman and Harmon 1986, 504]) are problem-solving *topoi* that the scientist can call upon, including experimental competence, observational competence, experimental replication, experimental originality, corroboration, explanatory power, predictive power, taxonomic power, qualitative precision, empirical adequacy, and significant anomaly (186–98). The second set of scientific *topoi* that Prelli identifies includes "evaluative *topoi*," those used in evaluating scientific arguments: accuracy, simplicity, scope, consistency, and empiricism, with elegance and fruitfulness as values also widely applied (199). The third set of *topoi* are "exemplary," which the scientist uses in explaining: scientific principles, examples, analogies, and metaphors (209–15). Such writing conventions, Prelli argues, often suggest, whether consciously or unconsciously, "the fixity of conventional knowledge and the objectivity of inquiry" (104). These conventions imply use of the inductive method, and when used with other stylistic conventions such as the use of the passive voice and the elimination of the first person, they suggest that the evidence presented is based on "impartial investigation of phenomena that have independent, objective, and undeniable existence"—in essence, the basic positivist program (104).

As part of the discourse of science, Eiseley's scientific rhetoric clearly calls upon rhetorical conventions such as these. *Darwin's Century* is a history of scientific investigations of phenomena. Though Eiseley analyzes the influences that prevent the mythical "impartiality," he clearly attempts impartiality in his own telling of the story. Throughout most of the book he avoids the first person— seeming to discover its power occasionally but then to avoid it when he needs most to assume the role of the reliable scientist. In examining causes, he relies on conventions such as enumeration and eliminates the emotionally charged adjectives that appear in the "other" rhetoric. For example, he enumerates developments that had occurred by Cuvier's time, concluding, "The rock formations of the Paris Basin were being quarried extensively in the days of the First Empire. There were strata interspersed with others containing fresh water forms" (*DC*, 83). The neutral prose and the distant narrator contrast sharply with the less detached language of Darwin's predecessor Cuvier (1769–1832), whom Eiseley quotes: "'I found myself as if placed in a charnel house . . . surrounded by mutilated fragments of many hundred skeletons of more than twenty kinds of animals, piled confusedly around me. The task assigned me was to restore them all to their original positions. At the voice of comparative anatomy every bone and fragment of bone resumed its place'" (cited in p. 84). Cuvier, Eiseley notes, "had a deep sense of drama" (84), though the comment, in another context, might well apply to Eiseley himself.

At crucial moments in the explication, Eiseley appeals to *topoi* such as experimental and observational competence or experimental replication or corroboration. In evaluating changes in thinking he calls on evaluative *topoi* such as accuracy, simplicity, scope, and consistency. Writing of Darwin's reluctance to extend his observations of the animal world to the dominant animal, humankind, Eiseley considers the intellectual climate of the time and makes his own restrained comment, before inserting Sir Charles Lyell's wry comment on the human position in the evolutionary system:

In the judgment of the present writer there can be no doubt, considering the temper of the times, that Darwin's caution was well justified, and probably had the salutary effect of broaching what was then an unpleasant topic by successive doses which were found assimilable rather than, as Lyell was accustomed to saying, "going the whole orang" all at once. (*DC*, 256)

Eiseley's "plain" or "scientific" style is painstakingly detailed and clearly organized to indicate the sequential development of evolutionary thinking and the ways that one development led to another. This style echoes, as noted in chapter 2, what Alan Gross describes as the customary "social imperative" of the scientific article, which focuses on the causal relationship of "physical objects and events," rather than the person describing them, as the "causal center" (1991, 934–36). In his emphasis on cause and sequence, for example, when Eiseley comes to the Comte de Buffon in the mid-eighteenth century, he pauses to enumerate patterns of contemporary thought that prepared the way for Buffon: notions and theories of cosmic evolution; recognition of fossils as historical and causal artifacts; exploration of reproductive and developmental stages with the microscope; in France, the growing interest in human population in relation to food production; and evidence in gardens and hothouses that supported Linnaeus's observations (*DC*, 35–38). In causal, reductionist analysis, Eiseley treats each thinker who contributed to evolution, organizing them according to understandings of geological change, of time and organic change, of the emerging awareness that the human being was part of the pattern; and he hierarchizes the contributions of those who "anticipated" evolutionary thinking, the "minor evolutionists," and the "giants." For the most part, he remains detached from the subject, relying on the third person and often on the passive voice.

But in addition to appealing to scientific reasonableness and to making causal relationships of objects and events central to the discourse, Eiseley turns frequently to the "other" rhetoric that will become his hallmark in the popular essays. As noted in chapter 2, this rhetoric centers on Eiseley's concern for the human side of science, relationship of the human being—as scientist or as layperson—to the web of nature and the web of culture. In centering on the human rather than on the relationships of objects and events, Eiseley's turns to conscious manipulation of narrative, to involvement of both author and audience in the material, to literary tutor texts and tutor structures, to reexamination of metaphors, and to deliberate creation or manipulation of metaphors.

Eiseley begins, almost hesitantly in *Darwin's Century,* to draw on personal observations and reactions. In describing de Candolle's anticipation of evolutionary thought, he breaks into the scholarly tone with the first person: "I noted that Darwin gave no direct source and remembering Darwin's own admission that he did not read French with facility, I was curious as to where he had found this reference" (101). As if his style were freed by the personal pronoun, he describes his own reading in Lyell to confirm the suspicion that Darwin's understanding of de Candolle came indirectly, through Lyell. Shortly thereafter, Eiseley uses his newfound freedom to defend Lyell against charges of equivocation: "I think that this charge of timid vacillation is in some degree unjust to the man who marshaled the evidence and took the stand which eventually destroyed the catastrophic doctrine" (105). From this point until the conclusion, however, the first person is suppressed. When he returns to it in the last pages of the conclusion (which I shall examine separately), he does so with the freedom to incorporate the first person, personal observations, and the personal tale as essential to the quest for human, in addition to scientific, understanding.

Narrative/Fictionalizing/Dramatizing

Eiseley's method relies heavily on effective use of narrative— whether the narrative of the history of science, the history of the American West, personal anecdote, or a tale of a dream. And often he dramatizes an event, inviting the reader into the scene as a participant. *Darwin's Century,* as an exploration of the development of evolutionary thinking, is, of course, heavily dependent on narrative. But the Eiseleyan fictionalizing of events emerges. Earlier I noted the fictionalized and dramatized accounts of Mendel's work and his audience's lack of reaction. Equally powerful is his fictionalized, dramatized account in which he attempts to place Darwin among those nineteenth-century scientists (not the majority) who did not always hold to notions of racial hierarchy. He notes that the young Darwin realized the closeness of the "savage" and the "modern," and he turns for evidence to Darwin's account of Jemmy Button's farewell to his European friends. The account, Eiseley writes, signaling his own rhetorical shift, "contains the pathos of great literature," rather than the hierarchical notion that Darwin embraced in later years. In the fictionalized narrative, Eiseley brings the reader into Darwin's experience as he watched Jemmy's last signal fire. The dramatic narrative provides a sense of participation that serves as a powerful emotional transition to an attack on those who try to see in living human beings the anatomical features of a vanished species:

> The fire from the dark headland stings the eyes a little even now, and Jemmy Button's wistful, forgotten face is an eternal reproach to those who persist in projecting upon the bodies of living men the shadow of an unknown vanished ape. Moreover, even the form of that ancestral ape is illusory. The long arms, the bandy legs, the pendulous belly, the semi-erect posture are conceived in imitation of the apes of today. (*DC*, 265)

The immediacy of smoke in the eyes brings the reader into the experience and underscores the lack of understanding that has led some to label living contemporaries as "missing links."

Synthesizing/Dialogizing

The condensing, synthesizing rhetorical perspective is especially evident in *Darwin's Century,* because of Eiseley's purpose of drawing together the links in evolutionary thinking. The narrator becomes meta-observer and synthesizer, for example, in interpreting the recurrent demise of civilizations. Going beyond simple narrative of the decay of Mayan culture, which presumably grew top-heavy under the burden of an increasingly large priestly class, Eiseley seems to predict death to any civilization, while the natural forces, "the hungry rootlets," continue: "In the Guatemalan jungles the mathematical computations of the Mayan astronomer-priests lay lost and unread beneath the hungry rootlets of the rain forest" (*DC*, 301). This synthesizing perspective recurs in the conclusion as Eiseley, having traced the emergence of evolutionary thinking, turns to what he perceives as similarities in the way that both organisms and societies "ingest" and "reject" the materials that come to them and, using this analogy, urges the need for assessment of the "new powers and limitations" of humankind:

> It must be remembered that in geological terms we are living perhaps at the very dawn of complex human society and this is most unfortunate because man, in coming to understand his genetic history, continues to look toward the past. This is the burden which science, and particularly evolutionary biology, has placed upon man's shoulders even as it has tried to free him from the shackles of superstition. (344–45)

This "other" rhetoric involves both interpretation and ethos, the attempt to project and in projecting to create a humane image that science can try to achieve instead of the dehumanized image that science has come to project. It is an ethos that will recur throughout Eiseley's texts—an ethos that combines the simple anthropological fact of the human hand's ability to reach (intertextual with Browning's "reach [that] must exceed his grasp") with the tendency of all life to reach to be more than it presently is and specifically with the human ability to reach beyond itself, to extend vision by seeing through another's eye or to stretch out the hand in a gesture of community.

Tutor Texts

A significant technique that begins as early as *Darwin's Century* is the use of a tutor text, from literature or philosophy or even another scientist who writes. In *Darwin's Century* some of the tutor texts that Eiseley uses are texts that were also Darwin's tutor texts, especially the notion of nature itself as a text or hieroglyph. Eiseley notes that a whole school of philosophy arose in Germany that perceived the world, in the words of Gode-von Aesch, "'as a gigantic system of hieroglyphics, as the language of God or the book of nature'" (cited in *DC*, 95). Such a system naively accepted the fixity of the book's meaning, and early uses of the root metaphor of the world as text all carry the notion of decoding or deciphering the book to discover the exact meaning God intended. Such notions of fixity in the text parallel the Cartesian paradigm's universe of laws, which incorporates the Judaeo-Christian concept of the Divine Lawgiver who created and set in operation the text of the world, with its unchanging *scala naturae,* or chain of being. The human role was that of puzzle-solver; once the text was read, the meaning was clear. In this view meaning, like the world itself, did not change.

Evolution, however, radically shifted the textual metaphor in a manner that anticipates postmodern notions of the text that never quite stands still for analysis, of the text that is created only as the reader participates in its creation. Eiseley frequently draws upon the textual metaphor, but he undermines notions of textual fixity in favor of the indeterminism unleashed by Darwin even though Darwin himself seemed still caught in the deterministic outlook of the nineteenth century. "Nature's hieroglyphs" emerge as a model for writing and understanding: existing before humankind, they are like oracles, ambiguous enough to be read in terms of the reader's needs. (For further analysis of the parallels of indeterminacy and textuality in Eiseley's texts, see chap. 7.)

Pursuing the text of nature through Lyell's "readings" of the geological record, Eiseley is able to contextualize Darwin's observations. Darwin went on board the *Beagle* "with a mind fresh from the European geological and biological controversies of his just completed student days" (*DC*, 158). The epigraph for this chapter is Darwin's own comment: "The force of impressions generally depends on preconceived ideas" (141). Thus Eiseley establishes a cultural framework for reading Darwin's texts, one that is strengthened when we learn that Darwin received in Montevideo in November 1832 a copy of the second volume of Lyell's *Principles of Geology*, the volume that treated biology (160). With this framework of Darwin's knowledge of contemporary thinking, Eiseley traces through Darwin's notes the evidences of extinction of species, of geographical variation, of the sequence of paleontological forms, and finally the possibility of modification of species through organic change. Dramatically, he then focuses on Darwin's observations in the Galapagos as the source of his realization that natural selection is the mechanism. "These burnt-out volcanic chimneys, parched and blackened as an iron foundry, made a profound impression upon Darwin," Eiseley writes, and "his arrival could not have been better timed to impress upon his mind a series of facts, both geological and biological, which were necessary to the formulation of his theories" (167).

Eiseley also draws on scientific pronouncements of Darwin's contemporaries as tutor texts, quoting them, commenting, and then relating them to both Darwin's position and his own. For example, Lord Kelvin's prediction that the universe would gradually run down becomes a tutor text that Eiseley criticizes and uses to make the point that science itself is constantly changing. He notes Darwin's attempts to meet such criticism, but he concludes that, "for all their striving, the physicists would be unable to coax the monster back into the bottle" (*DC*, 242–44). Eiseley uses the inaccurate predictions of the future made by nineteenth-century physicists such as Kelvin to conclude that, though the sciences that are most mathematically precise are often considered also the most reliable sciences, the geologists, "who appeared to their physicist colleagues as bumbling amateurs expressing themselves only in vague hunches," were actually "remarkably right" (234).

Another tutor text, also important to the intellectual movement of *Darwin's Century* but further significant in allowing Eiseley to establish his own position, is Alfred Russel Wallace's *Quarterly Review* article, which argues that the human brain developed rapidly and to a degree beyond what natural selection alone would have explained. This brain having once developed, ideas, language, and social interaction having emerged, the human brain would no longer be subject to the mechanics of natural selection (*DC,* 311–23). With the human brain, in Wallace's words, "'an instrument has been developed in advance of the needs of its possessor'" (cited in p. 311). Eiseley does not accept Wallace's later theological leanings, but Wallace's emphasis on the potential embodied in the human brain and on the role culture may play in shaping human-kind provides a powerful antithesis to what Darwin's followers (and sometimes Darwin himself) tended, in Eiseley's view, to overemphasize—the role of struggle and the survival of the fittest. Darwin could look briefly to human potential, writing in *The Descent of Man* that "with man we can see no definite limit to the continued development of the brain and mental faculties, as far as advantage is concerned" (quoted in p. 322), yet so powerful was the deterministic mind-set that his hint of the potential of the human brain is quickly overshadowed by the limits of natural selection. For Darwin, "the moment passed," and "across the pages of the selfsame book march struggle and habit, the war of tribe with tribe" (322). For Eiseley, the struggle cannot be overlooked, but neither can the element of cooperation, whether at the cellular or the societal level, if we are to understand evolution. The early evolutionists' "success with the concept of struggle in the exterior environment" is expressed in terms of a figure that will recur in the popular essays: their success "had led them to see everything through this set of spectacles" (336).

But Darwin did not simply observe and shout "Eureka!" Over a period of years he wrote to Wallace and others about his experiences, and these experiences Eiseley dramatizes by inviting the reader to see through the eyes of a young naturalist who was "already impressed by the similarity of the extinct armored glyptodonts to their living relative, the armadillo," who "had seen the slow variation in the form of related species as one moved along the great distances of the South American coasts, or passed from one side of the great Andean mountain barrier to the other," who had seen similarities in the structure of creatures both remote in time and presently living, and who "had stared at a penguin's wing and had perceived that by certain modifications a wing could be made to beat its way through either water or air" (*DC,* 168). Shortly afterward, Darwin wrote to his cousin of the impact of Lyell's views and added that he was "'tempted to carry parts to a greater extent even than . . . [Lyell] does'" (cited in p. 169). Later Darwin would piece together his synthesis because he had observed different species of turtles on islands barely separated from one another, but also because he had read Lyell and Humboldt. "The origins of his thought were as diverse as the fragments of the puzzle which he at last fitted together," Eiseley maintains (174). Eiseley's pursuit of the tutor texts that Darwin used is necessary to fitting together the pieces of the evolutionary "pirate chart," but it becomes also the foundation for a literary technique that moves from Darwin's journals to Thoreau's, to Melville's *Moby-Dick,* to Keats, and to C. P. Snow, among others.

One of Eiseley's most poignant uses of a tutor text comes on the last page of *Darwin's Century,* as he turns to twenty-eight-year-old Charles Darwin's insight into the interrelationship of living forms. Although Eiseley is a humanist, he constantly undermines any notion of human superiority or mastery that may have become embedded in the humanist perspective. In Darwin's lines linking animals and humankind there is embodied, Eiseley contends, "his heritage from the parson-naturalists of England," for Darwin wrote:

> If we choose to let conjecture run wild, then animals, our fellow brethren in pain, disease, suffering and famine—our slaves in the most laborious works, our companions in our amusements—they may partake of our origin in one common ancestor—we may be all melted [later emended to *netted*] together. (*DC,* 352)

None of Darwin's forerunners, Eiseley contends, "saw, in a similar manner, the whole vista of life with quite such sweeping vision," and "none . . . spoke with the pity which infuses these lines" (352). Beginning with Darwin's tutor text, Eiseley establishes his own position: like Darwin, he embraces what he has called the "more gracious, humane tradition of the parson-naturalists," though he does not reject Baconian observation and experiment. The popular essays and, to an extent, *Darwin's Century* emerge out of his need to create a dialogue of these two sets of values.

Tutor Structures

Eiseley's use of tutor structures from literature also begins in *Darwin's Century.* Whether he draws on a type of historical narrative, such as a tale of the American West, or on the epic, Greek drama, the beast fable, travel lore, liturgical structures, or autobiography, Eiseley incorporates these structures in his essays so that they function as codes to which the reader responds as to any code to which the response is "always already" culturally embedded. In *Darwin's Century,* the tutor structure is Homer's *Odyssey.* The book echoes classical epic conventions—even including the twelve books of the classical epic. The hero is dual: he is humankind personified by Darwin, who does emerge in a heroic light at the end of the book as Eiseley describes his sympathy with life. This dual hero undertakes a journey—the evolutionary journey for humankind, the journey aboard the *Beagle* for Darwin, the journey of knowledge for both. The hero must do battle frequently, even with the supernatural, which is presented ironically in terms of impeding myth that complicates and slows the hero's quest. Eiseley's epic catalogue appears in the lists of early evolutionists and the catalogues of their contributions. Even a kind of epic boast appears in Patrick Matthew's distribution of printed cards that proclaimed himself the first to discover natural selection. Darwin's own more modest version was his comment in a letter to Wallace that Matthew's explanation was "'a most complete case of anticipation'" (cited in *DC,* 126).

Even the epic simile has its equivalent in *Darwin's Century* in the sweeping comparisons of evolution to a monument or a boulder, which Eiseley introduces when he writes that the "mass of accumulated evidence had the weight of a boulder. Criticism flowed around and over it but the boulder in all its impenetrable strength remained" (*DC,* 195). The simile becomes pivotal in the conclusion, as Eiseley's method increasingly anticipates the method of the more "literary" texts. In the opening paragraph of the conclusion Eiseley introduces the boulder as the controlling metaphor for evolution, arguing that "ideas, like the disintegrating face of Hutton's planet, evolve, erode, and change" (325). Although some ideas vanish quickly, others, like rocky projections from the physical landscape, "may last for ages protruding, gaunt, bare, and uncompromising, from the soft sward of later beliefs" (325). Like its vehicle, the idea appears fixed but is nevertheless subject to erosion, to evolution, to constant and creative change. This "rock" is subject to the process that Eiseley calls "transmutation" through the kaleidoscopic effect of shifting light. The interaction of cloud and landscape produces the effect of shifting colors in the elaborate image:

Sometimes, in the clouds that pass over the formless landscape of time, [ideas] will seem to shift and catch new lights, become transmuted into something other than what they were, grow dull, or glisten with a kind of sunset color reflected from the human mind itself. Of such a nature is that vast monument to human thinking which is now called evolution. (325)

Evolution is established, then, as a rocky formation, a "monument," and a "structure" that is "ever vaster and more impenetrable" (325)—characteristics worthy of treatment in epic proportions. Landscape metaphors recur frequently in Eiseley's texts, serving to spatialize concepts and provide a point of reference for the reader, but the elaborately devised simile of evolution as a rocky formation anticipates the complexity of these comparisons. Because of its pivotal position in Eiseley's development of his *genre,* I turn now to a specific consideration of the conclusion.

Developing a Method: The Conclusion to Darwin's Century

Sensitively and elegantly written, the conclusion to *Darwin's Century* summarizes the major themes that later appear in the popular essays and employs the methods of the popular essays (some concurrently developed in *The Immense Journey*) and thus may be seen as the prototype of the popular essays. In the conclusion Eiseley establishes four of the points to which he will return in the popular essays: (1) the need to assess human powers, (2) the need to understand the human role in "indetermination," (3) undermining of metaphors that perpetuate emphasis on struggle, and (4) primitive humankind and "primitive" thinking as models for a culture dominated by a scientific-industrial-military institution that is accepted and even worshipped by many people. The conclusion is a combination of scientific information, which becomes the vehicle for the recurrent Eiseleyan themes, with metaphor and analogue, quotation, interpretation, and personal experience. In the later essays, the element of personal experience often plays a larger role, but here it adds a powerful affective dimension; and this essay, which summarizes the main points treated in *Darwin's Century* and looks toward their cultural and human implications, is a structural model for the later work.

Like most of Eiseley's essays, the conclusion begins with an epigraph that abstracts one of the main themes—in this case the theme of time and its human interpretation: "Life can only be understood backward but it must be lived forward" (*DC,* 325). The quotation foregrounds Eiseley's concern with the individual human being and her relationship to an evolving world. In this line from Kierkegaard, Eiseley pinpoints the problematic of the human relationship to time, which becomes the basis for organization of the essay into five sections:

I.   Time: Cyclic and Historic
II. The Pre-Darwinian Era
III. The Struggle of the Parts
IV. Evolution and Human Culture
V. The Role of Indeterminism

Each section treats the human awareness of time from one perspective or another, and the focus is on the impact that the prevailing concept of time has on human culture.

Not only is the conclusion about time and the relationship of human cultures to the perception of time, but in the research on Darwin as well as on pre- and post-Darwinian evolution, Eiseley has become acutely aware that popularization of evolution involves more than explaining natural selection and the "community of descent." Psychological and ethical implications are inextricably interwoven in his study, especially in the conclusion. In Eiseley's words, "the man of blood" and "the man of peace" have both utilized the arguments of evolution (from the age of the earth, to the gradual emergence of geological and animate forms, to the great quantity of individual variations and their roles in creating species, to the decline of acceptance of the world as a balanced machine [*DC*, chaps. 1–8; succinctly stated in *FT*, 70–71]). What is needed, however, is "a long second look at the history of this concept and at its moral implications" (*DC*, 326). The conclusion, then, not only summarizes the main points of the book, but also provides the beginning of this "long second look"—a "look" that leads Eiseley and the reader on the epistemological journey of the popular essays.

Opening with metaphor of evolution as a boulder (detailed in the previous section), Eiseley links the idea—the very word is related to "seeing," to the Old English *wit*, "knowledge, intelligence," to Latin *videre*, "to see," to the Greek *eidos*, "form, shape" (all dependent on the visual perception of form and boundaries)—to geological formations. Fittingly, it was understanding of gradual change in geological formations that led to understanding of change in life forms. Evolution is established, then, as a rocky formation, a "monument," and a "structure" that is "ever vaster and more impenetrable" (*DC*, 325). This "monument" is linked both to the atom's mysteries, which introduce the unpredictable and indeterminate in nature, and to human consciousness, which cannot be explained entirely by "the soft dust that flies up from a summer road" (326). Drawing on the biblical code of the human as dust as well as on the indeterminacy of the atom, Eiseley plays with the cloud, which gives shifting colors to a landscape that is equally but more slowly shifting. A cloud, like the dust from the road, consists of fine particles, and its formations are unpredictable. Evolution as an idea is linked to the human consciousness, which cannot be totally explained, as the cloud cannot be totally explained or predicted. Both are constantly shifting.

In addition, in all of Eiseley's cloud and dust metaphors there lurks a defamiliarization of the Judeo-Christian notion of humankind as supernaturally compounded of dust; Eiseley omits the supernatural but retains the shifting and diaphanous cloud of dust or moisture (in a Bachelardian sense, the ancient elements of fire, air, earth, and water recur in various guises in Eiseley's texts, as he notes in *The Immense Journey*. It is relevant that he concludes this first section of the essay with a speculation that between Sedgwick's supernaturally oriented "progressionism" and modern scientific "complacency" in the scientific ability to explain all things, there may remain "other mysteries as great as those that intrigued Darwin" (326).

Consciousness, then, seems in one sense like the cloud of dust, yet "no one has quite succeeded in identifying" the two. What we do not know is also what we do know—that contingency and equivocity are inescapable. In the text's uncertainty or equivocity, there is also strategy: the reader is allowed to experience and to assert the unexpected. Further, the equivocity of consciousness and its role in the evolutionary process is echoed in the equivocity of the history of evolution, for some have seen evolution as reducing humankind to a bestial condition, while others have viewed it in terms of human potential to be other than what we are. It is this tendency, above all, that leads Eiseley to posit the need for "a long second look" at evolution and its history, which becomes a look at evolution not only from the perspective of its social and moral ramifications, but also at evolution as itself a "figure" of thought; the notion is one that Eiseley will explicitly address in the later essay "The Illusion of the Two Cultures" (1964) when he says that evolution has become a root metaphor. With evolution as a rock taking on new colors as a cloud diffuses shifting colors onto it (later Eiseley will cite an essay by the unknown medieval mystic who wrote "The Cloud of Unknowing," as he wrestles with the "unknowing" that inevitably accompanies knowing), Eiseley introduces his consideration of the "social and moral ramifications."

This concluding chapter is prototypical of his concern with the social implications of evolution, with the problem of the individual in relation to evolution, and with the individual's quest for knowledge of the human place in an evolutionary world. And it is prototypical of the Eiseleyan method of organization, which employs (not in a fixed order) metaphor, quotation, personal reading of the quotation, and personal experience or anecdote—all in explication of or reflection on a point of science or of science in its cultural context.

The second part of the essay reviews and summarizes the pre-Darwinian nineteenth century, with its growing geological understanding, interest in morphological biology, realization of the enormity of time, and theories of "progressionism" that suggested a successive development of life forms from simple to complex through successive creations rather than actual phylogenetic descent of forms. The notion of progress in evolution, Eiseley will note in many essays, became embedded in nineteenth-century understandings of even Darwinian evolution; both "progressionism" and progressive evolution are results of the shifting light from Eiseley's "cloud of unknowing."

In this part of the essay Eiseley also emphasizes the notion of the past as unreturning and undermines Lyell's "safe," cyclical time predicated on the Newtonian machine. As astronomy and paleontology discovered an increasingly lengthy past, "the historic ever-changing, irreversible, on-flowing continuum of events was being linked to galaxies and suns and worlds" (*DC*, 330). Gradually the clues emerged that would present time itself as creative and thus would emphasize both cosmic and organic novelty. Even Malthus, for example, with his suggestion of the "balances" on population, was a product of eighteenth-century notions of "balance" inherited from Newton. What Darwin and Alfred Russel Wallace left to the next century was a new sense of time, which brought with it a new, continuously and infinitely creative world, but a world that was also far more indeterminate than Darwin himself, given his frame of seeing, could admit.

The stylistic strategy with which Eiseley makes the point is a metaphor as wild and uncontrolled and teeming with creative life as the Newtonian metaphor of the universal machine is fixed and controlled, not creative but representative: time, no longer subject to a Newtonian "natural government" (the word *natural* provides great tension in Eiseley's texts because humankind seems to accept as "given" anything that comes to be accepted as "natural"), has become "instead a vast chaotic Amazon pouring through unimaginable wildernesses its burden of 'houses and bones and gardens, cooks and clocks'" (*DC,* 332). Time, then, is metaphorized in terms that anticipate the interest, thirty years later, of scientists from widely disparate fields in the dynamics of flow and change, of the chaotic that is also orderly, the order that is also chaotic.[5] As a wild, uncharted, and romantic river rather than a machine, time carries forward the synecdochic elements of human life, in the phrase quoted from Wallace.

Eiseley turns next to the invention of the spectroscope, which led to astrophysics and to confirmation of Newton's contention that the sun's rays are a combination of colors separable by means of the lens. The solar spectrum provides metaphor and model in the popular essays. *Spectrum*—meaning "appearance," "image," "form"—emerges from the Latin *specere,* "to look at," which is related to other terms from the life sciences, including *speculate, specimen, spectacle, speculum,* as well as to other "sight" words that are relevant to Eiseley's reexamination of scientific "observation."

The emergence of astrophysics is treated with a metaphor that carries forward the archetypal notion of the human quest for fire; humans could not "dip a ladle into the glaring furnace of the sun and stars" (*DC,* 333). The spectroscope thus becomes the means of breaking light and of bringing heat closer, not of touching the "glaring furnace," but of using a characteristically human tool to dip into it. Traditionally, light is knowledge, the philosopheme of sight. The metaphor is further related to the metaphor of the journey in its relationship to the "shining" of *dies,* the Latin root of *journey.* The first chapters of *Darwin's Century* begin with the Renaissance voyagers of discovery, who were followed by "time voyagers," such as the Comte de Buffon and Erasmus Darwin, who journeyed into the depths of time, and with reference to the "pirate chart" of the early explorers of time. The spectroscope brought what Eiseley calls a "sidereal chemistry"—a chemistry whose crucible (either the vessel used for melting materials or the bottom of a furnace where molten metal collects) can be dipped into by means of a ladle of broken or splintered light (Shelley's "splintered radiance" of consciousness, from *Adonais,* and Eiseley's shattered mirror become insistent intertexts in later essays). Through spectroscopic investigations, the "sidereal chemistry" led one observer to write in 1869 "'that the whole visible material universe *is an evolution of things'*" (Pritchard, cited in p. 333).

The chemical metaphors lead, then, to an understanding that the elements of the "visible universe" are also the elements identifiable in our solar system (*DC,* 333). Organic and inorganic are related rather than separate. The visual and "chemical" metaphors dependent on fiery furnaces and gasses are related to the "cloud" of consciousness that breaks the light and scatters varying colors on the rocky monument of evolution, and to the prevalence of fire as an energy source. But the chemical metaphors also have imaginative import that foreshadows Eiseley's later attention to the human movement up the "fiery ladder" ("The World Eaters," *IP,* 63), which is the "energy ladder," and his observation that links *The Immense Journey* to the four ancient elements ("The Slit," *IJ,* 13). In the words of Northrop Frye, "earth, air, water and fire are still the four elements of imaginative experience, and always will be" (1968, vii). Science and alchemy, chemistry and imagination are essential mixtures for the scientist who insists on the unity of all creative insights.

Ending the second section of the conclusion, Eiseley returns to what the spectroscope confirmed—that time was not the circularity that it had been thought to be, that it was unreturning, that "it was a loneliness, an on-going," although "the silver thread of genetic continuity" can be traced in "the always looming and unknown future" (*DC,* 334). This silver thread is as cool as the chemical metaphors are hot. It is a winding thread., and the winding will appear again later in Eiseley's texts—for example, in his description of the illusion of a trunk snuffling, growling, winding behind a fresh-faced student ("How Natural Is Natural?" *FT,* 167–68). And the winding thread of genetic continuity repeats the motion of the model of life, the spiral of the DNA molecule, which Eiseley explores in his later texts, drawing on the code and examining its possibilities, subtly undermining deterministic interpretations of it by scientists whose insights, he insists, should be more complex.

The third section of the conclusion treats the emergence of evolutionary thinking and its rejection of preconceived Platonic "ideal forms." As he will write in various ways in the later essays, Eiseley contends that "the fixed taxonomy of life is an illusion born of our limited experience" (*DC,* 335). He compares our limited perception of the "writhing" of one form into another with a slowed-down motion picture; if the frames were speeded up, we would be aware of the illusion produced by our tempo of living. But evolutionary thinking produced the root metaphor of the struggle within the tangled bank, the "war of nature," which becomes, Eiseley says, "an apt illustration of the way in which a successful theory may be carried to excess" (335). As a countermodel Eiseley posits that "the co-operative aspects of bodily organization, the vast intricacy of hormonic interplay, of cellular chemistry, remained to a considerable degree uninvestigated" (335). Even in the embryo the early evolutionists perceived a battle; figuratively, "in a fertilized cell the very ancestors were struggling as to which might emerge once more into the light" (336). Eiseley's concern in this section of the essay is the human tendency, to which he returns throughout the later essays, to "minimize" cooperation among animal forms and to omit emphasis on "internal stability and harmony" (336) within the individual organism as well as within society. This is the approach for which the Malthusian model became—unconsciously for the Darwinian observer—the shaper of observation. Metaphorically, Eiseley concludes that the notion of struggle provided a "set of spectacles" for a generation of Darwinians (336). Recalling the *spectrum* and associated "sight" words, then Eiseley's spectacles reverberate with what language "knows" about sight, for spectacles frame, alter, and shape sight. The specular metaphors, including spectacles, mirrors, and kaleidoscopes, become important codes in the later texts.

Section IV of the conclusion centers on the chief concern of all of the later popular essays—the relationship of the concept of evolution to human culture and the role of the individual in an unfixed, nonteleological world in which not only geological and living forms lack stability, but knowledge itself is equally unstable. Here Eiseley establishes the dangers of the notion of humankind as nothing more than "animals" subject to struggle and survival of the fittest. His silent dialogue with Herbert Spencer and "social Darwinism" will run throughout the popular essays as he attempts to establish that continuous creation through evolution gives humankind—as indeed it has always given life itself—the ability to become more than it has been. Eiseley attempts to present the positive aspect of evolutionary becoming, though his efforts often end in a skepticism concerning human use of its potential, as embodied in the frequent references to his poem "The Maya," which evokes the astounding mathematical and astronomical achievements of the Maya, who used the zero and calculated time in eons, though they never knew the wheel, and who came, as all civilizations have, to nothingness:

> Behind, nothing,
> before, nothing.
> Worship it the zero. . . .
>
> ("The Maya," *AKA*, 23)

As the efforts to stress the positive capabilities bestowed by evolution falter, Eiseley draws the reader into his skepticism for the future of human culture; the stability of culture is defamiliarized as the reader, too, confronts cultural and epistemological instability. Yet in his dialogue with Darwin, Spencer, Freud, and other determinists, Eiseley always seeks evidence of the human freedom to defy the deterministic evolutionary constraints, and his dialogue expands to include knowledge and its shifting, knowing and not-knowing, which always returns to the potential evoked from nothingness (this point is most effectively made in the essay titled "The Star Thrower," which I examine in detail in chap. 7).

In this section of the conclusion, titled "Evolution and Human Culture," Eiseley discusses the parallels of biological and anthropological thinking and the mutual influence of the two. Mistakes in developing the biological theory were, he argues, duplicated in social thinking. Despite biblical literalism, he suggests, the Christian world had absorbed enough of the Aristotelian ladder of taxonomy to provides "the seeds of speculation" (337). Biological notions of "progress" and similar social notions seem to have emerged concurrently once the earth's antiquity was known; students of literature and science will note, with Hayles, that ideas are embedded in culture and hence are "isomorphic." Biological notions of stratification among human beings led to confusion of actual cultural levels with biological potential. Only with rejection of the unilinear in biology could cultural unilinearity be discarded; only then did it become clear that every culture is unique and that no observable "primitive" culture is an antecedent of others.

Similarly, both biology and anthropology were concerned with the external world to the exclusion of the specific organism. In anthropology, there was a period of studying the widespread "diffusion" of cultural characteristics over scattered geographical locations, of studying cultures as made of "shreds and patches" combined from varied sources, rather than examining the combining of these cultural traits into societies that were shaped, as it were, "by inner organizing forces" (*DC*, 343). The tendency toward partition thus functioned in both biological and cultural studies until cultures began to take on "individual personalities," as studied in the twentieth century. Eiseley's movement away from partition is evident as he suggests "that the holistic, organismic approach which finally emerged in biology when the intricacy of inner co-ordination and adjustment began to be realized has, once more, its analogue in the social field" (344).

Thus is the way prepared for an important analogue for social change: societies emerge as organisms, which "ingest or reject materials which come to them" and assimilate what they take in so that "when it reappears as part of the social body it has been molded to fit a purpose other than what was envisaged in another time and place where the trait arose" (*DC*, 344). Often, too, a given cultural or psychological trait will survive its political or technological usefulness, just as physical traits sometimes outlast their functions; as in the physical organism, there is an "inner cohesiveness" in the "social mind" (344). In the larger context of Eiseley's essays, one might think immediately of the ongoing effects of mechanistic metaphor—of a metaphorical concept that has survived beyond its usefulness and that Eiseley repeatedly attempts to undermine. Using the organismic model, Eiseley suggests that the emergence of the human brain makes possible a social tradition that is, of course, not hereditary, but that can be seen as paralleling heredity. Because humankind is still at the beginning of complex social organization, the organismic model underscores the human capacity for cooperation.

Yet in the attempt to free humankind from superstition, Eiseley continues, evolutionary thinking brings a burden because it may lead us to focus on the past and on our animalistic qualities rather than on our evolutionary potential. Used as a rationalization for bestial behavior, evolution becomes a danger, yet still Eiseley is at pains to demonstrate how it also reveals life's infinite creativity. He appropriates Henri Bergson's notion that "the role of life is to insert some *indetermination* into matter" (cited in *DC*, 347). For Eiseley, Wallace's contention that the brain has made unnecessary the further evolution of parts deserves more contemplation than does the notion of struggle in the tangled bank. No longer subject to determinism, the human mind has become "the arbiter of human destiny" (347). In Darwin, Eiseley finds personified the strange dialogue of continually shifting knowledge—the remarkable insights that led to indetermination versus the "blind spot" of deterministic emphasis on struggle and the Enframing of intellectual systems of his day (specifically, utilitarianism and Malthusianism).

Concluding this fourth section, Eiseley returns to the chemical metaphor established in the beginning, a metaphor that again provides a countermodel to the tangled bank: body chemistry depends on the cooperation at the subcellular and cellular levels, on cells joining and even sacrificing themselves for the larger being; the individual cell "is a laboratory where chemical processes are being carried on in an amazingly co-ordinated fashion" (*DC*, 349). As opposed to the nineteenth-century notion of ancestors warring for emergence in the embryo, Eiseley turns to Bergson's suggestion (in a phrase that echoes the *other* discourse of the nineteenth century, the romanticized discourse of family life that Queen Victoria herself would have approved) that, in the cellular chemistry, each generation "bends lovingly over the cradle of the next" (cited in p. 349). Thus are established the two tendencies of humankind to which Eiseley returns throughout his texts—aggressiveness and cooperation, the ape and the tiger that he calls "bad metaphors at best." And he emphasizes the human dilemma of dialogue between these tendencies by quoting from another popular writer who argued that, of the two tendencies, cooperation is both "the more elusive and the more important" (James R. Newman, cited in p. 349).

The last section of the conclusion centers on "The Role of Indeterminism." Here Eiseley most clearly foreshadows the themes and methods of the later popular essays, noting again Wallace's contribution to evolution in his emphasis on the insertion of human volition into nature and inserting an observation (which may seem a wandering but actually reinforces his earlier argument about ideas that outlive their time) about the human fascination with mechanisms and with "progress." Although humankind "represents the genuine triumph of volition, life's near evasion of the forces that have molded it," the species is threatened, Eiseley contends, by "confusion of the word 'progress' with the mechanical extensions which represent [human] triumph over the primeval wilderness of biological selection" (*DC,* 350). For Eiseley this confusion is a "reversion" that is manifested in enormous expenditures on industrialization and also on developing military implements that provide little more than a mechanical replay of the world of the dinosaurs.

To illustrate his contention that what the twentieth century considers "gracious living" is spuriously associated with amenities and machines, Eiseley inserts a personal example from his own field explorations. It is the only personal example in the scholarly *Darwin's Century,* but it exhibits the compassion and the use of individual experience as a means of thinking through larger problems that become hallmarks of the popular essays, and it gives a powerful affective dimension to the last pages of the book. In the passage, which is reminiscent of Claude Levi-Strauss's descriptions (and even his romanticization) of the Bororo tribe of South America, Eiseley recalls wandering with a companion, both men tired and lost, into the rude camp of a Mexican peon:

This man, whose wife and newborn child were sheltered in a little hovel of sticks into which one could only creep on hands and knees, supplied our needs graciously. To our amazement he gently refused any payment, and walked with us to the edge of his barren lands in order to set us on the right path. (350–51)

Having established the primitive conditions under which the family lived, Eiseley uses the family as an example of simplicity and dignity and as a contrast to modern North American notions of these qualities.

There was a dignified simplicity about this man and his wife, in their little nest of sticks, that was a total antithesis to gracious living in the great land to the north. It demanded no mechanical extensions, no stewards with shining trays. We had drunk from a common vessel. We had bowed and spoken as graciously as on the steps of a great house. I had looked into his eyes and seen there that transcendence of self is not to be sought in the outer world or in mechanical extensions. They can be used for human benefit if one recognizes them for what they are, but they must never be confused with that other interior kingdom in which man is forever free to be better than what he knows himself to be. (351)

From this simple example he draws the conclusion to an exhaustive work of scholarship that has traced the development of evolutionary thought from its roots in the Greek ladder of things through the *scala naturae* of medieval and Renaissance thinkers and through Darwin's departure from the scale, as juxtaposed to his and his followers' unconscious incorporation of it. The question that remains is one of the individual's part in the post-Darwinian world. Does the individual, perceiving primarily the selfish struggle in nature, abstract from it a role of struggle, a continuing battle of selfishness? Or does the individual perceive in evolutionary nature and modern cell chemistry a model of cooperation, of the social "organism"? For Eiseley, the mechanical world "tickles man's fancy" (351) but does not lead to concentration on the unfolding future. The mind behind the machine provides its "color"—an echo of the spectrum that began the conclusion, of the colors into which light is broken, and of the breaking itself as ongoing process.

Despite his references to Darwin's "blind spot," the concluding paragraphs defend Darwin in terms of art as well as science. Darwin, Eiseley argues, was "a master artist" who "entered sympathetically into life" (*DC,* 351). At age twenty-eight Darwin had an insight that none of his predecessors had, for he wrote in his journal of "the whole vista of life," providing a model that will be recurrent in Eiseley's texts (352). He suggests that few young people in mid-twentieth century would "pause, coming from a biology class, to finger a yellow flower or poke in friendly fashion at a sunning turtle on the edge of the campus pond" or would write that we are "all melted together" (352). With this detail of the individual's response to the natural world he concludes that Darwin's impact—his "shadow" (another reverberation of the "spectrum" of sight) is due not solely to the exact and complex reasoning that produced the *Origin,* but also his sympathy with life, which is "his heritage from the parson-naturalists of England" (352), who influenced his participation in life.

The conclusion, then, not only attempts to place in perspective the difficult path of development of the idea of evolution, but it marks the concerns that will inform Eiseley's popular essays. Early in *Darwin's Century* he notes the impact of the popularizer as "a very significant figure in the earlier centuries of science," whose "work might plant the germ of new ideas in other, more systematic minds" and whose popularity can indicate "the ideas which were beginning to intrigue the public imagination" (*DC,* 30). The conclusion marks the emergence of most of Eiseley's literary strategies: the epigraph from literature or philosophy; the statement of focus followed by simple and specific explanation of a scientific fact and by metaphors that condense or model the idea being explained; quotation from philosophical or literary texts rather than from scientific treatises, with personal rather than formal readings; and—what is perhaps most memorable in Eiseley's texts—use of personal narration to think through the scientific material that is the vehicle of reflection.

5
The Genesis of Method II:
The Immense Journey

While recovering from a respiratory illness that led to temporary deafness, Eiseley began writing the essays that were incorporated into *The Immense Journey*. As he describes the experience in *All the Strange Hours,* the experience of deafness so profoundly moved him (he had been haunted since childhood by his mother's deafness and mental instability) that he turned his writing talents in the direction he had always wanted to pursue, developing his "concealed essay" as a vehicle for communicating with a larger public. Some of the essays incorporated in *The Immense Journey* were thus composed concurrently with *Darwin's Century.* The essays in this first collection, however, are short and anecdotal, and the anecdotes are not extended for exploration of their literary possibilities as are the later essays. Yet striking parallels in insights, even in syntax, emerge as we study the two books, and the *themata* and *topoi* established in *Darwin's Century* reappear in the companion book, though its approach is more popular, more oriented toward narrative and social commentary.

In *The Immense Journey,* Eiseley confronts twentieth-century society's fascination with the machine, though he does not pursue this fascination in the context of the Cartesian-Newtonian paradigm so thoroughly as he will later. And in his basic explication of Darwinian themes Eiseley begins in *The Immense Journey* to examine the metaphoric sense of the organs through which we perceive and manipulate our surroundings, to explore the organs of the body in relation to knowledge; his pursuit of these metaphorical concepts or philosophemes will continue throughout the texts. In order to treat these metaphorical concepts and *themata,* I shall examine key essays in the collection.

A Crack in the Prairie: "The Slit"

Eiseley introduces the evolutionary journey of humankind as a controlling trope in "The Slit," the first essay in *The Immense*

Journey. The essay begins with one of Eiseley's dominant techniques—the contrast of landscapes and the experiences that landscapes evoke. The slit is a spatial model of the time dimension, an alternative to the well-known, humanly created ladder. The slit contains layers, but not *all* layers from all places; there is no single ladder. This spatialized image introduces Eiseley's recurrent themes of time and its uncanny relationship to life itself. The time dimension, he writes, is "denied to man" because each generation is tied to its own century ("The Slit," *IJ,* 11). The individual sees into the past and imagines the future, but is simply part of the "caravan," traveling as far as possible but never seeing or learning all that she would like to see or learn.

The framework narrative is a journey downward into a slit the width of the narrator's own body, until the sky itself becomes a mere slit. Eiseley's narrator-self chisels around an animal skull that links him to the time dimension, thinking of the human journey symbolized by the hand, but rejecting fixed "meaning":

> Perhaps there is no meaning in it at all . . . save that of journey itself. . . . It has altered with the chances of life, and the chances brought us here; but it was a good journey—long, perhaps—but a good journey under a pleasant sun. Do not look for the purpose. Think of the way we came and be a little proud. (6–7)

The chance that produced human life, continuing change, and the length and pleasantness of the sojourn will be ongoing themes. The journey itself is the meaning—a meaning that constantly changes and is constantly shaped and reshaped by the hand and the eye. Natural or supernatural *telos* and determinism are displaced, as well as the notion of fixity or of humankind as the purpose of the journey.

Chiseling his way, Eiseley's narrator notes "the cunning manipulability of the human fingers" and thus begins exploring the metaphor of the human hand: "It is not a bad symbol of that long wandering . . . the human hand that has been fin and scaly reptile foot and furry paw" ("The Slit," *IJ,* 6). Within the framework narrative in this essay is a flashback to another journey—the journey of an archaeologist to "prairie-dog town," where he watched a "shabby little Paleocene rat, eternal tramp and world wanderer, father of all mankind" (8). Though not a close relative to humankind, the prairie dog is part of a rodent line that succeeded in living on the ground, while ancestors of humankind took to the trees, where they developed the "cunning hands and the eyes that the tree world gave us" (10). The similar unpredictability of the human future leads Eiseley to ask, "Down how many roads among the stars must man propel himself in search of the final secret?" (12). The human journey toward knowledge, like the journey of life, "is difficult, immense, at times impossible" (12), yet it is one that, being human, we must make.

Eiseley explains this first book of essays as a "guide" on this journey, yet also "a somewhat unconventional record of the prowlings of one mind" ("The Slit," *IJ,* 12–13), which records a personal journey and an epistemological synecdoche:

> Forward and backward I have gone, and for me it has been an immense journey. Those who accompany me need not look for science in the usual sense, though I have done all in my power to avoid errors in fact. I have given the record of what one man thought as he pursued research and pressed his hands against the confining walls of scientific method in his time. It is not, I must confess at the outset, an account of discovery so much as a confession of ignorance and of the final illumination that sometimes comes to a man when he is no longer careful of his pride. . . . I can at best report from my own wilderness. (13)

"Forward and backward"—the very definition of *discursus,* or discourse—prepares the reader for a random journey that speaks to the randomness of knowledge, and for a text that seems to move at random from explication or critique to narrative and back again. The instability of knowledge underlies the instability of the text. In this key passage, Eiseley delineates the self-limiting nature of knowledge and the limitations that science imposes upon the "seeing" that it claims as transparent or unmediated reflection. The confining walls suggest the "framing" or "encircling" that troubles Eiseley and anticipates the image of "dancers in the ring" (*ASH,* chap. 18). The "illumination" that he records is not the unmediated knowledge that the Western scientific method espouses, but an illumination that "sometimes comes," that emerges from a dialogic interplay of ways of knowing, and that rejects closure and fixity, or "science in the usual sense," in favor of more intuitive knowledge. The passage echoes the biblical code of wandering in the wilderness—the sidetrack that is itself important in science and in the epistemological quest.

The evolutionary journey, which begins as synecdoche with one man's journey, is a journey of knowing that is undermined by not-knowing. The individual is often lost, but nevertheless capable of a unique vision: "On the world island we are all castaways, so that what is seen by one may often be dark or obscure to another"

("The Slit," *IJ*, 14). Eiseley reminds the reader that his personal journey is not "in Baconian terms, a true, or even a consistent model of the universe" (13). Here Eiseley's personal rhetoric becomes dominant. The Baconian epistemology is not enough; in Eiseley's quest, self must blend with science, intuition with observation.

Though the journey is incomplete and humankind is often a castaway, Eiseley later finds security in the very incompleteness of the journey, knowing that nature is "still busy with experiments" ("The Snout," *IJ*, 47). This dynamic quality of the evolutionary journey leads to a sense of sharing beyond the boundaries of human and animal form—whether in the narrator's encounter with a western jack rabbit or in Argos's welcoming Odysseus. Eiseley ends with the uniqueness of the individual's view—the view which, throughout these essays, will add a reflective, often self-reflexive, dimension to his own journey as a scientist who observes and as a poet who synthesizes.

The Magic of "Organization": "The Flow of the River"

"The Flow of the River" traces another journey—this time not downward into the earth's crust but following a river's journey that is also the journey of "magic," or "organization." Both have, for Eiseley, an unexpected, inexplicable quality that is the essence of the poetic. It is this unexpected "magic" that leads to the "eureka" moment in science, to the awe that great scientists have felt in their awareness of the universe, and to the synthesizing that makes even mundane moments poetic. "If there is magic on this planet," he begins, "it is contained in water" (*IJ*, 15). This "magic" is what he will call "organization," the movement of life toward form whose history evolutionary thought can trace, but whose reason for order, like the order of the crystal in the poem "Notes of an Alchemist," no scientist can explain (*NA*, 15).

This journey begins in the watershed of the Platte River, and Eiseley traces it over the prairie on its way to the sea. But he also recalls his personal "adventure" of drifting in the shallow water and identifying with "the unspeakable alchemies that gestate and take shape in water" ("The Flow of the River," *IJ*, 19), the "organization" that is life in any form. His personal journey brings reflection on human beings as "little detached ponds" that are, in a sense, "a way that water has of going about beyond the reach of rivers" (20). The river's journey and his own drifting are combined in the tale of "a yellow-green, mud-grubbing, evil-tempered inhabitant of floods and droughts and cyclones" (23), a catfish that he once took home and kept in a basement tank through one winter until instinct prompted the fish to leap from the shallows into the main channel and there was no channel. In the fish Eiseley sees a "kind of lost archaic glory that comes from the water brotherhood" and is aware of "the momentary shape" that is his own (24).

Form, then, is unstable, and Eiseley reflects on the distinctive noises of katydids, crickets, a rabbit's thumping, and human noise, finding "in all of them a grave pleasure" that ceases when no sound is heard along a frozen stream and "the enormous mindlessness of space settles down upon the soul" ("The Flow of the River," *IJ*, 25). The most intriguing principle of all, he concludes, is "organization," which is "not strictly the produce of life, nor of selection" (26). There is no "logical reason," he concludes, for the "organization" that produces snowflakes or evolution. Both are without causal explanation; they are moments of order whose explanation is as unstable as their form.

Life's "Reaching Out": "The Great Deeps" and "The Snout"

From the mystery of form to the mystery of life's "reaching" to be more than it presently is, Eiseley turns in "The Great Deeps" to a tale of a nineteenth-century exploration of the North Atlantic sea bed that dredged up a red sea urchin that seemed to pant in the hand of the expedition's leader. Embedded in the essay is the tale of an expedition that was expected to confirm the notion that the ocean depths contained living fossils that had escaped the disasters of shallower seas; a second expectation, that the deeps would yield the "Urschleim," a kind of half-living matter representing the transition between nonliving and living, also motivated the expedition. In the story of the quest for a primal ooze, Eiseley reflects on the depths of the sea as "the sole world on the planet which we can enter only by a great act of the imagination"—an act transcended only by nineteenth-century biologists' imaginative hope of finding "life in the process of becoming" on the floor of the sea (34). The "Urschleim," the "unindividualized ooze," represented, in Eiseley's view, the ultimate hope of mechanistically oriented scientists. "Man was mud and mud was man," in this theory that projected "an evolutionary family tree upon existing organisms" (*IJ*, 36–37). What such thinkers had to learn was that remnants of life were not neatly layered or mechanistically ordered, but that life crosses boundaries: life moved from the sea and also back to the sea. The abysses yielded a mixture of antique forms and more recent forms, the ancient forms representing migrations from above (41).

The essay combines the sea urchin's journey toward becoming more than it is with a unique human journey, the ability to project itself beyond its own form. Here the hand takes on a new metaphorical sense as the root metaphor for "reaching" or "groping" beyond the boundaries of form. As a living fossil, the sea urchin signifies the movement of life in the deeps, life's continuous "reaching" to be more than it presently is, "its eternal dissatisfaction with what is" ("The Great Deeps," *IJ,* 37). What the North Atlantic expedition revealed was that life did not begin in the deeps, but made its way there. "It was the reaching out," Eiseley writes, that brought ancient life forms ashore. This "reaching out" is a "magnificent and agelong groping that only life . . . can continue to endure and prolong" (43).

The essay concludes with the synaesthesia of the tactile reaching and the "reaching out" of human vision, which is "the most enormous extension of vision of which life is capable: the projection of itself into other lives" ("The Great Deeps," *IJ,* 46). "This is," Eiseley concludes, "the lonely, magnificent power of humanity. It is, far more than any spatial adventure, the supreme epitome of the reaching out" (46).

From studies of life in the abysses of the sea, Eiseley continues the journey of life in "The Snout," which focuses on the journey of a fish that was not quite a fish, an unsuccessful fish that was forced from the mud onto land, and the journey of "two bubbles" at the end of the Snout's brain, bubbles that became the cerebral hemispheres (*IJ,* 52). Exploring the Snout's journey, which is the unique product of the interaction of natural and genetic forces, Eiseley concludes that our worldview "is still Ptolemaic" because we see ourselves as the end of evolution (57). Yet Eiseley as narrator expresses "confidence to see nature still busy with experiments, still dynamic, and not through nor satisfied because a Devonian fish managed to end as a two-legged character with a straw hat" (47–48). Life, he concludes, "has no image except Life," which "is multitudinous and emergent in the stream of time," and the human being is "one of many appearances" of life itself (59). Here he plays on the literal stream of water and the metaphorical image of time. The link between water and life is "organization" and time bears a special relationship to life that it does not bear with inanimate objects.

"Reaching" beyond the Sea:
"How Flowers Changed the World"

From water to land again, Eiseley pursues the journey of life through the cooperation that made possible the emergence of herbivores because seed-bearing plants emerged, and of carnivores because herbivores evolved. "How Flowers Changed the World" focuses on the migration of life to land and on the continued "reaching out" of life on land, as Eiseley follows the "wandering fingers of green [that] had crept upward along the meanderings of river systems and fringed the gravels of forgotten lakes" (*IJ,* 62). In the evolution of plants on land—specifically flowering plants—he finds a model for the interdependence of life that we tend to overlook in our emphasis on dominance. This model of cooperation is, in fact, responsible for our very existence. It is a model, too, that traces the journey of plant life as it developed the capability of travel and moved into unfamiliar environments where natural selection could operate on nature's vast emerging variety. As always, Eiseley writes with a view to the past, choosing the perspective of looking from space at the earth (62). His thesis, "Flowers changed the face of the planet," is expanded with a reference to Francis Thompson's line that "one could not pluck a flower without troubling a star" and the comment that Thompson intuitively grasped "the enormous interlinked complexity of life" (63).

In painstaking detail Eiseley traces the emergence of creeping green life and the link of higher metabolic rates to the appearance of flowering plants. Flowers brought encased seeds that had the ability to travel, and these seeds provided a new energy source. In the midst of this consideration he inserts a small personal mystery tale about an exploding wisteria pod that disturbed him in the night. (Significantly, each species retains its old habits, though flowers and humans coexist even indoors.) As he studies the seeds, he reminds himself that "somewhere in here . . . was once man himself" ("How Flowers Changed," *IJ,* 70). With the emergence of flowering plants—embryonic plants in little boxes filled with food—plant life traveled and adapted: "All over the world, like hot corn in a popper, these incredible elaborations of the flowering plants kept exploding" (72). Concurrently, specialized insects emerged to feed on the flowers and to pollinate them. The great herbivores appeared, providing food for the carnivores that fed on them.

Humankind itself, Eiseley continues, is a product of this interdependent chain: "Apes were to become men, in the inscrutable wisdom of nature, because flowers had produced seeds and fruits in such tremendous quantities that a new and totally different store of energy had become available in concentrated form" ("How Flowers Changed," *IJ*, 75). When the great herds diminished, the hand (always the symbol both of the best human achievements and of human destructiveness) of a human-like creature "would pluck a handful of grass seed and hold it contemplatively" in a moment that held the future of the species (76). Without flowers, Eiseley suggests, "man might still be a nocturnal insectivore gnawing a roach in the dark. The weight of a petal has changed the face of the world and made it ours" (77). Eiseley's model is complex, involving the interlocked energy requirements of herbivores and carnivores and concluding with the synecdoche of "the weight of a petal." But it is a model that undermines the deeply embedded notion of human dominance and emphasizes the complex interdependence of human and other life forms.

Interdependence in Life and Society: "The Real Secret of Piltdown"

Eiseley's next step in this first collection of essays is to link the interdependence of life forms to interdependence of humankind upon its self-created society. He begins by focusing on the emergence of the human brain, drawing on the 1953 discovery that "Piltdown man" was actually a hoax. The sixth of thirteen in the collection, this essay treats what Eiseley considers the greatest mystery of all, the human brain and its journey toward understanding the relationship of the human to the natural world. Returning to a theme carefully treated in *Darwin's Century*, Eiseley explains that Darwin, in his explanation of natural selection, relied on his century's emphasis on struggle, which seemed the only way to continue the chain of natural selection. Yet Wallace, with his experience among primitive peoples, came to see them as more than inferior human beings: he came to see their mental capacity as "far in excess of what they really needed to carry on the simple food-gathering techniques by which they survived" ("The Real Secret of Piltdown," *IJ*, 83). The perpetrator of the Piltdown hoax, who attached the jaw of an ape to a fragment of a human skull, constructed a creature that supported Darwin's idea of early man as a kind of ape-man who emerged in a pre–Ice Age environment. The more recent understanding is that the actual human is much younger, and that the human brain emerged rapidly, even explosively. This understanding leads Eiseley once again to Wallace's notion that with the emergence of the human brain came the ability to transfer to tools the physical evolution that continues among creatures who lack this ability. In a sense, "it is man's ideas that have evolved and changed the world about him" (89).

The insight that Eiseley pursues is that the human being "is totally dependent on society," having created "an invisible world of ideas, beliefs, habits, and customs," an "invisible universe" that replaces instinct ("The Real Secret of Piltdown," *IJ* 92). Eiseley argues that "the profound shock of the leap from animal to human status is echoing still in the depths of our subconscious minds" (92), and he ponders the self-defining ability that humans alone possess, an ability that enables them to define themselves in terms of the social cooperation that they require or the struggle that they have perceived in nature. The "real secret," then, was not the hoax of Piltdown man, but its forcing a reconsideration of the human brain itself.

The Brain's Potential for "Reaching Out": "The Maze" and "The Dream Animal"

The next essays, "The Maze" and "The Dream Animal," further pursue arguments about the emergence of the human brain, and both provide a chance to explore Wallace's emphasis on the potential of the brain in even apparently "primitive" human beings. In "The Maze," Eiseley treats a debate between those who sought "forms which contained only the *possibility* of development" into the human ("The Maze," *IJ*, 99) and those who looked for evidences of "little men," concluding that we must admit that multiple lines—not a single line—seem to lead to the human. In "The Dream Animal," Eiseley argues that the "maze" through which evolutionary thinking leads us underscores that "little man" theories fail to account for the most important characteristic of humankind, the brain, which grows at an accelerated rate after birth, reaching triple its birth size during the first year (109), and he traces the argument that the human brain emerged rapidly in fairly recent evolutionary history (*IJ*, 110–18).

Such rapid growth of the brain leads to less emphasis on "struggle" than the early evolutionists suggested. Though he emphasizes that he does not reject natural selection (119), Eiseley pursues the notion of the "invisible" world, the sociocultural world that humankind created once language emerged (and the linguistic ability of "primitive" humans, as Wallace realized, is no less than our own linguistic ability). This world is "invisible" because it is created out of a network of symbolic communication, which also makes possible awareness of present and future. This "invisible" or "secret" universe is self-created, and it is even now "becoming merely a unit in the vast social brain" ("The Dream Animal," *IJ*, 125). Eiseley concludes in terms of the biblical story of creation. For this "dream animal," he writes,

> The story of Eden is a greater allegory than man has ever guessed. For it was truly man who, walking memoryless through bars of sunlight and shade in the morning of the world, sat down and passed a wondering hand across his heavy forehead. Time and darkness, knowledge of good and evil, have walked with him ever since. . . . In just that interval a new world of terror and loneliness appears to have been created in the soul of man. (125)

Linking language with awareness of time, culture with the interdependence of individuals, and the bane with the blessing of being human, Eiseley also underscores the link of hand and tongue and brain. His sense of wonder at the emergence of the "dream animal" is tempered by the subtext of the duality of human nature.

Reaching Forward: "Man of the Future"

Moving "forward and backward"—from the emergence of the human brain to predictions of what the "man of the future" will be, Eiseley extends his defamiliarizing of evolutionary myths (especially the myth of racial hierarchy) to a further consideration of pedomorphism (already introduced in "The Dream Animal") and the apparent direction that evolution is taking us. Noting the tendency to speculate that the "man of the future" will have an extraordinarily large brain, he concludes that this "future" being has already existed. His brain was bigger than ours, his face was childlike, and he was not white. South African fossils of large-brained humans long extinct give evidence that the "game" of foetalization and brain size "had already been played out before written history began" (*IJ*, 133). Why these people became extinct is unknown, but Eiseley speculates that they may have been poorly equipped to compete with "more ferocious and less foetalized [retaining characteristics that were once infantile, especially the ratio of brain size to facial size] folk" (137). From the evidence that "the man of the future" (in journalistic parlance) has already lived and become extinct, Eiseley reminds us of the "eternal flickering of forms . . . whose meaning forever escapes us" (138). The "social brain" is important now, not the need for a "bigger" brain.

The essay concludes with Eiseley's recollection of watching a peasant finger a submachine gun while Eiseley sat as prisoner. At the sound of a female voice, the man exchanged a look with his prisoner, "a male smile that ran all the way back to the Ice Age" (141). It was a smile that, Eiseley concludes, he always "weigh[s] . . . mentally against the future whenever one of those delicate forgotten skulls is placed upon my desk" ("Man of the Future," *IJ*, 141). In that look was embodied the dual capability of the human brain, the twin capacities that nineteenth-century Darwinians except for Wallace tended to de-emphasize—the fully *human* capacities for both cooperation and destruction, and the fully human ability to *choose*. The choice is one that Eiseley continually emphasizes; it characterizes the human, at least as much as the "encephalized forelimb" or the linguistic ability that the brain/forelimb link translates into writing. This choice is part of that "invisible" world that abstract thought and language have created, and it is a choice demanded by Eiseley's "extension of vision" or "projection of [vision] into other lives" ("The Great Deeps," *IJ*, 46).

Reaching into Space: "Little Men and Flying Saucers"

Linked to the human attempt to journey into the future by predicting "the man of the future" is the attempt to journey into space, the subject of the next essay in *The Immense Journey* and later of what Eiseley called his "rocket book," *The Invisible Pyramid*. In "Little Men and Flying Saucers" he ruminates on our fascination with the possibility of life beyond the earth, noting that, whether in folk tales or tales of flying saucers, the extraterrestrial form is that of the "little man" to whom he has alluded in the previous essays. This fascination evokes Eiseley's tale of his first encounter with the "little man," when a rancher brought to an archaeological camp a box containing a "mummy" that he offered to the archaeological party on his way to a carnival in town. The thing was bizarre to a practiced bone hunter—"an anomalous mummified stillbirth with an undeveloped brain" (*IJ,* 147), the kind of artifact that appears periodically in Eiseley's tales, circulating on the fringes of civilized society in carnival booths and eerie tales.

For Eiseley, the "little man" tells a story, but it is a story of evolution and of those, like Darwin, who wonder about living structures. The "little man" thus sparks a brief history of how we came to understand the shifting of living forms, and of how the "little man"—the archetypal anthropomorphic form that some nineteenth-century thinkers thought was prophesied in geological formations—has now shifted from prophesying the coming of "man" to serving as a figure from outer space. "If man is regarded as a good production here, he must be found in endless duplication throughout the worlds" ("Little Men," *IJ,* 157), Eiseley writes. In space stories, he continues, one may encounter "cabbage men and bird men," "lizard men and . . . tree men"—all of whom are nevertheless earthlings (158). Yet humankind, he argues as he returns to the point of the essay, "is a solitary and peculiar development" (158), and "every creature alive is the product of a unique history" whose "statistical probability of . . . precise reduplication on another planet is so small as to be meaningless" (160). Human form "is the evolutionary product of a strange, long wandering" (161), though we "torture ourselves" in a quest for the meaning of "the vagaries of the road" (162). "Little men," then, serve as a carnivalesque means of undermining the human self-reflexivity. They result from human projection of human form into the construction of a *telos* and a prophecy of things human; the wonder is that we evolved at all.

Beyond the Boundaries of Form: "The Judgment of the Birds" and "The Bird and the Machine"

After ending on the note of the evolutionary isolation, the "loneliness," of the human, Eiseley begins the next essay, "The Judgment of the Birds," by noting that even primitive seekers of visions have traditionally lived in the wilderness (this book, we may recall, began with an assertion that the narrator "can at best report only from my own wilderness," that "the important thing is that each man possess such a wilderness and that he consider what marvels are to be observed there" ("The Slit," 13). Though he disavows any "prophecies," Eiseley sets down in this essay, out of the isolation imposed on the naturalist, "a matter of pigeons, a flight of chemicals, and a judgment of birds, in the hope that they will come to the eye of those who have retained a true taste for the marvelous, and who are capable of discerning in the flow of ordinary events the point at which the mundane world gives way to quite another dimension" ("The Judgment of the Birds," *IJ,* 164). These tales exemplify that extension of vision that he suggested in "The Great Deeps," the "projection of itself into other lives" that "is the lonely, magnificent power of humanity" and epitomizes "the reaching out" (46).

The "matter" (an echo of the "Matter" of Rome, of France, and of Britain presented in terms of one individual's seemingly mundane encounters) of pigeons involves a man in a Manhattan hotel room who leaned out in the early morning light to experience the city "from a strange inverted angle" and to realize "it was not really his at all" ("The Judgment of the Birds," *IJ,* 167). The epiphany displaces the human from his self-appointed centrality and brings the reader into complicity in knowledge that, Eiseley suggests, "is better kept to oneself" (167)—like the knowledge of the poet who, in Shelley's "Kubla Khan," would be scorned as mad if he did build with words the "sunny domes" and "caves of ice" that he saw in a dream. The next episode tells of an enormous crow lost on a foggy morning and confronting, as the man confronted the pigeons, a man on the human level; it similarly tells about an inversion of vision, in which two "worlds have interpenetrated" (169). The "flight of chemicals" involves a flock of migrating songbirds swooping over a moonlit landscape in the Badlands, "a place of dry bones in what once was a place of life," where "carbon . . . ran blackly in the eroding stone" (170, 172). The man, standing amid the bones of fifty million years, with "the chemicals of all that vanished age" (171) surrounding him, lifts a fistful of the chemical-bearing earth, standing as a living intermediary between the dead chemicals and the same chemicals embodied in the life of a "wild flight of south-bound warblers" that "hurtled over [him] into the oncoming dark" (172). Amidst death, he experiences life, an "incredible miracle" that made the "vast waste . . . glow under the rising moon" (173). Like the previous epiphanies, this, too, has an epic quality, though the "matter" is chemical.

And the "judgment," again seen by Eiseley the narrator as wanderer over a wild Western landscape, was one made by a flock of birds that watched as a raven devoured a nestling. The birds "cried there in some instinctive common misery," but then the sigh died and a single "crystal note of a song sparrow lifted hesitantly in the hush," until others took the song and collectively the birds made "the judgment of life against death," forgetting violence ("The

Judgment of the Birds," *IJ,* 174–75). It is a sound that he recalls even as he knows that "eventually darkness and subtleties [will] ring me round once more" (176). The epiphany is renewed in another embedded tale of a spider spinning her web on a cold autumn evening, "warmly arranged among her guy ropes attached to the lamp supports—a great black and yellow embodiment of the life force, not giving up to either frost or stepladders" (176).

The epiphany elicited by the spider brings back the sound of the birds and the strength of life itself—what Eiseley calls "a kind of heroism" in "a world where even a spider refuses to lie down and die if a rope can still be spun on to a star" ("The Judgment of the Birds," *IJ,* 177). He hopes that the human being will fight as well when the time comes, but his final awareness is self-reflexive. It is the awareness that the human mind is able to devise "a kind of courage by looking at a spider in a street lamp" (178). Yet he concludes that the awareness retrieved from these wild things should only be recorded, without an attempt "to define its meaning" (178). In a self-reflexivity that combines the human brain, language, the sound of the birds, and the visual image of the undaunted spider—the brain, language, and the philosophemes by which we understand—Eiseley concludes that the recorded but undefined comment can "go echoing on through the minds of men, each grasping at that beyond out of which the miracles emerge, and which, once defined, ceases to satisfy the human need for symbols" (178). Not only the brain, but its symbols and its need for symbols, continue to reach out—beyond the boundaries of form, beyond the boundaries of understanding—toward new ways of understanding.

"Reaching Out" to Companionship: "The Bird and the Machine"

From his tale of "judgment" Eiseley turns to a tale of a single wild bird juxtaposed with the quintessential image of the mechanistic worldview in "The Bird and the Machine." Reflecting on an article about increasingly "smart" machines, including a mechanical mouse that outperforms a live mouse, he asserts, "It's life I believe in, not machines," and traces the human fascination with the machine, particularly in the eighteenth century, when "little automatons" and "clocks described as little worlds" were taken on tours (*IJ,* 181). At that time, human beings conceived of themselves in terms of their own tools, considering themselves made like puppets, so that the human being "was only a more clever model made by a greater designer" (181). He continues this history of the mechanistic view through nineteenth-century discoveries of the cell and finally its dissolution "into an abstract chemical machine— and that into some intangible, inexpressible flow of energy," with the "wheels" of the machine increasingly smaller (182). In opposition to this deterministic, mechanistic perspective, epitomized in a tale of a mechanical mouse outperforming a real mouse in a maze, Eiseley focuses on "the possible shape of the future brooding in mice, just as it brooded once in a rather ordinary mousy insectivore who became a man," and he concludes that "it leaves a nice fine indeterminate sense of wonder that even an electronic brain hasn't got" (182).

This sense of wonder marks his turn to another recollection from his bone-hunting days in the West, a tale that, as Gale Christianson has noted, was almost totally revised in Eiseley's memory from the trapping and release of a troublesome hawk (1991, 110–11) into a tale involving the narrator's epiphany on encountering the mystery of devotion when the captured hawk was released. As Eiseley tells the tale, he first had to learn a lesson about time when he was alone in a desert; there he discovered "that time is a series of planes existing superficially in the same universe" and that tempo is "a human illusion, a subjective clock ticking in our own kind of protoplasm" ("The Bird and the Machine," *IJ*, 183). The tale is of time slowed immeasurably, of an isolated research party looking for evidence of postglacial humans, wandering in dry canyons of the Great Plains, only occasionally finding a skull or other artifact. Part of the mission was to take back present wildlife, and in a deserted cabin the party found a pair of sparrow hawks. With some difficulty the narrator captured the male, which Eiseley anthropomorphizes by saying he had diverted his captor's attention to save the female. They captured the male, which remained in a box overnight until the next morning, when the narrator secretly took the limp bird into his hands and felt the heart pounding while the bird only looked upward until, after a long minute "he was gone straight into that towering emptiness of light and crystal that my eyes could scarcely bear to penetrate" (191). From far above the narrator heard the mate's cry, and he looked up to see the two meet "in a great soaring gyre that turned to a whirling circle and a dance of wings" as they cried "in a harsh wild medley of question and response" (192).

Laying aside the story of the mechanical mouse, Eiseley reflects, "Ah, . . . on the other hand the machine does not bleed, ache, hang for hours in the empty sky in a torment of hope to learn the fate of another machine, nor does it cry out with joy nor dance in the air with the fierce passion of a bird" ("The Bird and the Machine," *IJ*, 193). Through the elaborately fictionalized—and anthropomorphized—story of the hawks that danced and cried with passion, Eiseley thus reasserts his sense of wonder and indeterminism into a world that he sees as caught up in mechanism—even in the body as mechanism—and in technocracy.

Mystery and Indeterminism: "The Secret of Life"

The final transition is from unexplained life and power in the hawk to "The Secret of Life," which begins with the narrator's attention to seeds that attach themselves to his coat or socks as he walks in the woods. "I am sure now," Eiseley writes in a foreshadowing of his next book, the small masterpiece titled *The Firmament of Time,* "that life is not what it is purported to be and that nature, in the canny words of a Scotch theologue, 'is not as natural as it looks'" ("The Secret of Life," *IJ*, 197). He uses the folk belief in spontaneous generation of mice from old clothes to open his explication of what is known about life. Though evidence indicates that life has arisen from matter, Eiseley again invites the reader to make a journey backward into time and still to feel the impossibility of explaining life itself.

Despite magazine articles on topics such as "The Spark of Life" and predictions that scientists are close to creating life, he argues that science "is not the answer to the grasshopper's leg, brown and black and saw-toothed here in my hand, nor the answer to the seeds still clinging tenaciously to my coat, nor to the subtle essences of memory, delight, and wistfulness moving among the thin wires of my brain" ("The Secret of Life," *IJ*, 207). If science does manage to create life in the laboratory, he concludes, "we shall have great need of humbleness," and few will ponder "whether the desire to link life to matter may not have blinded us to the more remarkable characteristics of both" (208).

The point of the essay is that matter, life as matter, and life itself are subjects of wonder and respect. The "greatest missing link of all" is "the link between living and dead matter" ("The Secret of Life," *IJ*, 201), a link suggested earlier in the image of Eiseley as narrator holding a fistful of soil containing the chemicals of extinct animals while, living and energized in the form of birds, the same chemicals flew overhead. Eiseley is concerned with the need for awareness of the interlocked web of matter and life, with the human as a product of nature, not as "master" of a machine-like universe. Such a conception is as dead as the mechanical mouse that "out-performs" the real one.

The real "secret of life," he contends, is very likely not subject "to the kind of analysis our science is capable of making" ("The Secret of Life," *IJ* 202). Science itself—analytic knowledge—may not be able to plumb the mystery. Having analyzed the difficulties of exact scientific explanation, he concludes that "this long descent down the ladder of life, beautiful and instructive though it may be, will not lead us to the final secret" (202). The ultimate mystery is epistemological and existential, subject to the synthesizing under-standing of poetry and myth. "My memory holds the past," he reflects, "and yet paradoxically knows, at the same time, that the past is gone and will never come again" (208). Linking diverse living creatures with the human ability to wonder and the long tradition of human mythmaking, he concludes that if "dead" matter has produced "this curious landscape of fiddling crickets, song sparrows, and wondering men, it must be plain even to the most devoted materialist that the matter of which he speaks contains amazing, if not dreadful powers" and may even be, in Hardy's phrase, "'but one mask of many worn by the Great Face be-hind'" (210).

The concluding chapter of *The Immense Journey* is not the con-clusion of the rhetorical journey, nor the conclusion to the episte-mological one. It is an open-ended conclusion that science cannot contain—in the multiple sense of holding and holding back—the whole journey. There is no return to the past, but life is latency; life is, in Eiseley's recurrent phrase, "of the future." Like the root metaphors of the body, life itself has a dual nature—struggle and cooperation. What is unique about the human is the ability to choose. Often viewed as a traditional humanist, Eiseley clearly participates in the humanistic view, but he also repeatedly focuses on humankind, particularly through the Cartesian-Newtonian para-digm in which humanist thought is embedded, as the destructive force, the force that has the chance to choose, but has repeatedly chosen destruction. It is this paradoxically gloomy and yet almost hopeful perspective that emerges, along with his rhetorical method, in the first two of the major popular texts. Eiseley's ironic view has emerged; it will be formalized and lyricized in his next book.

6

The Move toward an Ironic View:
The Firmament of Time

> "[T]he natural world . . . is not natural
> but a queer event created
> in minds still queerer."
> —Eiseley, "The Bats"
>
> "'Natural' is a magician's word—and . . . should be used
> cautiously."
> —Eiseley, "How Natural Is 'Natural'?"

Based on a series of lectures delivered when Eiseley was visiting professor of the philosophy of science at the University of Cincinnati College of Medicine in the fall of 1959, *The Firmament of Time* is a small masterpiece that Angyal calls "lyrical and meditative" (1983, 57). Whereas the essays in *The Immense Journey* were composed over a period of several years while Eiseley was casting about for a *genre* through which to explore the relationship of science and culture and partly while he was working on *Darwin's Century, The Firmament of Time* marks his avowed ironic and deliberately literary "turn." Here he directly examines the role of metaphor in scientific thought, the anticipation of ideas (including scientific ideas) by literary figures, the role of the human brain in creating the world that we perceive, the implicit assumptions embedded in the common terms *human* and *natural* that help to shape science itself. The focal point of the essays is the code *natural*, which Eiseley reexamines and defamiliarizes. As part of the development of evolutionary understanding, certain unquestioned concepts had to be seen as "natural": first "the world" itself had to be seen as natural, then death, then life itself, then "man." After exploring these concepts, Eiseley ends with two essays that raise the central questions of encoded meanings: "How Human Is Man?" and "How Natural Is 'Natural'?" Throughout the essays he explores both literary tutor texts and extended personal narratives for their epistemological implications. As the pivotal text in his movement from popularization in *The Immense Journey* to reflective, meditative essays on nature and the human relation to it, *The Firmament of Time* anticipates some of the techniques employed in contemporary literary studies:[1] (1) examination of the essential metaphoricity of understanding, (2) "decentering" what has been viewed as central, (3) reversal of hierarchies—especially man/nature and scientific knowledge/poetic understanding, and (4) recognizing the linguistic basis of concepts, including that of "natural."

Tutor Texts and the Sequence of the Essays

The Firmament of Time takes its title from *Adonais,* Percy Shelley's elegy for John Keats, which transcends the limits of mourning for an individual to defend the sensitive individual writing in a harsh world: "The splendours of the firmament of time / May be eclipsed, but are extinguished not" (cited in *FT,* viii). Shelley's "splendours" are great writers, and he continues,

> . . . When lofty thought
> Lifts a young heart above its mortal lair,
> And love and life contend in it . . . ,
>   . . . the dead live there
> And move like winds of light on dark and stormy air.
>
> (cited in *FT,* viii)

Whether these "splendours" are major writers or major ideas, the passage suggests that human continuity comes through thought, specifically through writing—a concept basic to all of Eiseley's texts. Later in *The Firmament of Time* he writes of "a great mystery," which "opens" in the human brain: "that life and time bear some curious relationship to each other that is not shared by inanimate things" ("How Natural?" *FT,* 169). And part of this mystery is that, through social and written memory, humankind transcends the limited world of the present to shape a future out of an incredibly vast potential.

The book begins with two sentences that draw on this human transcendence of time through other tutor texts—through myth (social memory) and language: "Man is at heart a romantic. He believes in thunder, the destruction of worlds, the voice out of the whirlwind" ("How the World," *FT,* 3). These three items of faith are encoded in myth—the primitive myths of thunder, Judaeo-Christian notions of the end of the world, the voice that Job heard in the whirlwind—and all three evoke explanations of occurrences that either are or once were inexplicable. Such myths evolve because the brevity of human generations makes it difficult for us to comprehend the vastness of time. It was only two hundred years ago, in fact, when "a few wary pioneers began to suspect" that the literal chronology that estimated the earth's beginning at 4004 B.C. could not be accurate (3).

In this context of reexamining myths, Eiseley begins the first essay, "How the World Became Natural," in which he explores the notion of *natural* in the primitive mind and the modern mind. The notion of "natural," as it has become a code, has taken on meanings that we do not consider. Thus nature is usually viewed as (1) observable, accepted and used, thus (2) known and explored, and thus (3) expected. Eiseley reminds us, however, that what we call "nature" contains more than we normally grant, for nature itself contains the potentialities that underlie continual change, continual creation. Because primitive humankind added "a shimmering haze of magic" to an "everyday, observable world" that was "utilized" for survival ("How the World Became Natural," *FT,* 4), the natural world in its earliest sense is a compromise of reality and dream. Eventually, however, the natural world was "tamed" and labeled "'known and explored,' as if it had little in the way of surprises yet in store for us" (7–8). Eiseley states his purpose as exploration of the evolution of science in relation to the changing intellectual climate and reexamination of what we call "natural," combining the empiricist view and a more primitive, evocative view that "takes into account that sense of awe and marvel which is part of man's primitive heritage, and without which" we would not be human (8).

In examining the history of science, Eiseley notes that "a given way of looking at things, a kind of unconscious conformity" ("How the World Became Natural," *FT,* 6) may affect this history, for scientists are human beings who, like everyone else, "are capable of prejudice" (6). Eiseley thus openly embraces an ironic journey, asserting that the individual "who learns how difficult it is to step outside the intellectual climate of his or any age has taken the first step on the road to emancipation, to world citizenship of a high order" (7). And if the meta-view is necessary to emancipation, it is also necessary to individual growth. In exploring how "even the scientific atmosphere evolves and changes with the society of which it is a part," Eiseley expects to find an accompanying "extension" of the individual's "own horizon as a human being" (7).

In thus "treating simply" of large matters, Eiseley focuses on the two common terms, *natural* and *human,* which science must use and whose meanings science assumes, as he leads the reader through a series of historical perspectives that made familiar or "natural" what we today do not question. The first chapter, "How the World Became Natural," focuses on perceptions of the "natural" world from primitive to modern. Being "both pragmatist and mystic" (*FT,* 4), Eiseley argues, humankind saw the "natural" world as useful and real, yet for the primitive "unseen spirits moved in the wood" (4). Outlining in broad strokes the emergence of evolution, Eiseley traces the "retirement" of the magical or the supernatural and the emergence of a world that, after Darwin, finally became "completely natural" (29), a world of unreturning time and unfixed form.

In the second chapter, "How Death Became Natural," Eiseley looks at the primitive assumption that life was natural and death was somehow "unnatural," "the result of malice or mistake, the after-message of the gods, or, in the Christian world, the result of the Fall from the Garden" (*FT,* 33). Before evolution could be understood, species death had to be established. Thus both death and creation became natural; individuals and species passed from existence, yet nature was continuously creative.

The third essay, "How Life Became Natural," examines life itself as a code culturally linked to the supernatural but forced to undergo a shift. The essay opens with both a tutor text and a comment on the role of literary figures in anticipating ideas: "Great literary geniuses often possess an ear or a sensitivity for things in the process of becoming, for ideas which are just about to be born" (*FT,* 61). Coleridge in 1819 referred to an idea that "'has become quite common even among Christian people, that the human race arose from a state of savagery and then gradually from a monkey came up through various states to be man'" (cited in p. 61). Although Eiseley notes that "Coleridge was not an evolutionist" (61), he pursues Coleridge's statement because of the writer's sensitivity to a new notion that was widespread before Darwin's *Origin* and because Coleridge further "observes in a very shrewd fashion . . . the way in which the intellectual climate of a given period may unconsciously retard or limit the theoretical ventures of an exploring scientist" (61).

In Coleridge, as in Bacon earlier, Eiseley finds the realization of what he calls in *Darwin's Century* the "frame," in *Francis Bacon and the Modern Dilemma* and later in the autobiography "the dancers in the ring," and elsewhere the "conservatism" of any professionalized group, the tendency of the group to hold its members within a circumscribed set of understandings and practices (rather like Kuhn's "paradigm" or "disciplinary matrix"). Looking at the history of philosophy during the two or three previous centuries, Coleridge writes that one "'cannot but admit, that there appears to have existed a sort of secret and tacit compact among the learned, not to pass beyond a certain limit in speculative science. The privilege of free thought so highly extolled, has at no time been held valid in actual practice, except within this limit'" (*Philosophical Lectures,* cited in p. 62).

Using Coleridge's passage as tutor text, Eiseley contends that major scientific change is not often swift, but is held back by its intellectual and cultural milieu. In Darwin's time, Eiseley argues, "it was not really new facts that were needed so much as a new way of looking at the world from an old set of data" ("How Life Became," *FT,* 72). The remainder of the essay traces those figures who contributed to the concept of evolution through natural selection, with some emphasis on Edward Blyth, whom Eiseley saw as both anticipating and influencing Darwin's theory. He pursues Darwin's debt to Blyth as a matter of undermining "unsophisticated hero worship" and understanding "how an idea . . . was altered by slow degrees into something which set the world of life adrift in an unfixed wilderness" (69–70). (Eiseley's work on Blyth was largely dismissed in professional circles and may have hurt him professionally.)

After geological change was linked to time and after the death of both individual and species was "naturalized," life itself required reexamination. Life did not become "natural" in a day or in a generation, but in the course of generations the notion of a "fixed, static, immovable" chain of being ("How Life Became," *FT,* 65) was transmuted from "a stabilizing factor making for providential control of the living world" into an understanding of evolution through natural selection (81). Eiseley concludes the essay with an image of unity for life as "natural": "An order of life is like a diamond of many reflecting surfaces, each with its own pinpoint of light contributing to the total effect" (85). As human beings, we are "one flashing and evanescent facet" of "the many-faced animal" that is continuously emerging in time (86). As part of the larger animal, the human being, like life, is "natural."

From the image of the human among other forms of life, Eiseley turns to "How Man Became Natural," the chapter in which he traces nineteenth-century biologists' quest to understand human-kind as part of nature, a quest in which sometimes they placed living humans and living animals closer than the evidence would support. Defining the human as "natural" opens a new dimension—the human capability for self-definition—as an animal or as more than an animal ("How Man Became Natural," *FT,* 111–13). Because freedom from determinism has given us the chance to explore human potentiality, the question of the "naturalness" of humankind is also a double question of the "humanness" of this product of nature that has evolved to the point of defining its own "nature." The use of the same word in both contexts is significant, providing another example of what Lakoff and Johnson call the basic conceptual metaphors by which we understand experience; thus *nature* in the scientific sense is itself a metaphorical concept that structures understanding in terms of "nature" and "humankind" as separate entities, and in *human nature* the metaphor redoubles the separation of entities to apply the metaphor to that which defines humankind. A *human* being, as the origin of the term from the Latin *humanus* (from *humus,* "earth") suggests, is a creature of the earth. The "human" was originally part of the earth, not master of it as in the traditional Western hierarchy. Closeness to the earth, not domination over it, is characteristic of the primitive and, Eiseley argues, worthy of emulation by modern humankind. The key to the question "How Human Is Man?" is, for Eiseley, the human concept of the future, which we "invite" (*FT,* 117)—that is, we create—through the human imagination. The future exists only in language, and awareness of this linguistic self-referentiality is crucial in Eiseley's reexamination of the "human" versus the "natural."

Humankind has become natural, Eiseley contends, but without understanding this "naturalness." Possibly, he speculates, sounding more like an existentialist alone in a universe that has doomed him to freedom than like a traditional humanist who would place the human at the top of a hierarchy, this "human freedom has left him the difficult choice of determining what it is in his nature to be" ("How Man Became Natural," *FT,* 114). And possibly, he concludes, the nineteenth-century argument over whether the human being is warrior or weakling is simply a reiteration of the old duality, the warfare within the human being. Eiseley sees the Darwinian duality as "only an evolutionary version of man's ancient warfare with himself—a drama as great in its hidden fashion as the story of the Garden and the Fall" (101, 114).

From ancient and evolutionary versions of the human duality,

Eiseley turns to focus in the last two chapters on two great questions, "How Human Is Man?" and "How Natural Is 'Natural'?" Because of the centrality of these two questions to the Eiseleyan project, I shall examine in detail these essays and their reexamination of the terms *human* and *natural.* Eiseley's exploration of the term *human* and his defamiliarizing of *natural* are basic to his project of reexamining and defamiliarizing common understandings that are the underpinnings for a worldview, coded meanings that we do not consciously consider because they are part of the "always already" of human culture. Because nature itself has come to mean something external, accepted, known, and used, we forget the element of surprise in nature.

Thus Eiseley pursues the potentialities in nature, whether genetic potential in life or the potential for unpredictable geological change. In his reexamination of "natural" Eiseley subverts positivist views as he consciously examines the sense of wonder/wondering which, tied to language, distinguishes the human and is the founding of philosophy.

The Metaphoricity of Thought

Earlier, I have noted Eiseley's view of the role of metaphor. It is in *The Firmament of Time* that he argues for the role of root metaphor—"the hook of metaphor" ("How the World Became Natural," *FT,* 20)—in extending the domain of science. Even though the analogy may be faulty, he notes,

> yet so potent is its effect upon a whole generation of scientific thinking that it may lie buried in the lowest stratum of accepted thought, or color unconsciously the thinking of entire generations. While proceeding with what is called "empirical research" and "experiment," the scientist will almost inevitably fit such experiments into an existing comprehensive framework, an integrative formula, until such time as that principle gives way to another. (20)

This awareness (Kuhn's "paradigm shift" framed in terms of metaphor) underlies all of Eiseley's oeuvre. Fascinated by "the flow of ideas from one field into another," he explores the "curious and ambivalent paths" that flow may take, often through the vehicle of metaphor (17).

In "How the World Became Natural," Eiseley traces Newton's metaphor of the universe as a machine and its extension into geology, an extension that "was both advantageous and, paradoxically, retarding" (*FT,* 17). James Hutton of Edinburgh, a physician, took the old notion that the microcosm reflects the macrocosm and applied it to the earth, which he saw in the physician's terms of circulation and metabolism—in dynamic rather than static terms (18–19). Hutton's view, however dynamic, nevertheless treated the earth as "a completed mechanism—whether we regard that mechanism as organic in essence or mechanical" (19).

Because the individual looks first to forces that can be observed within his own life span, events outside that span are often explained in terms of "myths incorporating outright violence on a gigantic scale" ("How the World," *FT,* 21). Hutton, however, differed from his contemporaries when, influenced by Newton's concept of continuity, he suggested that small events might combine to produce major changes; geological changes, he suggested, might be wrought by "raindrops and aerial erosion" (22). Hutton, analogizing with the organic, wrote that "'the earth, . . . like the body of an animal, is wasted at the same time that it is repaired'" (quoted in p. 24). In Hutton's dynamic view, "'Nature . . . lives in motion'" (quoted in p. 25).

Hutton's eighteenth-century deistic view combined the machine and the organism, but the machine triumphed in that he saw nature as operating on "self-balancing" principles rather than experiencing continual change and adaptation ("How the World," *FT,* 28). The eighteenth-century world remained of divine origin specifically designed for human beings. The world would become "natural" only when the supernatural was "retired from the earthly scene"; only then would the microcosm fail to reflect the macrocosm and the "celestial clocks" fail to "chime in perfect order" (29). Hutton analogized from his own experience and in his own time, but his blending of the organic analogy with the mechanistic helped to focus understanding of the "natural" in a way that both anticipated and retarded evolutionary insights. The "hook of analogy," the "root metaphor" is a major point of Eiseley's analysis.

Decentering "Man": "How Human Is Man?"

In addressing the two great questions with which he concludes the book, Eiseley moves from the past tense to the present, from the human relationship with the past to the human relationship with the present and especially with the future. The first essay, "How Human Is Man?" opens with a tutor text and then strategically uses a remotivated dramatic metaphor to develop the essay. The tutor text is from Kierkegaard—a passage that Eiseley admits is "cryptic" ("How Human?" *FT,* 117). Kierkegaard writes of the human relationship to the future: "He who fights the future has a dangerous enemy. The future is not, it borrows its strength from the man himself, and when it has tricked him out of this, then it appears outside of him as the enemy he must meet" (cited in p. 117). Eiseley draws on Kierkegaard's comment to argue that in the West we "have rushed eagerly to embrace the future" and in so doing we have "invited," strengthened, and created that future (117). Kierkegaard's conclusion, "Through the eternal, we can conquer the future" (cited in p. 117), becomes the basis for Eiseley's analysis of our relationship to time, past and future, and to the problems of "within" and "without" that have more recently troubled other thinkers.

To approach this problem he invokes, defamiliarizes, and remotivates the drama metaphor, speculating that we can best understand the human "predicament" through "the simple mechanics of the theatre," including the time allotted for the play, the stage, and the apparent plot. The drama, Eiseley suggests, may indeed be an essential root metaphor. "It may well be," he writes,

> that we can see our history in no other terms, being mentally structured to look within as well as without, and to be influenced within by what we consider the nature of the "without" to be. It is for this reason that the "without," and our modes of apprehending it, assume so pressing an importance. Nor is it fully possible to understand the human drama, the drama of the great stage, without a historical knowledge of how the characters have interpreted their parts in the play, and in doing so perhaps affected the nature of the plot itself.
>
> This, in brief, epitomizes the role of the human mind in history. ("How Human Is Man?" *FT,* 119–20)

The theater's "simple mechanics" thus take on an epistemological dimension as the dramatic metaphor provides images for the varying appearances with which human understanding has clothed reality. For Eiseley the dramatic metaphor echoes the Western philosopheme of "within/without" and thus serves as a vehicle for examining the relationship of the two, the way that our perceptions of the "without" affect our understanding of ourselves and nature, the linguistic "creation" of both past and future. The "simple mechanics of the theatre" provide a model for his reexamination of the human predicament—the time, the stage, and "what appears to be the plot" (119), though this appearance is illusory. As the essential metaphor of prescientific Christianity, which viewed the earth as the stage for a "short-lived drama" ("How the World Became Natural," *FT,* 11), the dramatic metaphor is forced to undergo changes. Eiseley's "world stage . . . has the skeletons of dead actors under the floor boards, and the dusty scenery of forgotten dramas lies abandoned in the wings" ("How Human Is Man?" *FT,* 118). Now the ultimate displacing question confronts us: "What if we are not playing on the center stage? What if the Great Spectacle has no terminus and no meaning?" (118–19).

Not only does the metaphorical question decenter the human, but it also displaces teleology. As individuals "have interpreted their parts in the play," they have "perhaps affected the nature of the plot itself" ("How Human Is Man?" *FT,* 120). The artistic metaphor effectively demonstrates the relationship of human understanding of the role to the actual working-out of that role. In terms of the doubling effect of our ability to change the nature that produced us, the metaphor of the drama takes on a new dimension. Eiseley's remotivated play is not the traditional teleologically oriented play of the Greeks or the medieval religious thinkers. It resembles instead an Elizabethan play within a play, for it introduces the element of "making"; later Eiseley will draw heavily on the Elizabethan notion of an impromptu play that the characters work out as they go. No play is ever the same twice, and the impromptu play is especially subject to modification. There is no script—or there is only a probabilistic genetic script, itself subject to modification through chance and the influence of the environment. The metaphor of human life as a drama projects the image of man as a "shape-shifter" and a "changeling" whose forms and "realities" are never fixed, never what we may wish to think they are, changing costumes and styles of knowledge from one age to another.

The metaphorical drama is also one of Eiseley's many "spectacles" (from the Latin root for seeing)—a play on the framing effect of worldviews, and on the drama as spectacle. The point is both scientific and linguistic:

> Each time the world has appeared real, and the plot has been played accordingly. Strange colorings have been given to reality and the colors have come mostly from within. As science extends itself, the colors, and through them the nature of reality, continue to change. The "within" and "without" are in some strange fashion intermingled. ("How Human Is Man?" *FT,* 120)

In the renewed dramatic metaphor, self-defining, self-creating humankind makes the plot. "Observation" is more than the nineteenth-century notion of objective observation, for observation depends upon the frame through which we see. If the drama provides an effective vehicle, it does so because it is metadramatic: through the drama we look at our own role in shaping the outcome. Eiseley has earlier demonstrated that the "natural" could emerge only when the supernatural was removed from its domain ("How Death Became Natural," *FT,* 33–58). But there remains the sense of the uncanny or the unreal that human beings have always experienced in nature. A primitive sense of participation in the environment is reintroduced. "Perhaps, in a sense," Eiseley suggests, "the great play is actually a great magic, and we, the players, are a part of the illusion, making and transforming the plot as we go" ("How Human Is Man?" *FT,* 120). This participatory form of "observation," which is part creation, is privileged, in Eiseley's epistemology, as a means to help humankind learn to survive.

Eiseley not only uses the dramatic metaphor to decenter the human being, but he also decenters the human as master—a concept that he traces to Judeo-Greek origins and to the dominance of machine metaphors in the last two centuries. From the outset, he demonstrates the anthropocentrism in which the metaphor originated and the assumptions of human centrality that it encourages. He also links the theatrical metaphor to the scientific view; science itself, he contends, emerged out of the Judeo-Christian quest for a return to the original garden and the physical and mental voyages of exploration that characterized the Renaissance. Bacon's followers departed from the Baconian argument that we need to follow an "inner" path to the Garden, substituting the quest for technological mastery of the earth ("How Human Is Man?" *FT,* 126). Thus science seeks the "earthly Paradise" through technological mastery of nature rather than through inward reflection (126–27). The dramatic metaphor blends with science in Darwin's speculations on the question of the human being as "warrior" or weakling.

Nature itself, before human tampering, was in a sense "steadfast and continuous," having a kind of "'expectedness'" ("How Human Is Man?" *FT,* 122–23). Yet with the human creation of "another nature" (clearly a play on the polysemy of the word *nature*), which is culture, then "the new order imposed by cultural discipline, became the 'nature' of human society" (124). Modern humankind, however, in the quest for mastery of nature, threatens nature itself—to "the point where we must look deep into the whirlpool of the modern age," which threatens nature's stability (128). The whirlpool is a metaphor for a cultural centripetal force, for a self-reflexivity that ignores self-reflection. By focusing on "things, and assum[ing] that the good life would follow" (130), human beings, Eiseley contends, "increasingly are the victims of what they themselves have created" (131).

Primitive societies found the divine in nature, but in modern society the divine has been displaced, so that humankind strives against itself ("How Human Is Man?" *FT,* 132). The whirlpool is an image of the technological revolution, which, Eiseley contends, has brought three conditions: (1) a social environment that changes so rapidly that individuals cannot adjust to it; (2) direction of human attention to machines that leave little time for solitude or daydreams or reflection; and (3) through "this outward projection of attention" and "the rise of a science whose powers and creations seem awe-inspiringly remote, as if above both man and nature," the creation of a human being "who is not human," who "no longer thinks in the old terms," who "has ceased to have a conscience" and has become "an instrument of power" (134–35). This individual fails to realize that scientific power "partakes of human freedom. It is no longer safely *within* nature; it has become violent, sharing in human ambivalence and moral uncertainty" (135). The technological whirlpool, then, has displaced human and scientific concern with *ethos,* for creating the image of the self, whether of the scientist or of the followers of science.

Thus, Eiseley argues, is created "asphalt man" (136), who follows a group ethic amidst fragmentation of society into small groups. The future, in such a world, "is no more than the running of the whirlpool"; it is characterized by a "secularized conception of progress" and by "the mass loss of personal ethic as distinguished from group ethic" ("How Human Is Man?" *IJ,* 137). Faith in "progress," which Eiseley linked to the Judaeo-Christian tradition, has not diminished, but "progress" has become "secularized"; it has become "the increasing whirlpool of goods, cannon, bodies and yielded-up souls that an outward concentration upon the mastery of material nature was sure to bring" (138). For Eiseley, "the roar of the whirlpool" threatens to engulf the "second nature," which is custom (139). Even though this "progress" has a dual nature, for science has also relieved pain and opened the universe for exploration (139–40), Eiseley's concern is for the individual's relationship to the whirlpool; the individual will not find "the secret of the Garden" in inventions or in exploration, but within herself, in deciding the future by deciding what she wants to be (140–41).

Within the reexamination that the tutor text and the remotivated drama metaphor bring, Eiseley notes the changes in the Western ethic "toward conformity in exterior observance and, at the same time, toward confusion and uncertainty in deep personal relations" ("How Human Is Man?" *FT,* 121). In looking for "the meaning of Kierkegaard's faith in the eternal as the only way of achieving victory against the corrosive power of the human future" (121), Eiseley's model is clearly an interactive one. He has clearly delineated his dramatic metaphor as interactive; the stage, in a sense, sets the stage for an interactive examination of the intertext of nature and culture.

Eiseley's synecdoche and symbol of the human predicament is a single individual, a panhandling derelict whose "clever neo-modern, post-Freudian . . . lingo" centered on the statement, "'I can't help myself'" ("How Human Is Man?" *FT,* 143). Though the transient ambiguity called the human soul or spirit "merely came," with no help from the human being, it made possible the human quest to be more than it presently is. This "spirit," however defined, is what Eiseley interprets as the vehicle through which "Kierkegaard glimpsed the eternal, the way of the heart, the way of love which is not of today, but is of the whole journey and may lead us at last to the end" (145). Eiseley's difficulty with the term is evident; the notion of soul, spirit, consciousness, "inner light" is a traditional concept that he does not reexamine but attempts to define as a materialist and an evolutionist. The result is an emphasis on self-definition. Eiseley concludes that we can conquer the whirlpool "only by placing it in proper perspective," by retaining self-awareness (146).

To make this point, Eiseley tells a tale of a physicist, a man who contributed to the making of the atomic bomb, who found in the woods a tortoise that he picked up to take to his children. After a few seconds, however, he returned to where the tortoise had been and set it down, commenting that "'perhaps, for one man, I have tampered enough with the universe'" ("How Human Is Man?" *FT,* 148). The tale exemplifies the self-reflection and self-definition that can help to make a positive future, what Eiseley calls "a growing self-awareness, a sense of responsibility about the universe" that is the opposite of the whirlpool (148). The scientist's act "was not a denial of science. It was a final recognition that science is not enough" (148–49).

Human "nature," Eiseley argues, is neither totally identified with external nature nor totally separated from it. The question of "humanity" is a question of self-creation, directly related to the human

view of the self and of the environment. Only the sense of reflection, of "inward" looking and awareness of responsibility to the
nature that produced us, can make us "truly human" ("How Human Is Man?" *FT*, 149). In terms of Kierkegaard's tutor text, only
this awareness can link the present with the eternal and prevent
the future from becoming an enemy. We will be "truly human"—
that is, we will have defined the *human*—only when we realize our
participation in, rather than separation from and mastery of, the
natural world.

"How Human Is Man?" marks one of Eiseley's greatest quarrels
with Christianity, which, he argues, displaced divinity from nature
and placed the human above nature (138), but it also marks his
quarrel with traditional humanism, which has collaborated in separating humankind from external nature and placing the human atop
a hierarchy. The conclusion of the essay is a call for what he sometimes calls a "revivified humanism"; *revivified* reintroduces the
sense of the Latin root *vivus*, "life," to the concept of humanism.
Eiseley's "revivified humanism" would require not only denial of
fragmentation and assumed hierarchical superiority of the human,
but an awareness that science and technology alone cannot serve
as the route to understanding, that the human being must find his
own way (a word that inevitably has epistemological overtones).
And that "way" to the "Garden" must place the human in the larger
perspective of a participatory relationship with the natural world.

Scientific Knowledge/Poetic Knowledge: Inverting the Hierarchy

Eiseley inverts the epistemological hierarchy of scientific obser-
vation/poetic insight to reexamine the term "natural" and to explore how we know. Earlier I have noted his contention that the
literary figure often has "an ear or a sensitivity for things in the
process of becoming" ("How Life Became Natural," *FT*, 61). For
Eiseley, as we have seen, the scientific view falters when not accompanied by poetic insight. In *The Firmament of Time*, he reexamines the two kinds of knowledge and inverts the hierarchy that
makes science the dominant path to knowledge. In the foreword,
Eiseley openly embraces the task of combining scientific knowledge and humane understanding as he writes that the lectures were
designed "to promote among both students and the general public
a better understanding of the role of science as its own evolution
permeates and controls the thought of men through the centuries"

(*FT,* v). Perhaps nowhere in Eiseley's texts is his synthesizing technique more evident than in this book. As meta-observer he looks at encompassing terms that fix the understandings by which we live, acknowledging the scale of his treatment as a "survey" of elements of the evolutionary "drama." Eiseley's secondary purpose (a fulfillment of his declaration as a high-school student that he wanted to be "a nature writer") is a defamiliarization of a different kind; he hopes to "direct the thought of an increasingly urban populace toward nature and the mystery of the human emergence" (v). Of his merging of the simple and the complex Eiseley writes, "I make no apology for my attempt to treat simply of great matters, nor to promote that humane tolerance of mind which is a growing necessity for man's survival" (vi). In this text he not only combines science and the quest for humane understanding, but openly incorporates major literary texts—including those of Darwin, Freud, Agassiz, Coleridge, the Bible—that are encoded in our language and in our understanding. In what is his most traditional humanistic text, these encoded understandings become tutor texts for Eiseley's examinations of the human role in a post-Darwinian, post-Einsteinian world.

Eiseley's alternative to the positivist way of looking incorporates the poetic and emerges as holistic. With the retirement of machine or deity in favor of chance and change, reality has become an illusion. Creatures are many yet one, and Eiseley says, "it is almost as though somewhere outside, somewhere beyond the illusions, the several might be one" ("How Life Became Natural," *FT,* 82). The image suggests, as he will later explore, the view of the poet "attempting to see all" ("Science and the Sense," *ST,* 200), which is distinguished from the fragmented, particularized view of science. In humankind Eiseley finds "a curious wistful gentleness and courage," qualities that science alone cannot explain and that "have little to do with survival" ("How Human Is Man?" *FT,* 145). These qualities provide the key to "the way of the heart, the way of love which is not of today, but is of the whole journey" (145). This is the way of Heidegger's "poetic dwelling," of the poetic "evocative" that supplements the scientific "assertive," of the literary that yields the insight necessary to supplement scientific understanding.

Defamiliarizing "Natural": "Natural" as Linguistic/ "Nature" as Potential

In asking the second great question, "How Natural Is 'Natural'?" Eiseley focuses on the life-world that we call "natural"—examining scientific "observation" of this world, the effects of science on it, and the human relationship to it. He sees the history of all life as a history of reaching to be more than it presently is, and he sees the unique quality of the human being as a *conscious* reaching. "*Homo faber* the toolmaker is not enough," Eiseley writes, despite "the age-old primate addition to taking things apart" ("How Natural Is 'Natural'?" *FT,* 159). This reaching, linked to the hand as root metaphor, takes us beyond both physical "taking apart" and into intellectual analysis; the two are closely related. The earliest hefting of stones, Eiseley contends, is linked to the desire for "mastery over the materials of [the] environment" (158–59). In hunting cultures, James Bunn notes, hands were "intermediaries that magically symbolized a degree of control over animals, which were the measure of all things in that they supported hunting life" (1979, 49). Examination of the human "reaching" is part of Eiseley's re-evaluation of the human relationship to the world from which we emerged and also part of his quest for an epistemological position that does not deny science, but allows it to incorporate poetic insight. From "man the toolmaker" the essay moves to man the creature able figuratively to reach outside himself, to go beyond what he has previously determined to be "natural."

In "How Natural Is Natural?" Eiseley turns to "natural" as a linguistic construct rather than a representation of the real as the positivist epistemology would argue. The essay asks the question that underlies the human relationship with "external" nature and explores ways of remotivating the concept to effect the "uses" of all life, not just human. The three possible "natural" worlds that Eiseley describes reflect three perspectives on the natural environment that frame the holders' seeing. The first "world"—that of a physicist who, having helped to create the world that modern physics knows, in his old age became afraid of falling through the molecular space around him and wore oversized padded boots—is insubstantial, but underscores human self-definition. The second "world"—that of youths roaring about on a lake, secure in the power of their motorboats and unaware of the world that the physicist feared—is equally insubstantial. The third "world"—that of a young muskrat bringing ashore a bit of water foliage—may be considered "naive" and is certainly fragile. The narrator's response is a question: "In so many worlds, I thought, how natural is 'natural'—and is there anything we can call a natural world at all?" ("How Natural Is 'Natural'?" *FT,* 157).

However eccentric, the physicist's view, Eiseley writes, is close to the creation of worlds, to "the void where science ends" ("How

Natural?" *FT,* 158). In "The Star Thrower" in his next book, Eiseley will treat the void as the source of meaning, the Nothing from which the human mind creates meaning. Here Eiseley again examines the human quest for mastery of the environment, quoting Pascal's comment that "'there is nothing which we cannot make natural,'" and, further, "'there is nothing natural which we do not destroy'" (cited in *UU,* 159). Once we see something as "natural," as understandable in other than supernatural terms, we have the urge to control it; and mastery is the precursor of destruction. Both toolmaking and learning falter.

In the anecdotal climax of the essay, Eiseley tells of finding in a desert area a hen pheasant and a blacksnake intertwined—the bird probably having surprised the snake as it crept toward her eggs and the two having become so locked together that neither could free itself. The scientist might have waited to see what would happen, but the man intervened, letting the harmless serpent coil around his arm and taking it over the ridge. "The bird," he writes, "had contended for birds against the oncoming future; the serpent writhing into the bunch grass had contended just as desperately for serpents" ("How Natural Is 'Natural'?" *FT,* 175). The human role is unique—not the human as benefactor, but the human being "struggling" to understand himself: "I had struggled, I am now convinced, for a greater, more comprehensive version of myself" (176). Physically (through evolution) and metaphorically (as the one capable of intervening on the side of life), the man "contained . . . the serpent and the bird" (178). As traditional codes, the serpent and the bird signify the dual nature of humanity (good and evil, creative and destructive forces), as well as freedom from and bondage to earth. In Herman Hesse's context, the bird flies to Abraxas, the deity that "is God and Satan and . . . contains both the luminous and the dark world" (1970, 93). In "containing" both serpent and bird, Eiseley's narrator experiences the beginning and the end, the "eye of the poet attempting to see all." The experience leads the narrator into "another sphere of reality" ("How Natural Is 'Natural'?" *FT,* 178), which exists *within* nature and cannot be separated to be examined and "understood."

Thus Eiseley concludes, "I no longer believed that nature was either natural or unnatural, only that nature now appears natural to man" ("How Natural?" *FT,* 178). *Natural,* then, is a term for what we accept as reality and what we think we understand. It is linguistic, not "real," and positivist observation does not lead to a totalizing conception of nature. Eiseley finds more to the "natural" than we realize. As he writes in his next book, "there looms, inexplicably, in nature something above the role men give her" ("The Star Thrower," *UU,* 92). The "natural"—like the three "worlds" of the muskrat, the youths, and the physicist—is "unreal" because it is continuously changing ("How Natural?" *FT,* 178). The individual "who makes nature 'natural,'" Eiseley contends, "stands at the point where the miraculous comes into being, and after the event he calls it 'natural'" (179). As a scientist, Eiseley uses the word *miraculous* carefully; he stops to define it "as an event transcending the known laws of nature," which change continually as science changes, for the earth "is always just passing from the unnatural to the natural, from that Unseen which man has always reverenced to the small reality of the day" (171). The Unseen is the abyss of potentiality, the "domain of absolute zero" from which the aged star thrower will assert meaning ("The Star Thrower," *UU,* 87). It is the potential from which meaning is generated, "the reality of night and nothingness" ("How Natural?" *FT,* 171) from which the miracle emerges. Only the human imagination brings us close "to the world beyond the nature that we know" (179). And that world is the randomness of potentiality, the randomness that is rich in information, from which the changing reality emerges.

Not science, assertive knowledge, alone, but poetic knowledge, "the flashes of beauty and insight which trouble us so deeply," can suggest "what the race might achieve" ("How Natural?" *FT,* 179). If humankind destroys what it makes natural, then the last idol "is no less than man made natural" (180)—that is, the assumption that human manipulation is a "natural" part of an explicable, "usable" world. This "idol" is not technological, "not the outward powers of man the toolmaker" (180), but Heidegger's "Enframing" that "threatens man with the possibility that it could be denied to him to enter into a more original revealing and hence to experience the call of a more primal truth" (1977b, 28). What we profess to understand ignores the potentiality. The "last idol"—the Enframed view of the human as "natural," which explains and fixes the human as if it were unchanging and also encompasses human mastery over an externalized nature, human separation from a fragmented nature, human reduction to "reality," and hence destruction, of all things—is what Eiseley would displace for the primitive concept of a "nature, not of this age, but of the becoming," a nature of the meta-view of wholeness and potentiality, for we are "not totally compounded of the nature we profess to understand" ("How Natural?" *FT,* 180).

For Eiseley, then, positivism falters and the uncanny remains. We are "always partly of the future," which we help to determine

("How Natural?" *FT,* 180). That future exists in language, in dreams; it affects the "real" just as the observer affects the "nature" she observes. The word *natural* is indeed "a magician's word"; it is linked to a traditional code of magical words whose conjuring effect we should fear. The Enframing of technology extends to the language of concepts—of "human" and "natural." In contrast, the imagination and sympathy demonstrated by humble early humans who poured gifts into the grave could, Eiseley speculates, possibly "take man beyond the nature that he knows" (181).

Eiseley's defamiliarization of "natural" is essential to his reconsideration of the human relationship to the environment. To be truly "human"—to define ourselves in terms of the nature that produced us—will require an understanding that moves away from the Enframed view of nature as that which we can isolate, fragment, understand, manipulate, and eventually—as Pascal suggested—destroy. It will require an understanding that looks to dreams and imagination, to insight and sympathy, to potential and unpredictability, to a dialogic understanding that allows for not only the scientist's but also the poet's and the amateur's insights. It is this understanding that Eiseley seeks, though he cloaks it in a dramatic tale of a lonely rider who intervened in a battle between a serpent and a bird or in a recollection of a neurotic physicist who feared that he would fall through the interstices of the molecular world he had helped to create. The quest for this understanding leads Eiseley from scientific generalizations back to Heidegger's "here and now" and "little things" (1977b, 33) that provide the discursive "way" out of Enframing, which is the real danger of technology.

7
Reexamination of Science I:
The Unexpected Universe

If *The Firmament of Time* marks Eiseley's direct turn to an ironic perspective and his increasing experimentation with extended personal narrative and with literary tutor texts, the next four books represent Eiseley's literary and rhetorical maturity. The first two, *The Unexpected Universe* and *The Invisible Pyramid,* examine science as an institution, explore the personal in relation to the universal and to epistemology, further explore metaphor and personal narrative, and frankly address the uncanny. The essays in both books continue his program of defamiliarizing the cardinal myths of Western thinking—mechanism, determinism, positivism, reductionism, and fragmentation—as Eiseley continues his linguistic reexamination and exploration of metaphors.

Eiseley on the "Unexpected"

The Unexpected Universe contains some of Eiseley's best work—and some of his worst. Published in 1969, with three essays based on lectures given at Stanford University, it contains, in a departure from earlier dedications, a dedication to an unknown dog, Wolf—an indication of Eiseley's increasing retreat (evident in his later essays, but especially evident in his poetry) from the human world in favor of the animal world. In the acknowledgments Eiseley refers to the three Stanford lectures as "explorations of the unexpected universe," embracing the notion of exploration that is traditionally inherent in the essay. The epigraph (by now "expected") for the book is, fittingly, from Heraclitus, who maintained that the universe consists of change and flow, and it concerns the paradox of "unexpectedness": "If you do not expect it, you will not find the unexpected, for it is hard to find and difficult" (*UU,* ix). The epigraph echoes a recurrent theme in Eiseley's texts, a

theme given shape in the metaphor of the "dancers in the ring"—
the theme of the difficulty of breaking out of an accepted way of
looking at the world. As Eiseley discusses the theme, it is primarily
a characteristic of science, though the "narrow professionalism"
of science has its counterparts in any discourse community. Even
Heraclitus, who emphasized the role of flux and change and looked
to a returning cycle of time, recognized the difficulty of discerning
the "unexpected." In this book Eiseley gives greater rein to his
imagination in pursuing unexpectedness. The essay titles are metaphorical (or metonymic, as in "The Innocent Fox"), beginning with
"The Ghost Continent," in which he uses the Odyssean voyage in
literary versions from Homer to Kazantzakis to explore the ghostly
terrain of the individual psyche. In a complex text that interweaves
the Odyssean voyage of legend with voyages of science including
Cook's Antarctic voyage and Darwin's voyage, Eiseley incorporates elements of his own journey in science to explore the human
quest for knowledge, the quest for an epistemology that will unlock
the secrets of the journey. He concludes the essay with a reference
to his own voyage of knowing:

> I have listened belatedly to the warning of the great enchantress. I
> have cast, while there was yet time, my own oracles on the sun-washed
> deck. My attempt to read the results contains elements of autobiogra-
> phy. I set it down just as the surge begins to lift, towering and relentless,
> against the reefs of age. ("The Ghost Continent," *UU*, 25)

In this concluding paragraph of the first essay, Eiseley hints of
what is to come, of what I shall call his "re-nunciation"—a paradoxical simultaneous renouncing and re-announcing of his "scientific heritage" in one of his finest but most intricate essays, "The
Star Thrower." The re-announcing is especially powerful in its
positing the emergence of "value" from the "domain of absolute
zero" ("The Star Thrower," *UU*, 87), followed later in "The Invisible Island" by the notion of culture emerging from nothingness.
Both insights, as I shall treat in further detail, powerfully anticipate
the understandings of chaos studies that began to receive attention
more than a decade later. The paragraph hints, too, of Eiseley's
pursuit of the uncertainties of encoding and decoding that make
"The Golden Alphabet" a remarkably insightful exploration of the
textual principle and the reader's role in responding to the text. It
hints further of Eiseley's experience tumbling with a young fox in
"The Innocent Fox," when he realized, as painfully as a postmodernist can recognize, "There was a meaning and there was not a
meaning, and therein lay the agony" (*UU*, 211).

The Unexpected Universe is a collection of carefully wrought, powerful essays, which are often characteristic of Eiseley at his best. Eiseley's anticipation of the postmodern dilemma—that meaning is both present and absent—remains evident. He continues to interweave the topoi of the body and acquisition of knowledge, as well as the positive and negative, creative and destructive, sides of life's ability to change. And in this collection the theme of potential emerges strongly. The evolutionary ability to change, he writes in "the Star Thrower," "is both creative and destructive— a sinister gift, which, unrestricted, leads onward toward the formless and inchoate void of the possible" ("The Star Thrower," *UU*, 78). Such a "force can only be counterbalanced by an equal impulse toward specificity. Form, once arisen, clings to its identity" (78), creating a dialectic of stability and inevitable change.

In this chapter I shall examine Eiseley's treatment of themata and metaphors and their roles in his defamiliarization of the cardinal scientific mythemes, beginning with his treatment of games of chance in "The Unexpected Universe" and the Odyssean journey in "The Ghost Continent." Because of the centrality of its expression of "rejection" of Eiseley's "scientific heritage," I shall examine in detail both the structure and the thematic and metaphorical explorations of "The Star Thrower." Because of its penetration into the encoding and decoding of life itself, as well as of the human understanding, I shall examine "The Golden Alphabet." In examining "The Inner Galaxy" and "The Invisible Island," I shall note Eiseley's continuing exploration of the interactive theater and of human language. Finally, I shall examine Eiseley's use of a simple personal experience and the "magical" crossing of the boundaries of form between human and animal in "The Innocent Fox."

Magical Reality and Games of Chance:
"The Unexpected Universe"

Chance—in nature, in the course of an individual life, in the universe—is basic to Eiseley's cosmogony. From the earth's atmosphere to the emergence of humankind to genetic recombination, chance is dominant. The human brain's linguistic center has broadened chance by enlarging the possibilities of human social behavior ("The Unexpected Universe," *UU*, 38). In the title essay of this collection, Eiseley begins an ongoing dialogue with Einstein's widely quoted comment that "God does not play dice with the universe," as he explores metaphorical games of chance that undermine reductionism and determinism. In a random universe, the constant, unpredictable shifting of forms gives a magical effect, which Eiseley explores figuratively, taking advantage also of the link of early science with magic and the tendency today to ascribe almost magical powers to science.

In a verbal play on the meaning of experimentation, Eiseley takes the reader back to the first experimenters, the alchemists. As Jung and others have observed, science has its roots in alchemy and magic. "To a universe already suspected of being woven together by unseen forces," Eiseley writes, humankind has added "the organizing power of primitive magic" ("The Unexpected Universe," *UU,* 32). Reexamining a science that seems to perceive itself marching steadily toward ever greater knowledge, Eiseley returns to the "magical" origins of a science rooted in reductionism and to a mutability and a mystery that undermine the apparent certainty and epistemological stability of science in favor of fluidity and a poet's sense of wonder, which he finds the essential supplement (an addition, yet necessary) for science.

Magic provides a metaphorical undermining of reductionism and an underscoring of the role of contingency in evolution. As the primitive "organizing power," it is the result of the human's "first conscious abstraction from nature, his first attempt to link disparate objects by some unseen attraction between them" ("The Unexpected Universe," *UU,* 32). In primitive magic, Eiseley finds a kinship with his own message and method. Rejecting the nineteenth-century emphasis on primitive warfare, Eiseley chooses human sympathies as models for modern study. The primitive mind retains a sense of connection between the human and the environment, and it attempts the linking of disparate objects that Eiseley adopts as a hallmark of his method, which links disparate objects as well as the discourses of science and poetry in an attempt at dialogue and synthesis.

In *The Unexpected Universe,* Eiseley explores three possibilities of the notion of magic. First, *magic* is a term for the pull of entropy, the tendency toward disorder that affects even the biological world. "Black magic," he writes, "the magic of the primeval chaos, blots out or transmogrifies the trues form of things. . . . Instability lies at the heart of the world" ("The Star Thrower," *UU,* 78). Thus "black magic" is equated with chaos to emphasize the instability of living forms. The human body also appears as "a magical vessel" dependent on the opposite of chaos, on a symbiosis based on photosynthesis that only plants can accomplish—hence the "intricacy" of the human interrelationship with other life forms ("The Hidden

Teacher," *UU,* 52). Science, too, is explicitly "magical," having its origins in the primitive organizing power. Eiseley refers to the human being as "the self-fabricator who came across an ice age to look into the mirrors and the magic of science" (55). Like the mirror, science is magical: it both distorts vision and reflects the maker or the magician, for the scientist is reflected in her work. Magic is linked with both chaos and organization in a tension that echoes what Eiseley cites as the "universal tension" between the present and the potential (in current chaos studies the relationship between chaos and organization is revealed in its increasing complexity[1]). Magic thus suggests the unpredictability and instability of the universe, the interconnectedness of life, and the impossibility of reducing matter or life to simple terms or building blocks.

The epistemological dimension of magic as a metaphor—knowing as unpredictable, indeterminate, not necessarily or predictably causal, and not reductionist—is underscored by the meanings trailing behind the word itself. The Greek *magike* is *tekne,* referring to the sorcerer's art, related to the Persian *magus* and the Indo-European root *magh-',* meaning "to be able" or "to have power." Closely related is the root *magh-os-,* meaning "that which enables" and becoming the Doric *makhos,* "device," "machine," "mechanism." *Tekne,* as art or skill, is also related to weaving and textiles and to technology. *Magic* thus belongs to the family of words to which both *art* and *technology* belong. As Heidegger notes, there was a time when *tekne* incorporated art, not exclusively technology (1977b, 34). In *magic* the two come together, as technology and art, science and Heidegger's "saving power" of reflection. The linking of the two is especially important in Eiseley's quest for an altered epistemology. If language, as Heidegger suggests, indeed "knows" or carries with it more than we know or more than we realize, then the language that links science and magic suggests that we may indeed know poetically as well as conceptually.

As the epigraph to "The Innocent Fox," which treats the theme of a "magical" crossing of the boundaries of form between man and animal, Eiseley cites a passage from Peter Beagle: "Only to a magician is the world forever fluid, infinitely mutable and eternally new. Only he knows the secret of change, only he knows truly that all things are crouched in eagerness to become something else, and it is from this universal tension that he draws his power" (cited in *UU,* 194). As epigraph, this passage draws together magic, continuous change, and the recurrent emphasis on the potential—all essential, for Eiseley, to a fuller understanding of science itself, of the total journey.

In the metaphors of chance, his dialogue with Einstein is mostly silent, but at times explicit, as when he mentions Einstein's remark but contends,

> it would appear that in the phenomenal world the open-endedness of time is unexpectedly an essential element of His creation. Whenever an infant is born, the dice, in the shape of genes and enzymes and the intangibles of chance environment, are being rolled again. . . . Each one of us is a statistical impossibility around which hover a million other lives that were never destined to be born—but who, nevertheless, are being unmanifest, a lurking potential in the dark storehouse of the void. ("The Unexpected Universe," *UU,* 40)

Einstein never completely resolved his quarrel with quantum mechanics. Though he helped to create the world of randomness, he nevertheless clung to a causal view more characteristic of the thinking of the two previous centuries. Ironically, Eiseley, too, in the passage just quoted, perceives the randomness of the universe—Darwin's contribution, he repeatedly notes, was to make the emergence of life forms, including the human, a totally random occurrence. Yet his own words undermine the randomness he embraces; as he writes of the "lives that were never destined to be born" (40), the notion of *destiny* again injects a presumably unwanted deterministic perspective. The problem only heightens what the linguistic episteme has led us to understand: that we may unconsciously undermine the very meaning we attempt to convey.

As the title essay, "The Unexpected Universe" begins with an epigraph from John Donne: "Imagine God, as the Poet saith, *Ludere in Humanis,* to play but a game at Chesse with this world; to sport Himself with making little things great, and great things nothing; Imagine God to be at play with us, but a gamester" (cited in *UU,* p. 26). The essay opens with a tale of a train stalled by chance beside a burning city dump, with Eiseley the narrator talking with an attendant who forked material into the flames about what could be found there. As if in answer the attendant held up a fragment of an old radio, whose dangling wires reminded Eiseley of other times that he had, as an archaeologist, encountered disconnected mechanisms of communication, whether human or machine. From this opening image Eiseley draws two strands of scientific understanding that recur throughout the essay: perception through the human senses and perception within context, a given "society's moment of time" (30). Returning to Donne, Eiseley quotes the lines

"I am rebegot / of absence, darkness, death: / Things which are not" (30). These "things which are not" are focal points in Eiseley's later texts—marginal "things" of the past and the future as much as the present; of (symbolic) darkness rather than daylight (Eiseley "would speak . . . not as a wise man, with scientific certitude, but from a place outside, in the role . . . of a city-dump philosopher"); and of death because the archaeologist's task is to put "civilizations to bed" (28–29).

This "city-dump philosopher" moves at the margins of science and the margins of humanism, doubting both yet reaffirming both. This book, he argues, will further pursue the "ill-concealed heresy" introduced in "How Natural Is Natural?" ("The Unexpected Universe," *UU*, 31). His pursuit will engage him and his audience

> not in the denigration of science, but, rather, in a farther stretch of the imagination as we approach those distant and wooded boundaries of thought where, in the words of the old fairy tale, the fox and the hare say good night to each other. It is here that predictability ceases and the unimaginable begins—or, as a final heretical suspicion, we might ask ourselves whether our own little planetary fragment of the cosmos has all along concealed a mocking refusal to comply totally with human conceptions of order and secure prediction. (31)

In Donne's words Eiseley finds a recognition "that behind visible nature lurks an invisible and procreant void from whose incomprehensible magnitude we can only recoil" (31). The notion of the "procreant void" underlies Eiseley's treatment of potentiality throughout this text and the later ones. Eiseley emphasizes the range of "*possible* behavior" (38) of humankind and undermines notions of causality, fixed determinism, and teleology in favor of randomness that we can direct toward Bacon's "uses of life." "Great events," he writes in the *Lost Notebooks,* "are cheapened and made vulnerable simply by being isolated in a world made up of continuities and causal sequences" (*LN,* 191–92). Causality, in his view, can be considered only in view of the web of existence and in probabilistic terms.

In pursuing science as a dialogue of "senses" plus context (a dialogue in which the mechanism of communication is often broken, as it is in science itself), the first two parts of the essay trace the eighteenth- and nineteenth-century sense of security that arose from the concept of the world machine, which displaced primitive magic and reduced nature and life itself to lawful obedience. Part 3 treats Darwin's introduction of "absolute random novelty," what Alfred Russel Wallace called "'indefinite departure,'" and linear time ("The Unexpected Universe," *UU*, 36). The realization that chance itself evolved, aided in human life by the mechanism of sexual attraction, leads Eiseley to focus on the human brain, which could not have been predicted until it emerged. With the emergence of the brain, human physique could remain remarkably similar while thought could become increasingly diverse, leading to "such vast institutional involutions as the rise of modern science, with its intensified hold upon modern society" (39). And with the introduction of technology's "irrevocable forces," Eiseley turns to the "biological analogy" of "a single gigantic animal embod[ying] the only organic future of the world" (39). Still there remains an unexpected quality, a dual potential. The Roman Empire, in spite of its "political and military machine" lost its "lines of communication" (43). Today, science is similarly the center of a political and military machine and itself "is of human devising and manufacture"—a fact that Eiseley finds frightening because "contingency has escaped into human hands and flickers unseen behind every whirl of our machines, every pronouncement of political policy" (43–44). Contingency itself—and for this insight Eiseley refers to D'Arcy Wentworth Thompson—has evolved. The lines of communication are not always reliable.

The last section of the essay introduces another eerie image of desolate and unworking lines of communication as Eiseley tells of passing a bleak cemetery on whose borders were six nonworking telephone booths. Here he reiterates the randomness of the emergence of photosynthesis, which made possible oxygen-consuming fauna, and "the phenomenon of sex that causes the cards at the gaming table of life to be shuffled with increasing frequency and into ever more diverse combinations," to conclude that nature itself "contains the roiling unrest of a tornado," a metaphor that will recur in "The Star Thrower" as an extended metaphor for contingency ("The Unexpected Universe," *UU*, 46). Order in the human sense, he argues, "is at least partially an illusion," for our order is based on our time, which is infinitesimally small in relation to the cosmos (46). It is primitive philosophers such as the Hopi who have sensed that beneath our brief superimposed order, "lurks being unmanifest, whose range and number exceeds the real" (46). Thus the unexpected remains, in spite of human accomplishments. And, returning to the marshy city dump site, Eiseley links it to the ancient sense of chaos as unformed matter: the unexpected, he writes, "is why the half-formed chaos of the marsh moved me as profoundly as though a new prophetic shape induced by us had risen monstrously from dangling wire and crumpled cardboard" (46).

The sense of potentiality, of "latency," drives Eiseley's fascination with the future. "To the unexpected nature of the universe," he writes, we owe our existence, for we incorporate "unknowingly, the shapes and forms of an uncreated future," which will come from ourselves ("The Unexpected Universe," *UU,* 47). If the negative aspect of potentiality is the evolution of contingency itself, the positive aspect is that the human consciousness of existence has led us to ask "terrible questions" (47), questions that can lead to a knowledge of self and a human dignity, which Eiseley sees as a goal for humankind, though one that seems far away to this "city-dump philosopher" viewing his century from the margins of science and the margins of humanism. The metaphor is the recurrent one of "the Unseen Player in the void," rolling "his equally terrible dice" (47). In place of a causal or teleological view of evolution, Eiseley suggests, through the metaphorical concept of the dice, a universe shaped by chance, but a universe whose very indeterminate potentialities open human beings to the quest for knowledge that brings their greatest dignity. And the journey, we recall, is the human search for knowledge. The role of the dice functions not only at the macro level of the randomness of the universe, but also at the micro level of the genetic code, undermining teleological determinism for random determinism and fascination with potentiality. The dice thus symbolize both unrealized and realized potentialities, both chance and probabilities. For Eiseley, the significance of what *is not* echoes from behind what *is:* what is not remains the potential that is creative.

The chapter ends on the dual note of human potential for both positive and negative. "We are more dangerous than we seem," he writes, "and more potent in our ability to materialize the unexpected that is drawn from our own minds" ("The Unexpected Universe," *UU,* 46). A thema that shapes the growing sense of urgency in Eiseley's texts is this constructivist view, which he knows that few have grasped. Reiterating the mediating effect of language, he quotes Emerson, who referred to the human discovery of existence as the figurative "Fall of Man" and observed "'that we do not see directly'" (47). For Eiseley, "both the light we seek and the shadows that we fear are projected from within" (47). And in the "terrible questions" that we have learned to ask Eiseley finds a possible resolution of the human predicament, though the resolution reiterates his dialogue with Einstein: "Perhaps it is just for this that the Unseen Player in the void has rolled his equally terrible dice," for "the self-knowledge gained by putting dreadful questions" is the potential for human dignity (47).

The Odyssean Tutor Text: "The Ghost Continent"

With the Odyssean voyage, Eiseley draws together the individual, evolutionary, human, epistemological, scientific, and creative journeys that he began in *Darwin's Century* and pursued in the next two collections of essays. His most comprehensive treatment of the Odyssean voyage, which is an effective model of the random journey, appears in "The Ghost Continent." He divides the essay into three parts—the first establishing Odysseus and his journey as a model, the second treating the "Odyssean voyages" of eighteenth- and nineteenth-century science, and the third linking the Odyssean voyage with contemporary science through images from a Scandinavian myth contemporary with Homer.

Opening with the statement that we all contain "ghost continents" where we can encounter "strange shapes" that we would fear to expose to our fellows, Eiseley begins his personal record by joining "the Odyssean voyages of legend and science" ("The Ghost Continent," *UU,* 3), taking Homer's legend and modern variations of it as tutor texts and explicitly outlining his allegorical intent. Eiseley is willing to "claim no discoveries [but] . . . only the events of a life in science as they were transformed inwardly into something that was whispered to Odysseus long ago" (3). What Circe whispered to Odysseus, he suggests, was magic—the reassurance that magic could not touch him. The fragment of magical knowledge is linguistic knowledge, for Circe's reassurance enabled Odysseus to eschew magic as such and to rely on the magic of words. What Eiseley claims to report is the knowledge granted by the magical power of words.

Eiseley compares Odysseus's journey to the human journey among "the shape-shifting immortal monsters" of human legend, the magical figures that in the present world "assume [the] more sophisticated guises" of science and modern myth ("The Ghost Continent," *UU,* 4). The Odyssean journey explores the textual metaphor as the narrator describes the possibilities of the Odyssean allegory that "can be read as containing the ingredients of both an inward journey of reflection and an outwardly active adventure" (4). In contemporary culture Eiseley discovers "all the psychological elements of the Odyssey . . . present to excess: the driving will toward achievement, the technological cleverness crudely manifest in the blinding of Cyclops, the fierce rejection of the sleepy Lotus Isles, the violence between man and man" (5). Odysseus himself symbolizes not only the "human journey," but Eiseley's recurrent Faustian tension between desire for "illimitable knowledge" and yearning for "a lost world of inward tranquility" (7, 5).

In the context of this tension, Eiseley compares Odysseus to Cook and Darwin, scientific voyagers who found new worlds after journeys "as irretraceable and marvel-filled as any upon the lost sea charts of Odysseus" (9). Like Odysseus, Cook found a "ghost continent," the human "inner world" ("The Ghost Continent," *UU*, 3), and Darwin found "the whole Circean labyrinth of organic change" in the Encantadas (15). Eiseley suggests an allegory of evolutionary shape-shifting in Odysseus's experiences: "What had appeared to Odysseus as the trick of a goddess was, in actuality, the shape shifting of the incomprehensible universe itself" (15). He compares the effect of this instability of form on the human mind to the effect of Circe's tricks on Odysseus's crew, for Darwin found that all of life could be changed by "a power hidden in time and isolation . . . into wavering shadows" (15). The real difference, as Eiseley observes in his notebooks, is that earlier human beings thought in terms of immediate changes, whereas evolution introduces change over a long period of time (*LN*, 186).

And in the Odyssean analogue Eiseley also finds the tendency of the mind to turn "homeward, seeking surcease from outer triumphs" ("The Ghost Continent," *UU*, 18). For this analogue he turns to Giovanni Pascoli's "Ultimo Viaggio" (1904), in which Odysseus's return home is portrayed as an anticlimax. Pascoli's Odysseus represents, for Eiseley, "the hunger for lost time" (18). In Pascoli's conclusion, the waves take the dying Odysseus to Calypso, who hides him with her hair. Thus Odysseus, who had escaped the Cyclops by identifying himself as "No Man," "has come home to Nothingness" (18). This figure of "Nothingness" anticipates Eiseley's "domain of absolute zero" from which the human mind creates meaning, one of his repeated uses of a kind of semantic vacuum from which the human mind creates meaning (see "Science as Aporia," pp. 192–210). In the return to nothingness, Pascoli's Odysseus experiences not knowledge but the calm of sympathetic understanding. The Sirens do not sing for him, but he understands them, aware that "knowledge without sympathetic perception is barren" (18). Again the thema of competing epistemologies, always just below the surface, emerges in Odysseus's realization.

Eiseley turns next to another journey from folklore—a Scandinavian tale from a time approximately contemporary with Homer. This allegory involves "a strange crossroads religion"—a phrase that evokes associations of intersecting "paths," or ways of knowledge, of cultures. In the tale an earth goddess wanders through the forests in an enormous coach, pausing with drawn curtains at villages where the poor and ignorant live. Priests and followers come before the coach, but none can look behind the curtains or speak to the coachman or touch the oxen that draw the coach. After a ritual and a human sacrifice, the coach moves on. The allegory, Eiseley finds, is still powerful, for "the same awkward coach still lurches through the darker hours of our assembled scientific priesthood" ("The Ghost Continent," *UU*, 19–20). The coach is the vehicle of science itself (the vehicle of the dominant epistemology is itself vehicular), which has become our version of the primitive "crossroads religion," with all of its implications of intersecting cultures.

The narrator muses that in one age the coach bears "the masked face of Newton, his world machine ticking like a remorseless clock in the dead and confined air," or that in another age "Darwin lurks concealed behind the curtains, and all is wild uncertainty and change in the misty features of his company," or that Freud looks out at "a sea of leering goblin faces" ("The Ghost Continent," *UU*, 20). Again, he muses that the Abbe Lemaitre's followers "hear the alternate expansion and contraction of nature's pounding heart," or that, "silhouetted gigantically in the fierce rays of atomic light streaming from the carriage, four sinister horsemen trample impatiently" (20). Finally, he speculates that "the figure in the coach is a changeling and its true face is no face, as Odysseus was 'no one' until he shouted a vengeful name before the Cyclops," or that "behind the concealing drapery, hooded in a faceless cowl, there is caught only the swirling vapor of an untamed void whose vassals we are" (20).

The wanderer, then, confronts chaos, and science is its servant. In this "ancient car" (science itself, Eiseley suggests elsewhere, has its origins in questions as ancient as those of the voice in the whirlwind questioning Job ["The Hidden Teacher," *UU*, 48]), the face is "No Face"; and "we" are the acolytes, "toiling in a hundred laboratories with our secret visions of what is, or may not be, while the wild reality always eludes our grasp" ("The Ghost Continent," *UU*, 21). Eiseley's rhetoric clearly implies attitudes toward each thread of the dreamlike sequence: the mechanistic worldview "ticks like a remorseless clock" in "dead" air; the frightening world of the unconscious evokes "leering goblin faces"; the beginning of astrophysics evokes the "pounding heart" of nature; and the atomic era is imaged in terms of the four horsemen of the Apocalypse. The figure in the coach, Eiseley imagines finally, is a changeling whose face is "no face," just as Odysseus was "no one" until he declared an identity. Or perhaps "we"—followers of science—serve only the uncontrollable spiral of nothingness (whose "center cannot hold," in Yeats's words), in contrast to the centering and circling of traditional metaphysics. Yet out of this existential confrontation with nothingness, the human mind creates value. Out of the semantic void we create our own meanings. The point is, further, that science is not one method or one body of facts. Science shifts as constantly as the "natural" world that it both observes and shapes. Linking the folk tale to the *Odyssey,* Eiseley uses Homer's classic journey to assert both the shifting of reality and the human role in reading or creating that reality.

In the twentieth century, Eiseley notes, Odysseus has been portrayed "as a symbol of the knowledge-hungry scientist" ("The Ghost Continent," *UU,* 23), but this portrayal overlooks the epistemological implications of Circe's warning about magic. Scientists, he warns, often forget "the inward journey" (24). In promising Odysseus that magic would be unable to touch him, Circe gave Odysseus a magical self-fulfilling prophecy that would itself lead to protection. Epistemologically, Eiseley emphasizes this "magical" linguistic power, which he finds a necessary complement to scientific knowledge. If the Odyssean journey symbolizes "both [human] homelessness and . . . power" (24), then human beings need the "magic" of words. They need to verbalize Odysseus's wonder and his empathy with the dog Argos, an empathy that transcends the boundaries of form. The Odyssean voyage, then, involves at once the journey of life itself, the "inner" or psychological journey, the journey of knowledge, and the journey of culture, which includes science. As a model of a random journey, with byways and unexpected adventures, it displaces the fixity, determinism, and teleology embedded in Western thinking, especially since Newton's world machine and the nineteenth-century incorporation of it.

Science as Aporia: Epistemological Re-nunciation in "The Star Thrower"

> Poetry is the first and last of all knowledge—it is as immortal as the heart of man.
>
> Wordsworth, preface

Ultimately, all of Eiseley's texts are about knowledge—about science as *scientia.* "The Star Thrower" is not only a profound emotional renunciation of the narrator's "scientific heritage" in favor of the uniquely human sympathy with "the lost ones, the failures of the world" ("The Star Thrower," *UU*, 86), but a re-announcing, a "re-nunciation." The root of *renunciation, nuntiare,* "inform," from *nuntium,* "message," derives from the Indo-European form -*neu,* "to shout." The essay is at once a defamiliarization, an undermining, and a renewing of science as knowledge. In his renunciation, Eiseley is also re-announcing an epistemological position. Because "The Star Thrower" is the pivotal essay in *The Unexpected Universe* and a major text in Eiseley's epistemological quest, I shall trace its development in detail.

As a vehicle for understanding, "science," in Eiseley's view, must be reexamined, even undermined. The nineteenth-century notion of an unmediated vision is subverted by a new paradigm— that of the mediated vision of primitive or interactive reading. Eiseley leads the reader to encounter an aporia from which comes an insight. To draw the reader into this experience, he employs a tutor structure like that of classical drama, an alternation of episode (act, narration) with ode (comment, insight, "poetic" knowledge). Such "poetic" knowledge, for Eiseley, suggests the understanding or insight that cannot be measured or quantified, but is no less real; the intuitive or "uncanny" understanding that is peculiar to human beings; the empathy that enables human beings to identify with creatures other than the self; a Heideggerian "dwelling" rather than a Cartesian existence; and an awareness of the web that interlinks all things. The movement of the text is both literally and metaphorically a journey (the Indo-European -*ode* also indicates a way or path), in the sense of Maurice Blanchot's comment that the Sirens forced Odysseus "to undertake the successful, unsuccessful journey which is that of narration—that song no longer directly perceived but repeated and thus apparently harmless: an ode made episode" (1982b, 61). In this narrative movement of ode/ episode, Eiseley employs the reader's response to the text to reflect the epistemological problems treated. Science becomes aporia— not only an impasse that signals undecidability of meaning, but also a blindness, as well as an abyss or a "rift," which is nevertheless also a "joining." Confrontation of the aporia/rift, Eiseley contends, must come before understanding.

Although the important movement (journey) of the essay is epistemological, the alternation of episode and commentary forms a simple narrative or episodic framework for the essay. Part 1 tells of the fictionalized (autobiographical) narrator's coming to Costabel, encountering the madman who hurls "stars" back into the ocean (light and life archetypes united), and rejecting the man's action.

Part 2 gives a narrative flashback to the trickster who postured behind the priest of a primitive tribe many years earlier, followed by the narrator's return to his hotel, where he is haunted in the darkness by multiple eyes from the previous day and from days long past, the final eye seemingly coming from a torn photograph. Part 3 presents a flashback to the narrator's finding the photograph from which the torn eye came—a photograph that includes the narrator's mother as a child, her eyes speaking of the isolation of total deafness. As the narrator ponders a maelstrom of eyes, he experiences the climactic encounter in which he rejects both traditional Western culture and his own "scientific heritage." Only then does the eye from the photograph look at him "sadly" and depart. Finally, in Part 4, the narrator chooses to join the madman throwing "stars," acting out his rejection of the "way" (path, as well as method) of science and his choice of a primitive closeness to nature.

Alternating with episode, the commentary accomplishes the essay's epistemological movement—from the eye in the skull (the eye of science) to the abyss that the eye observes and to the confrontation with other eyes that leads to acceptance of poetic vision. This movement of the essay relies heavily on interrelated visual and spatial metaphors and parallels the four parts: (1) establishing the abyss or aporia as necessary to understanding; (2) undermining the stability of knowledge, especially scientific knowledge, in favor of the omnipresence of the abyss; (3) exploring the necessity of the abyss; and (4) establishing the answer, whose "way" is not the way of science, in another epistemology—one that incorporates both science and the supplement of poetic and humane knowledge, which is actually not outside nature, but in another "dimension" of nature.

i

In his epistemological exploration, Eiseley turns to a constellation of metaphors: the revolving eye in the skull that alternates with the eye from a torn photograph; the trickster from a primitive culture whose counterpart in nature is a tornado; the rift that is also a joining, the spatial image for what he calls the "discontinuities of an unexpected universe" ("The Star Thrower," *UU*, 91); and metaphors of textuality and reading. All the metaphors that inform "The Star Thrower" and provide its epistemological movement are related to reexploration of the visual philosopheme of light/sight/insight.[2] Plays on the word *star* repeatedly suggest light; even the star-*fish* is not only the product and vehicle of the archetypal source of life, the sea, but also the vehicle that leads to poetic insight. The journey as metaphor, which begins in *The Immense Journey* and informs this text, among others, is also linked to this basic root metaphor through the Latin *dies,* "day," and its association with luminosity, the traditional root metaphor for understanding. Western thinkers invented and enshrined the scientific method—use of the supposedly unbiased eye (the "innocent eye," to nineteenth-century painters and natural historians) to observe and record nature (assumed to be "present" and external to the observer), and from that observation to discover the patterns or laws by which nature operates. Central to Eiseley's project is a questioning of the "illumination" of science.

The visual act of reading is itself a journey. As Michel Serres comments, "To read and to journey are one and the same act" (*Jouvences,* 12, cited in Harari and Bell 1983, xxi). Metaphors of textuality and reading are inseparable from the Eiseleyan journey of exploration. But this reading is a random journey (Serres's *randoneé*), like that Roland Barthes suggests: "We read a text (of pleasure) the way a fly buzzes around a room: with sudden, deceptively decisive turns, fervent and futile" (1975, 31). Discourse is *discursus,* literally a running back and forth, and the word *essay* itself suggests exploring, attempting, experimenting. In Eiseley's hands, the essay is a journey that employs the reader's winding through a path, a movement that reflects but is less formalized than the "turning" of Greek drama. Not insignificantly, then, the journey is linked to the dominant philosopheme of light/sight/insight and thus to the metaphors of the eye, of the "rift" that is a spatializing of discontinuity to be perceived by the eye, and of textuality.

Epistemologically, Eiseley contends, an aporia must precede insight, and the aporia is translated into the spatial metaphor of landscape that accompanies the metaphoric eye. The landscape is literally what the eye sees, with the metaphoric sense an extension of that seeing (metaphor arising, as Umberto Eco suggests, from a subjacent metonymic relationship [1984, 68]). Eiseley establishes early that there is always an "apparent break, [a] rift in nature, before the insight comes" ("The Star Thrower," *UU,* 67). The Eiseleyan rift/abyss is both semiotic and semantic, both a figure of landscape and an image of the answer to the "terrible question" (67) that requires translation. The spatial metaphor, the "rift," is both a break and a joining, an aporia and a resolution. Eiseley explores the metaphoric vehicle in terms of landscape. The essay is itself an exploration and a "translation" of the rift of scientific observation and human understanding. Language itself, Heidegger finds, leads to an abyss and to darkness, for language is not grounded in something outside itself ("Language" 191). Fittingly, the abyss, the rift, and blindness/aporia are linked to a problem of language—translation of the "terrible question."

And this linguistic element of the abyss links the first two metaphors to the third—that of textuality, which Eiseley treats as the chief means to knowledge, which must be acquired interactively. Again he explores the vehicle of the trope. For Eiseley, the world is to be treated as text, read as text, with all that "reading" implies of the reader's participation in re-creating the text, in seeing in the text what the reader brings to it. "The Star Thrower" thus becomes not only a renunciation of an epistemological position, accomplished through the narration of one individual's growth, but a model of the Eiseleyan epistemological method.

Eiseley's textual exploration invokes what Barthes calls the "cultural codes," the always "already-written" cliches of the culture that appeal to "what has been written, i.e. to the Book (of culture, of life, of life as culture)" (1974, 20–21). The journey itself is a code—in the traditional motif of the conscious human journey, in the evolutionary sense of the journey of all life, in the notion of the individual's epistemological quest, in the ritual journey or the journey of initiation. Further, the journey is a code for literary method, so that the metaphor of the journey is also the journey of metaphor. The code of the primitive—at least as old as the eighteenth-century "cult of the noble savage"—is evoked in Eiseley's trickster and his reader of goose bones. The code of the Book of Nature survives from the sixteenth-century notion of nature as the one sure divine revelation "unclouded by human error and confusion" (Eiseley, *MWSTT,* 40). As Douglas Kneale has suggested, the question of "tropes of nature . . . is perhaps naive," for "enlightened minds no longer think that nature speaks or writes its own language, though they can conceive of a grammar of nature"; yet attributing to nature the power of speaking and writing suggests the poet's "ability to interpret the mind's linguistic projections as a continuous allegory" (1986, 352). Eiseley is acutely interested in exploring the allegorical role of the human mind's linguistic projections. In exploring the literal dimension of the metaphor, he uses the Book of Nature as he uses the other codes— as a means of breaking into the nonscientific consciousness at the level of a popular coded response. The Book of Nature is the starting point for reading a "text" to be demythologized, a text that undermines its own meaning, a text whose "reading" is as much a function of the reader as of what is read. Nature's text is as arbitrary as human texts; human reading gives semiotic meaning to nature, and reading is a re-creating.

Exploring the paradigm of textuality/reading, "The Star Thrower" suggests a mediated knowledge that is at least as important as scientific exploration. If, in reading, we help to shape the experience, then our experience and cultural conditioning, like the dead letters of the semantic text, mediate our response. Reading, then, becomes the paradigm of exploration as well as of understanding in Eiseley's epistemology. "Reading" underscores ambiguity, change, discontinuity (a key term in Eiseley's vision of the universe), as opposed to persistent notions of an unmediated vision. Reading is privileged over the voice of presence: science is reduced to nagging "whispers" while the earth's insistent semiotic writes itself.

The epistemological thrust is clear from the beginning of the essay, in which the narrator states that, though a teacher, he has been "taught surely by none" ("The Star Thrower," *UU,* 67). The opening note undermines the certainty of knowledge, the adequacy of any teacher. But the next sentence undermines any possibility of exact knowledge as it shifts from accepted "teachers" to texts whose meaning is uncertain: "There are times when I have thought to read lessons in the sky, or in books, or from the behavior of my fellows, but in the end my perceptions have frequently been inadequate or betrayed" (67). Reading, then, is not-reading; reading falters. And knowledge is not-knowledge, for it too falters. As Eiseley writes in "The Running Man," "to grow is a gain, an enlargement of life. . . . Yet," he continues—"almost deconstructively," as John Clifford has noted (1986, 8)—"it is also a departure. There is something lost that will not return" (*ASH,* 31). Growth and knowledge both speak of something that is gone. Knowledge is limited; it is not immediate, but a question of language, of translation. Before knowledge can come, there must be the translation of the "terrible question" into an existential awareness of "an even more terrifying freedom" ("The Star Thrower," *UU,* 67).

Eiseley's self-undermining knowledge, his "growth" and "departure," reflect the scientist's philosophical struggle to know. The struggle between the two eyes—the eye of science and the eye of human sympathy—is a tropical exploration of the struggle between the "truth," or supposedly im-mediate knowledge of science, and the mediated vision of textuality, which leads to poetic knowledge. This notion of "immediate knowledge" is especially important in

Eiseley's scientific context. What Nalimov calls the "stereotyped vision of the world," emerging from the language of determinism, causes us to see the world in terms of cause and effect, of logic, of human and natural "mechanisms," of hierarchies, of human separation from the environment. "Within this vision," Nalimov suggests, "a human being is nothing more than a block of matter which became so sophisticated that it managed to master the logic built into the foundation of existence" (1982, 3–4). The problem of textual "meaning" is linked to the "nineteenth-century, common-sense view," which "depicted an objective world which people knew through direct experience of their senses. In that dualistic empirical model, the words people used to describe the world corresponded to objects 'out there,'" and meaning was assumed to be based on correspondence between language and the world (Staton 1987, 3). Such positivist views of "truth," "reality," and "representation" are often dependent on what Eiseley calls "styles" of scientific thinking, and his exploration of the visual as philosopheme is directly related to his reexamination of scientific "observation."

ii

Because of the complexity and occasional faltering of "The Star Thrower," I turn now to an analysis of each "movement," with emphasis on the function of the three dominant figures. The first movement establishes these figures—-eye (root metaphor or philosopheme), rift, and textuality. In the epigraph from Seccho, the eye in the skull merges with the human quest for a path: "Who is the man walking in the Way? / An eye glaring in the skull" (*UU,* 67). The epigraph thus links the organ of the philsopheme of sight, which is embedded in the emblem of time and of emptiness, with the journey (itself related symbolically to light) and with the "Way" that has no fixed, teleological goal. In the first movement, the eye is the "beacon" eye of scientific observation—the "search beam" (68) that always observes but does not necessarily understand.

Thus the eye is also blindness—not knowing, or suppressing part of what is before it. Eiseley clearly links the eye with science as he writes of the revolving eye in the skull, "With such an eye, some have said, science looks upon the world" ("The Star Thrower," *UU,* 68). But this scientific "eye" is only partial knowledge, a "seeing" that does not see, for the narrator turns to a disclaimer of the efficacy of scientific observation: "I do not know. I know only that I was the skull of emptiness and the endlessly revolving light without pity" (68). The revolving beam, in a reversal of the

Shelleyan intertext of the desire of the moth for the flame, "consumes" the ideas that swarm toward it. The "light" is not the traditional code for generating ideas, but a consuming, destroying force, just as institutionalized science is a consuming, pitiless scanner unless it is balanced by art and sympathy—unless the scientist realizes, as Eiseley writes elsewhere, that the images of science are "as powerful as great literary symbolism and equally demanding upon the individual imagination," so that "one and the same creative act [occurs] in both domains" ("The Illusion," *ST,* 275). As he undermines the traditional symbol, Eiseley frustrates the reader's expectations. The reader is left uncertain, groping, as the authorial persona himself is groping, for "upon that shore meaning had ceased" ("The Star Thrower," *UU,* 68).

Also established in the first movement is the narrative of the man who comes to a landscape—a "wave-beaten coast at dawn"—where he encounters the "rift" in nature. The beach at Costabel is the beach for shipwreck, the place to which figuratively all must come ("The Star Thrower," *UU,* 69). To the revolving beam are added the nagging "whispers" of scientific observation; voice and science are subordinated to the text to be read, reduced to troublesome and obscure whispers in an empty skull. The juxtaposition of the eye of science and the eye of semantic or semiotic reading is crucial to the essay's focus on the struggle of science (positivism, "truth," "reality") and reading (the uncanny, the unexpected, the unreal that is more real than instruments can measure). In this first movement, the narrator asserts that the encounter with the abyss/aporia must come before insight can come, and he traces the beginning of his encounter to a shore of meaninglessness. This coast is a literal and metaphorical desert where the narrator juxtaposes the eye and the "edge of an invisible abyss" (73)—the organ of sight and the metaphorical "rift" that science cannot observe. The narrator, and with him the reader, must confront this abyss through the empty skull (itself an abyss) that contains the revolving beam.

Drawing the reader into the encounter with the abyss, the narrator calls attention to the present activity of reading: "If there is any meaning to this book, it began on the beaches of Costabel with just such a leap across an unknown abyss" ("The Star Thrower," *UU,* 67–68). The authorial intrusion shatters the reader's confidence, at least momentarily. If the author is uncertain about his meaning, how can the reader understand? The reader is thus brought into the uncertainty of knowledge that informs the essay. Neither writer nor reader can be sure of knowledge, but reading is the paradigm through which both will construct, create, or re-create meaning.

The vehicle that introduces semantic and semiotic textuality is the narrator's recollection of "a modern primitive" who "reads a goose bone for the weather" ("The Star Thrower," *UU,* 68). Al-though this "modern primitive" seems to have little connection with the context, the turn to folklore leads to a semiotic "reading" that is privileged over voice. The reader of goose bones lived in Costabel, where even a faintly buzzing voice (like the "whispers" of science) in the narrator's mind makes no sense. In this disjunc-tion the narrator's and the semantic reader's experiences are paral-lel, for the reader's expectations of clarity and understanding are shattered. Nothing in this place makes sense, and the narrator wonders if everyone is "destined at some time to arrive there as I did" (68), pointing to his own experience as synecdoche for that of all readers. The reader of goose bones, then, links Costabel to the recurrent textual metaphor.

iii

In Part 2, the narrative moves forward through a maelstrom of "eyes," which also blend the visual metaphor with the text's central epistemological statement or re-nunciation. The scientist's revolv-ing eye slowly becomes aware of a sequence of other eyes that symbolize the dialectical pull of negentropy and entropy, ending with an eye torn from an old photograph. Linking the eye as trope with the visual perspective of landscape, the "torn eye" looks through the narrator "as though it had already raced in vision up to the steep edge of nothingness and absorbed whatever terror lay in that abyss" ("The Star Thrower," *UU,* 80). As the eye of the narrator's long-dead mother, who, totally deaf, had known the abyss of nothingness in the absence of voice, the "torn eye" links narrator and reader to the abyss.

Both reader and narrator are uncertain of the maelstrom of "eyes." Both clarity and confusion are present—the clear image of the "revolving beam" that is the eye of science and the blurred image created as the narrator watches a sequence of eyes (of crea-tures on the beach, of creatures recollected from childhood, of the young girl in the torn photograph) changing as in a dream. To the limitations of the "eye" of science in Part I, Eiseley adds personal confusion. Science's confusion is also the individual's. The "eye" from the narrator's past becomes a powerful means of drawing the reader into the narrator's personal and epistemological confusion.

The text is muddled and, at this point, yields no clearer understanding than does detached scientific observation of nature's semiotic text.

The narrator begins this second movement with a seemingly disjunctive comment that underscores the visual vehicle of the metaphor and specifically links the rift to landscape as he notes the "difference in our human outlook, depending on whether we have been born upon level plains, where one step reasonably leads to another, or whether, by contrast, we have spent our lives amidst glacial crevasses and precipitous descents" ("The Star Thrower," *UU,* 73). The vehicle of the metaphor—vision or outlook in terms of landscape—has its epistemological tenor in the vista and perspective of human thinking. The apparent rhetorical disjunction actually extends the metaphor of the rift through a play on the vehicle of the metaphor and the philosopheme of sight. The difference in "outlook" (those who are accustomed to level plains and those who are accustomed to the "desperate leap over a chasm" necessarily "see" differently [73]) is analogous to early man's "crater lands and ice fields of self-generated ideas" and later man's "level plains of science" (74). With science as the level plain of cause and effect, of apparent order, as opposed to the terrain of desperate leaps from peak to peak, crossing rifts, Eiseley spatializes the two epistemologies. In the level plain each step seems to follow another, the universe seems orderly, and illusion seems to give way to "the enormous vistas of past and future time" (74). A *vista* suggests a broad awareness of past, present, or future happenings. The word is related to *wisdom; learning; eidos,* "form"; *wit,* "intelligence"—all related to *videre,* "to see." Space and time thus join in a metaphor of landscapes and perspective, linked to the dominant philosopheme of knowledge (light/vision) and to the journey—itself linked to the philosopheme of light. *Metaphor* (the word itself, as Derrida has established in "The Retrait of Metaphor" [1978a], suggests a vehicle and hence a journey) thus explores the epistemological dimension.

But accompanying the "level plain" of science is a force less evident than peaks and chasms. Contingency lurks in the level plains country in the form of the cyclonic "twister." The analogue in modern science is the struggle of "chaos versus form or antichaos"; thus "form, since the rise of the evolutionary philosophy, has itself taken on an illusory quality" ("The Star Thrower," *UU,* 75–76). Looking backward, we find only dissolving forms until we reach a physical and an epistemological void. Even the philosopheme of light/sight is shifting and unstable, for we are "changelings" whose identity dissolves into dream, so that "we are process, not reality, for reality is an illusion of the daylight—the light of our particular day" (76). Thus what we "know" is temporal and shifting, and any "transcendent lesson" is, like the meaning of the text, actually "preparing in the mind itself" (76).

Underlying this "abyss" of physical and epistemological instability is contingency, symbolized by the black-painted trickster from a primitive tribe that gesticulates behind man's back. In a doubling of the metaphor, contingency is the trickster, and the trickster is the cyclonic twister of the level plains of science. The effect of the doubled metaphor begins at the signifier—at the doubling of sounds in *trickster* and *twister*. But the trickster is literally and figuratively a twister—motioning, whirling—while the twister is also a trickster that surprises with its cyclonic whirl. The doubling of *trickster/twister* invokes a code of indeterminate whirling trickery embodied in the lore of whirlwinds, "dust devils," and "whirling dervishes." The trickster posturing behind solemnity is invisible to modern scientific man; like the unpredictable whirlwind that invades the level plains, however, it appears without warning and undermines the stability of the most confident science, even of knowledge itself. Although modern scientific man, with his emphasis on observation and measurement, seems to suppress the trickster, the figure's "posturing" underscores the contingency basic to scientific theory. Earlier societies incorporated it in their lore.

Associated with primitive societies and folklore, the trickster not only lurks on the level plains of science, but also links the reader of goose bones and the metaphor of textuality. As the trickster haunts science and undermines its certainty, it reminds the narrator "why man, even modern man, reads goose bones for the weather of his soul" ("The Star Thrower," *UU,* 77). In the absence of epistemological certainty, the figural "writing" of folklore (which "writes" as a mime or as the choreographic figures in Barthes's *A Lover's Discourse* [1983] write) becomes a kind of refuge. And Eiseley explicitly links the trickster to textuality as he admits that the figure breaks into, mocks, and undermines even his own writing process: "That mocking shadow looms over me as I write. It scrawls with a derisive pen and an exaggerated flourish" (77). Even the assumed certainty of dead letters is threatened by the posturing figure. The trickster contingency thus intrudes into the writer's act and into the reader's re-creation, though "in a quarter of a century it has never spoken" (78). The figure haunts the writer, displacing the Voice of Presence. Folklore, the oral Writing that is subject to contingency (reshaped and reinterpreted as it is told), has long known what nineteenth-century science seemed to discover: the instability of form, which reverberates epistemological instability. Knowing, then, is threatened by the Writing figure, who undermines it and reasserts the dance of contingency.

Eiseley's shift from the epistemology of science (of what Kuhn has called "normal science"), which can know only the ambiguous "creative and abolishing maelstrom" ("The Star Thrower," *UU*, 78) of form and formlessness, to an epistemology that will include poetic awareness is signaled by a shift in the semantic text itself as he blends scientific and Romantic intertexts to describe the dialectic of negentropy and entropy: "As the spinning galactic clouds hurl stars and worlds across the night, so life, equally impelled by the centrifugal powers lurking in the germ cell, scatters the splintered radiance of consciousness and sends it prowling and contending through the thickets of the world" (79). In this curious intertext of the Shelleyan "dome of many-coloured glass" and the Darwinian "tangled bank," Eiseley treats life as code ("impelled" by the "germ cell") and the reading of nature and life as an unstable self-undermining reading, for consciousness both "prowls" (explores, stalks) and "contends." Stars and worlds, germ and life, are linked to the semiotic dialectic of form and chaos in life and knowledge.

The textual metaphor continues as the desert coast evokes "a kind of litany": "As I came through the desert thus it was, as I came through the desert" ("The Star Thrower," *UU*, 73). The tropes of eye, abyss, and textuality are juxtaposed as the litany continues while with a "restless hand" the narrator "fingers" the edge of the "invisible abyss" and the "world-shriveling eye" revolves (73). The "litany" suggests the alternating reading of a liturgical prayer, but it is addressed to desolation. *Desert* is etymologically related to the Indo-European *ser-*, "to join (in speech)" or "to discuss"; the desert litany thus invokes the textual potentiality treated in the next movement.

iv

After the second movement has drawn together the reader's and the narrator's textual difficulties through the maelstrom of "eyes," the rift, and the unreliable text of nature, the third movement turns to the question why ships come to coasts that invite shipwreck. The question is literal in terms of the island of Costabel, but figurative in its epistemological dimension—the question why the abyss is necessary, why one seeking knowledge is left with nothing but

a revolving beam "whose light falls only upon disaster or the flot-
sam of the shore" ("The Star Thrower," *UU*, 80). The question
itself echoes Odysseus and the journey of knowledge. The path
(epistemological "way") to the answer, suggests the narrator, lies
"not across the level plains of science, for the science of remote
abysses no longer shelters man" (80). The abyss is now near, in
the spatial vehicle and the temporal tenor of the metaphor. Science,
unable to provide security or to evade the abyss, "reveals [human-
kind] in vaporous metamorphic succession as the homeless and
unspecified one, the creature of the magic flight" of folklore (80).
Science itself now is drawn into the abyss, and the landscape metaphor serves as vehicle for the history of science, which has taken
man from his security in "the satisfying supernatural world" of the
village to the "precipitous edges" of the world separated from man
and labeled as "natural." The journey has led to an existential
"region of terrible freedoms" (81), where man creates tools but
cannot control them, and the "huge shadows" of the trickster—
invisible to science but visible to poetic insight—leap "trium-
phantly" behind each scientific advance (81).

    Continuing the motif of the visual, the narrator attributes the
trickster's invisibility to modern science to "the heavy spell of the
natural" ("The Star Thrower," *UU*, 82), the mystified term that has
emerged as a result of human separation from and dominance over
a world assumed to be external, the term whose demystification
Eiseley undertook in *The Firmament of Time*. Primitive humankind was a part of the natural; the interrelated forces knew no line
of separation. Primitive humans worshiped spirits in nature, prayed
to the creatures they hunted, found safety in the imagined forces
surrounding their villages. Scientific humans, however, face a new
terror, itself linked to the philosopheme of light: "Humanity was
suddenly entranced by light and fancied it reflected light" (82).
Scientific humankind identifies itself with the mystified philosopheme of knowledge rather than with the world around him. Separated from nature, having internalized the dominant philosopheme
of knowing (in Derrida's terms, the sun is interiorized in Western
thinking [1978b, 113]), humankind controls forces for which, Eiseley argues, it has neither understanding nor feeling.

    In an attempt to examine the discoveries that brought human-
kind to this state of separateness from the "natural" and released
the "invisible" trickster's posturing, Eiseley turns to Darwin, Einstein, and Freud. But after a brief discussion of Darwin, he falters
from his ostensible purpose. With the mention of Freudian
"changelings," Eiseley abandons exposition and returns to the re-

volving eye in the empty skull. The intricately organized text falters in explicating science and turns (in a powerful textual strategy that seems unconscious) to a painful recollection of the photograph from which the "torn" eye seemed to come. Recollection of "the coast demanding shipwreck," the narrator adds, stems from "a dark impulse toward destruction lurking somewhere in the subconscious" ("The Star Thrower," *UU*, 84)—an impulse that ties the narrator to the encounter with multiple eyes. Disguised with sunglasses against the eye that asks for pity, he writes, "It was as though I, as man, was being asked to confront, in all its overbearing weight, the universe itself" (86). As he abandons scientific exposition, the narrator turns to the emotional experience that leads to poetic understanding, exploring his own confrontation with the self-created ghost that haunts him. The eye of observation/science, the notions of nature, man, and science embedded in Western culture, and the eye from the photograph are caught for an instant in confrontation in the text itself. The confrontation is a code, craftily drawing the reader into the archetypal experience of "confronting the universe," as Job confronted the unanswerable questions in his own whirlwind.

In this confrontation, the narrator recalls the biblical injunction, "Love not the world, . . . neither the things that are in the world" (cited in "The Star Thrower," *UU*, 86). But his is a rejection of Western culture's pattern of "rejection." As Paul de Man once commented, traditional "renouncing" may actually be a "ruse" that allows the subject "the possibility of enjoying the ethical value of an act of renunciation that reflects favorably on the person who performs it" (1985, 113). Protective though it may be, this renunciation suggests "the blindness of the subject to its own duplicity" (113). Eiseley's narrator turns from this Western notion of "rejection"—in this instance, rejection of the world of science—to reaffirm a primitive sense of identification with the living world.

At this moment, the narrator becomes aware that the "revolving beam" of science has ceased, as well as "the insect whisperings of the intellect" ("The Star Thrower," *UU*, 86). The eye of science and the Voice of Presence have ceased, while the eye from the photograph considers him. And in this climactic encounter the narrator responds, speaking himself in a whisper that displaces and defies science, culture, and any tradition that would displace him from the natural world: "But I *do* love the world. . . . I love its small ones, the things beaten in the strangling surf, the bird, singing, which flies and falls and is not seen again. . . . I love the lost ones, the failures of the world" (86). The narrator's whisper is both timid and defiant; it is the voice of victory and defeat, of gain and loss. It expresses the faltering of one sight and the triumph of the other, the privileging of the eye of mediated vision and humane concern, as an extension of the Darwinian "interlinked and evolving web of life" (83)—a poetic insight that transcends its empirical origin.

The choice is the climax of the narrative sequences of the essay; it is, Eiseley writes, "like the renunciation of my scientific heritage" ("The Star Thrower," *UU,* 86). The word *like* is a key word: the narrator renounces but does not renounce his heritage. In writing the phrase, he brings the notion of renunciation before the reader, but with the word *like,* he avoids the trap that snares Western "renunciation." On the brink of renouncing his heritage, he nevertheless retains it as the starting point for essays that blend science and poetry. What he makes, then, is a re-announcing: a statement of an epistemology that does not abandon the scientific method, but questions it and attempts to incorporate the epistemology of poetry. The narrator does renounce the present stage of human development, a stage of separation from and control of a world that humankind does not love, of releasing natural forces without understanding them. Only the star thrower, madman that he is, is the model for the narrator's awareness. The narrator chooses the supplement of poetic knowledge over empiricism, of interactive reading of nature's semiotic text over the supposed immediate vision of science, though the gain of sympathy is undermined in the loss of faith in science and modern man.

Rejection of the scientific way for the way of poetry might seem by now a commonplace. What Eiseley attempts, however, is not a rejection of the scientific way, but an incorporation of the way of poetry, a recognition of the relationship of creativity and science. Stephen Jay Gould writes that twenty-five years after Kuhn and others noted "the intricate interpenetrations of fact and theory, and of science and society," the "simplistic" view of science as the triumph of objective observation has become "bankrupt" (1981, 6). Eiseley, like Kuhn, questions the myths of scientific thinking. He asserts, through the encounter with a madman, the value of an epistemology that recognizes and incorporates poetic understanding. Thinking itself is a journey. At least as important as the end of the journey is the way, the method, the experience, of the journey. If the world is a text, then its Writing is the supplement of poetry to science, of sympathy to observation.

With the renunciation, the torn eye departs, completing the dramatic climax, and the "merciless beam" ceases its "traverse" ("The

Star Thrower," *UU*, 86–87). Only then can the narrator comprehend his blindness and his insight, which he expresses through the trope of the "rift." As the "torn eye" departs, the landscape metaphor returns in the statement, "I had come full upon one of the last great rifts in nature" (86–87). The "rift" is both a cleft and a joining, both aporia and understanding. Of the rift between World and Earth in the presencing of truth, Heidegger writes: "It is the intimacy with which opponents belong to each other. This rift carries the opponents into the source of their unity by virtue of their common ground. It is a basic design, an outline sketch." (1975b, 63). Eiseley's narrator discovers this paradox of conflict as he realizes that what he perceives is "not a rift but a joining: the expression of love projected beyond the species boundary by a creature born of Darwinian struggle, in the silent war under the tangled bank" ("The Star Thrower," *UU*, 87). The expression comes from a madman who has, in the continuing landscape metaphor, "moved to the utmost edge of natural being, if not across its boundaries," as if momentarily and hesitantly, natural and supernatural have touched (87). The narrator is concerned with the boundaries of being and of knowledge—boundaries that seem to lead to the touching of natural and that beyond "nature." And in this touching of natural and supernatural the rift joins landscape and eye metaphors with the recurrent metaphor of textuality.

Textually, the narrator's insight arises "from the domain of absolute zero" (87). As Nalimov suggests, Nothing is "the *semantic vacuum*," "the absolute lack of semantic manifestation (or, in physical terms, an unobservable state) which must be complemented by the idea of potentiality as a source of semantic manifestation" (1982, 77). This notion of the human potential for creating out of "Nothing" the semantic insight that forms the climax of the essay's dramatic movement is what concerns Eiseley, who has said earlier that the experience described in the essay has given "a fugitive glimpse" "of what man may be" (67). Out of the rift comes insight; out of the abyss, potentiality.

The semantic Nothing that is semiotic potentiality joins the trickster and other tropes as the narrator recognizes the human ability (itself, like the trickster, unpredictable) to derive value from nothingness: "The nothing had miraculously gazed upon the nothing and was not content. . . . A little whirlwind of commingling molecules had succeeded in confronting its own universe" (Eiseley, "The Star Thrower," *UU*, 87). The whirlwind—the human being—confronts the larger whirlwind—nature—that Job confronted. The act of confrontation is central; it is a creation of semantic value created from semiotic reading: "The act was, in short, an assertion of value arisen from the domain of absolute zero" (87). "Contemporary physics," writes Nalimov, "has come close to the concept of *non-existence* (the unobservable state) as a potential basis for *reality* (the observable state)" (1982, 75). Out of the Nothing, identified by the gnostics with silence (79), comes an "assertion of value" that assumes the force of a text.

The act itself, like the act of giving central to Heidegger's "The Thing," the outpouring of a libation that, gushing, "means to offer in sacrifice" (1975d, 173), is the turning point: the rift "beyond Darwin's tangled bank" is identified with the human action of reaching out in pity (Eiseley, "The Star Thrower, *UU*, 87–88). "In the gift of the outpouring," Heidegger writes, "earth and sky, divinities and mortals, dwell *together all at once*" (Heidegger 1975d, 173–74; italics in original). Eiseley's outpouring vessel is the human form; in the act, the vessel becomes "the twofold containing, the container, the void, and the outpouring as donation" (173–74). The rift that is a joining is the staying of Eiseley's natural and supernatural.

For Eiseley, the word *supernatural* requires defamiliarizing, for "there looms . . . in nature something above the role men give her" ("The Star Thrower," *UU*, 92). *Super-* denotes dominance in degree, size, or quality; the supernatural is the greater natural that denotes human potential, symbolized by the star-thrower's projection of affection across the boundaries of form—the act that, like the libation, brings value out of the domain of zero. The moment when natural and supernatural are "as if" touching is the moment when "thought mediated by the eye," which is "one of nature's infinite disguises" (88), asserts the value arisen from the "domain" of nothingness. The supernatural is, then, the intensely *natural* that the eye's mediation, its interactive semiotic reading, creates out of the semantic void, in the absence of voice. (Traditional science, having been reduced to "whispers," has withdrawn.) The eye's mediation returns us to the text of mediation, to the *eye/I* of the text that "looked out upon what I can only call itself" and, not content with nothingness, asserts value (87). In an essay whose dominant metaphor is the eye—in the form of the eye of science and the eye of humanity—the narrator clearly embraces what to the scientific eye would be blindness, but to the poet is the rift/joining of insight.

v

The essay's fourth movement opens with an "ode," a comment referring to humankind as a narrative, "a tale of desolations," but the movement then turns to "episode" that reinforces the third movement's "renunciation." In this final movement—overt recognition that the answer to the "terrible question" derives from poetic knowledge as a supplement—the narrator follows his insight with action as he sets out to find the mad thrower of "stars." The individual, says the narrator in a Sartrean reflection, possesses "the awesome freedom to choose . . . beyond the narrowly circumscribed circle that delimits the animal being" (88). It is this freedom that makes possible the journey of knowledge or the narrator's quest for the madman. To the madman he explains only, "I understand. . . . Call me another thrower" ("The Star Thrower," *UU*, 89), as the two men working futilely become "part of the rainbow—an unexplained projection into the natural" (89).

The two throwers' motions are like "the sowing of life on an infinitely gigantic scale" ("The Star Thrower," *UU*, 90). Theirs is a positive motion amidst the dialectic of negentropy and entropy, for "all about us roared the insatiable waters of death" (90). The sensitive human being thus participates in a throwing into life (compare Heidegger's "thrownness into Being") through the act of flinging or of reaching out, which is for Eiseley the role ("destiny") of the human, who may have lost his sense of direction, but nevertheless remembers "the perfect circle of compassion from life to death and back again to life" (90). Modern humankind retains only the memory, but Eiseley privileges the primitive's semiotic reading, which placed the human as a being in nature, over the modern "unmediated" vision of science, which divides the human from the natural.

In the motion of throwing, the narrator participates in a previously "unknown dimension of existence" ("The Star Thrower," *UU*, 91). The old man signals the self-undermining of science, for he represents the rift/joining, the "subtle cleft in nature before which biological thinking had faltered": "From Darwin's tangled bank of unceasing struggle, selfishness, and death, had arisen, incomprehensibly, the thrower who loved not man, but life" (91). Primitive and modern, like the Heideggerian divinity and human, join in this hurling that is also a flinging of the self as archetypal "forfeit" (91). This identity in the web of nature is a result of a knowledge founded on "the discontinuities of the unexpected universe" (91).

Awareness and desolation, being and nothingness, human and nature are joined, then, through the act of throwing, which reveals a rift that contains a "hint" that there is to be read in nature's text "something above the role men give her" (Eiseley, "The Star

Thrower," *UU,* 92). Rainbow, thrower, and the shipwreck-inviting beaches of desolation at Costabel, like the changeling shadows of the Freudian unconscious, combine to unveil a difference that is a joining, a discontinuity that is a link, a knowledge that is not a knowledge, a supernatural that is a heightening of nature through the reader who is more than detached observer. The essay is a moving insight into the human relationship to the natural world, but it is also an exploration of the human philosopheme of sight— whether for its own mediating power or mediated by metaphors of landscape or of reading. The essay is an exploration of the instability of knowledge and a re-announcing of an epistemological position that incorporates the supplement of poetic understanding in what has often been considered the unmediated vision of science.

If Blanchot is correct in asserting that narrative is a "successful, unsuccessful journey," a "song no longer directly perceived but repeated and thus apparently harmless" (1982b, 61), then the narrative journey, incorporated into the genre of the essay, provides a dual means of epistemological exploration, "an ode made episode." "The Star Thrower" exemplifies Eiseley's method and reveals the adaptability of Eiseley's unique genre to this dual epistemological exploration—as well as to incorporation of elements from classical drama and myth, and to exploration of metaphor and philosopheme in the quest to understand how we know at all.

The Dialogic Understanding of Indeterminacy/Textuality: "The Hidden Teacher" and "The Golden Alphabet"

> . . . the noblest written words are commonly as far behind or
> above the fleeting spoken language as the firmament with its
> stars is behind the clouds.
> —Henry David Thoreau, *Walden*

Both "The Hidden Teacher" and "The Golden Alphabet" treat the code or the alphabet as metaphor, which Eiseley explores both in terms of linguistic and genetic evolving and in terms of the dialogue between the code's stability and the forces of change, between human creation of meaning and the tendency to attempt to fix meanings once they have emerged. In exploring these codes, Eiseley draws on two of the dominant paradigms of the twentieth century—indeterminacy and textuality. Emphasizing human potential, Eiseley explores the essential dialogue between stability and instability, determinacy and indeterminacy. Eiseley's "unexpected" is essential to the evolutionary view, and textuality provides a complementary trope.

As a paradigm, indeterminacy emerged in earnest early in the twentieth century with the discovery of particle physics (Born's indeterminacy, Heisenberg's uncertainty relations, Bohr's complementarity principle that defines light as both wave and particle, depending on the experiment being conducted). Transferred to the realm of scientific or artistic creativity, which Eiseley later treats as "one and the same creative act" ("The Illusion," *ST,* 275), the "indeterminate" or the "unexpected" continues to evolve. Both artist and scientist play their roles in shaping, even creating, humanity. The unexpected and the discontinuities basic to the Darwinian view and to twentieth-century physics are interwoven in Eiseley's view not only with scientific understanding, but with creativity as well.

In literary studies, the parallel of indeterminacy as a scientific paradigm is the notion of textuality that has emerged since structuralism; the notion is embedded in Saussure's concept of the indeterminate relationship of signifier and signified. Because we know that language is not fixed, we know also that the text cannot be considered determinate. Authorial intention cannot fix potential readings of the text. The structure of a particular text may impose certain limitations, but readings may vary. A caveat is necessary, however, concerning analogizing between scientific theory and other fields. Umberto Eco notes Niels Bohr's warning against naive comparisons of physical theory with other fields—equating indeterminacy with moral freedom, for example. But, writes Eco, "indeterminacy, complementarity, non-causality are not *modes of being* in the physical world, but *systems for describing* it in a convenient way" (1984, 66, n. 9). A system of description in one field may well have an impact on systems of description in other fields. Eco finds a "relationship between a *scientific methodology* and a *poetics*" (66, n. 9). A student of anthropology and linguistics and a reader of Claude Lévi-Strauss before literary theorists discovered him, Eiseley demonstrates an uncanny linking of textuality and indeterminacy, while retaining awareness that both are systems for describing, not reflections of reality.

Textuality is for Eiseley a major metaphor for the means to knowledge. The world is to be read as text, with the reader participating in re-creating the text, seeing in the text what she brings to it. Eiseley pursues the ordering of nature itself according to a system of codes like hieroglyphics or alphabets—a system that provides order but nevertheless retains a fundamental indeterminacy in the communication process such as Derrida stresses with the metaphor of the post card (we take for granted that the message sent will actually arrive, though often it does not).[3] Thus "natural" and "unnatural" codes, physical and cultural codes, share similar characteristics and provide a heuristic of coding that has fascinated modern thinkers from the structuralists (in whatever field) of the 1960s to students of information science and chaos in the 1990s. The paradigm of textuality is basic to Eiseley's epistemology because it makes possible a dialectic of constraint and freedom such as one sees in the growth pattern of a plant or in the genetic ramifications of an embryo. Even science is translated into the terms of textuality—a mediated form of knowledge, as opposed to the nineteenth-century myth of im-mediated knowledge.

Clearly Eiseley is aware of the problematic of textuality as a metaphor. In *Darwin's Century* and *The Firmament of Time,* he has traced the notion of the Book of Nature from the sixteenth- and seventeenth-century notion of nature as "the second Book of Revelation" (*FT,* 10). For the English progressionists (who saw life as developing through series of forms yet not through phylogenetic descent), "the fossils were true hieroglyphs, signs from earlier ages as to God's intention and design" (*DC,* 96). The teleological thrust of such notions is, of course, problematic to the Darwinian view. Still, the established use of writing and the book as figures provides a cultural code that can be defamiliarized and renewed—so long as, in E. R. Curtius's term, the subject is "value-charged" (1953, 303), as is any human consideration of nature. Employing the book, Eiseley undermines book culture as he returns to primitive notions and approaches the "reading" of texts and codes other than the book: of nature itself, of civilizations, of humankind as a palimpsest of what and where we have been. Drawing on and adapting for his purposes the historical tradition of the world as hieroglyph—in later centuries, the world as book—he emphasizes the epistemological uncertainty of the search for meaning in nature.

For his model, however, Eiseley turns more specifically to the metaphor of "reading" nature among primitives, especially among the Native Americans of North and Central America, whom he studied in depth. Thus he both builds on and departs from the coded assumptions of post-Renaissance thinkers. Eiseley's world is a book, but only in the primitive (and contemporary) sense of a text that the reader creates. Eiseley's metaphor of the hieroglyph is close to that expressed by Coleridge and practiced by the Eskimo or by our primitive forebears. He argues that humankind, since the emergence of the "symbol-making mind, has sought to read the map of that same universe," for the human being is "an oracular animal" who, with only minimal dependence on instincts, "must search constantly for meanings" ("The Golden Alphabet," *UU,* 144). Eiseley reminds us that, like children, we began to read before we learned to write. Early humans read

> what Coleridge once called the mighty alphabet of the universe. Long ago, our forerunners knew, as the Eskimo still know, that there is an instruction hidden in the storm or dancing in auroral fires. The future can be invoked by the pictures impressed on a cave wall or in the cracks interpreted by a shaman on the incinerated shoulder blade of a hare. . . .
>
> But the messages, like all the messages in the universe, are elusive. (144–45)

Significantly, in turning to the primitive for insight and for method, Eiseley also invokes the insight of the poet—of "poetic" as opposed to "scientific" knowledge—for support of his understanding; and the poet "invokes" the future, for reading is itself a creation of reality. Although eighteenth-and nineteenth-century "readings" of the book of nature depended on answers already in the mind, on underlying assumptions of teleology and onto-theological "meaning" in nature's book, Eiseley attempts to displace such notions in favor of the individual's interaction with the book or the alphabet, arguing that each individual "deciphers from the ancient alphabets of nature only those secrets that his own deeps possess the power to endow with meaning" (146). Among such messages, he notes, some will be unreadable. Still, the basic human need is to try.

Eiseley's text or alphabet or hieroglyph as a symbol and a model is thus never fixed, just as reality is never fixed and never interpreted teleologically in Eiseley's texts, but is "read" interactively according to the purpose and insight of the individual reader. The reader, then, is shaman, poet, and conjuror at once. Eiseley is less interested in hermeneutic investigation of the "hieroglyphs" of nature—since interpretation for him is largely individual—than in exploring their potential as metaphors and in employing the individual's involvement in the text as a means of stressing individual involvement with the environment, both for its own sake and for the sake of understanding how we came to where we are. The Eiseleyan "hieroglyphs" of nature are writings in the process of becoming; dialogue with them is as much the individual's task as reading a text is. As Marshall McLuhan once noted, since the invention of printing, a "reader" is one who scans, whereas earlier, "a reader was one who discerned and probed riddles" (1982, 344). The reader as the primitive who probes riddles is the model that Eiseley employs in reading nature, civilizations, and humankind itself as texts or "hieroglyphs."

The term *hieroglyph,* which Eiseley often uses almost interchangeably with *alphabet* or with the notion of reading nature's text, brings the reader to a metaphor more ancient than the Book of Nature—one to which Eiseley brings the understandings of modern linguistics and anthropology. For the student of language and of writing, a distinction exists between "hieroglyphic" and "alphabetic" writing. A hieroglyph is a symbol that may be read as ideograph or pictogram, or as a phonetic symbol—hence many of the difficulties in interpreting the Egyptian hieroglyphs, some of which functioned pictographically while others functioned phonetically. A true alphabet requires, ideally, a single symbol for a single sound, or at least (as in the case of the Roman alphabet and the English language) an approximation thereof.[4]

In "The Hidden Teacher," Eiseley pursues notion of the reader as puzzle-solver, or at least one who contemplates "formidable riddles." He recalls the voice that spoke out of a whirlwind to the biblical Job, asking "pitiless questions . . . that have, in truth, precisely the ring of modern science. For the Lord asked of Job by whose wisdom the hawk soars, and who had fathered the rain, or entered the storehouses of the snow" ("The Hidden Teacher," *UU,* 48). Such riddles are common to both art and science, to the individual and the culture. One such riddle is that of the genetic code and what Eiseley considers its parallel in human culture. For Eiseley, a "filamentous seed" is "one of the jumbled alphabets of life," and the human body itself is produced by "an alphabet we are only beginning dimly to discern" (56–57, 58–59). The code controls, for life "*instructs* its way" into form, and species change as the genetic code itself changes. The key term *instructs*—from the Latin *instruere,* "to build, prepare, teach"—carries the sense of building or teaching, of leading not forcing, of what seems on the surface a deterministic code but actually allows freedom for the individual gene. Always, he suggests, the genetic codes resemble human language codes, for "like genuine languages, [they] ramify and evolve along unreturning pathways" (59). In Eiseley's analogy, the individual vanishes, and institutions evolve "in invisible flux not too dissimilar from that persisting in the stream of genetic continuity" (61). Civilization subsumes the individual as hieroglyphs subsume individual sounds. A dead pharaoh may unwittingly have conveyed an image that has outlasted his culture. Only the figure is readable, and only in the sense that its meaning requires active participation of the reader.

The hieroglyphs of civilizations are not only ambiguous, but partake of other qualities of the text—transitoriness, the death or effacement of the author, and shifts in meaning: "The little scrabbled tablets in perished cities carry the seeds of human thought across the deserts of millennia. . . . [T]he minds that wrought the miracle efface themselves amidst the jostling torrent of messages, which, like the genetic code, are shuffled and reshuffled as they hurry through eternity" ("The Hidden Teacher," *UU,* 59–60). As in information theory, the macro level is what endures, shaped by the coding of information. The individual vanishes, and the institutions evolve. Yet Eiseley insists that the human individual must sense this relationship with the culture. The human mind, he contends, can "unconsciously seize upon the principles of [the genetic code] to pass its own societal memory forward into time" (61). "The Hidden Teacher" ends with a forecast that is both terrifying and hopeful: that we may be near the end of "that wild being who had mastered the fire and the lightning" (66), for evolution now is more cultural than physical. Preserving life itself requires "transcending" the human image, for the human being shapes his own image, and the final form of that image is still unknown (66).

The human need for messages, which actually are self-created, is the subject of "The Golden Alphabet," in which Eiseley explores at length the epistemological dimension of nature perceived as a remotivated hieroglyph. The essay is flawed by sentimentality and an overextended metaphor of human "spectacles" (another word that suggests the "framing" effect of ways of seeing and derives from the Latin *specere,* "to see") borrowed from *The Wizard of Oz,* but it is nevertheless memorable in its heuristic dimensions. Invoking the primitive as reader of the universe, Eiseley quotes the Eskimo's proverb: "Be not afraid of the universe." Lacking instinct, he contends, humankind searches for meanings, attributing to the universe the qualities of an alphabet. Darwin and Thoreau, scientist and poet, sought to "read" as their experience and learning prepared them to read. A recurrent thematic duality emerges as Eiseley explores the two kinds of knowledge—poetic and scientific. Treating Thoreau and Darwin as comparable yet contrasting "readers" of nature, Eiseley argues that their experiences as scientific observer and as poet taught them to "read."

In the shell that gives the essay its name, Eiseley finds an allegory for the human reading of nature's pictograms, for "golden characters like Chinese hieroglyphs ran in symmetrical lines around the cone of the shell," as if it bore some message from the sea ("The Golden Alphabet," *UU,* 145). Although a shell dealer quickly classified it as *"Conus spurius atlanticus, . . .* otherwise known as the alphabet shell" (145), Eiseley rejects the word *spurious* and affirms the human impulse to discover—which is to create—some message in the shell, though not a fixed message or what one might expect in the eighteenth-century moralistic sense. "We *live* by messages," he writes, "—all true scientists, all lovers of the arts" (146). The point is that the quest for the message is a quest for knowledge, though the knowledge may ultimately come from the self and lead simply to engaging the imagination. The human lack of instinct is also a freedom to explore, to learn, which is to create and to explore both natural and human codes. The involvement of the self in the message is essential in Eiseley's view—for scientist or artist, for the human being.

Eiseley's shell is as precious, he suggests, as "the tablets of a lost civilization," which we could not specifically decipher. Like the dream or the unconscious, the natural alphabet has no fixed or generalized meaning, but is to be untangled according to the individual context. For Eiseley, "the golden alphabet, in whatever shape it chooses to reveal itself, is never spurious. From its inscrutable lettering is created man and all the streaming cloudland of his dreams" ("The Golden Alphabet," *UU,* 146). Humankind creates itself—its dreams, culture, art, science, and future. Eiseley's metaphor is informed with the notion of interaction with nature and exploration of the imagination rather than with any sense of fixity or determinism in the message, which shifts as quickly as nature or the human imagination (which includes any "true" science) can shift. In this view, the alphabet, or hieroglyphics, create the future. The code creates the reality. This is the major insight of Eiseley's textual exploration—not that we observe and encode "reality" but that we create reality by encoding it, or by ascribing a code to it.

Eiseley's text or alphabet or hieroglyph is never fixed, just as reality is never fixed and never interpreted teleologically in Eiseley's texts, but is "read" interactively according to the purpose and insight of the individual reader. In the context of textuality, the attempt in literary studies "to restore a sense of the 'unnaturalness' of the signs by which men and women live" (Eagleton 1983, 14) is anticipated by Eiseley's attempts to restore a sense of the "unnaturalness" and "unexpectedness" not only of linguistic signs, but also of the universe. We had to learn to see the world as "natural," he insists ("How the World Became Natural," *FT,* 3–30). As he plays with the notion of "reading" the universe, Eiseley's emphasis is on the unnatural and the unexpected. His metaphors of hieroglyphs are a means of achieving a nonphonetic reading (mediated by figures reminiscent of Barthes's choreographic figures in *A Lover's Discourse* [1983], not by voice or presence) of his unexpected universe.

Eiseley thus suggests the ambiguous textuality of nature itself, of civilizations, and of the natural history of humankind. He draws on a traditional code—nature as a book—but defamiliarizes and remotivates it in seeking messages that bypass sound—hieroglyphic messages. The metaphor of textuality continues the notion of "meaning" in nature even as it undermines the notion of nature as a book—that is, nature framed, closed and enclosed, teleologically prepared for the "right" reading. For Eiseley's readings, like readings of texts or of dreams, depend on the individual. The hieroglyph itself connotes ambiguity and indeterminacy, phonetic and nonphonetic meaning, life and death. For Eiseley, textuality and indeterminacy are complementary; they suggest the ambiguities of a discontinuous, unexpected universe that cannot be read as the traditional book, with a movement toward closure and "truth," but as an unexpected and unfinished unfolding of ambiguities requiring interaction with the individual. And it is the probing that is worth doing, for humankind is by nature "oracular," seeking messages. To seek these messages through a close relationship with the nature that produced us is, in a sense, to depart from dispassionate science and to return to a primitive mode of observation. In Eiseley's view, it is the return that may save humankind from itself.

The Tongue's World: "The Invisible Island"

> Him, haply, slumbering on the Norway foam,
> The pilot of some small night-foundered skiff,
> Deeming some island, oft, as seamen tell,
> With fixed anchor in his scaly rind,
> Moors by his side under the lee, while night
> Invests the sea, and wished morn delays.
> So stretched out huge in length the arch-fiend lay,
> Chained on the burning lake. . . .
>
> —John Milton, *Paradise Lost*

In "The Invisible Island" Eiseley explores what he calls the "web" or "screen" of all life—a net or screen "that keeps things firmly in place, a place called now" ("The Invisible Island," *UU,* 152); the "unexpected" is the rent or tear in the screen. The screen and the rent in the screen that allows the unexpected produce an ongoing dialectic, the pull of the present species to maintain the status quo versus the pull of the unexpected—genetic anomaly or change in an environment. In this screen "the living constitute a subtle, though not totally inescapable, barrier to any newly emergent creature" (154). Against this web or screen Eiseley sets the human removal, through language, from the struggle within nature. Eiseley analogizes the human relationship to the earth in terms of the water strider, an insect that dances on the surface tension of ponds and brooks. "The insubstantial film" on which these insects tread "resembles the surface tension of the living screen of life, in which every organism, like the forces in the atom, exerts an enormous hold, directly or indirectly, upon every other living thing" (153). Humankind, he continues, "has similarly defeated and diverted the entire web of life and dances, dimpling, over it" (153). The human being, like the water strider, "possesses the freedom of a dangerous element," but the human "dances upon shadows, the shadows in his brain" (153–54).

The essay's beginning is overly dense but interesting in its play of intertexts. An almost playful intertext of Milton's powerful if mistaken epic simile that depicts the fallen Satan's size and deceptiveness underlies the essay. The epigraph is from Shakespeare's Caliban, whose fish-like form befits Eiseley's theme of the failed fish that became a land creature in his speculations on failures as successes, of which the human species is one. As a witch's child, Caliban symbolizes the unexpected aspect of the original elements of earth and water. And Caliban asserts his own original ownership of the earth and the magical means by which Prospero, representing humankind, acquired the island: "I say by Sorcery he got this Isle;/From me he got it" (cited in "The Invisible Island," *UU,* 147).

Beginning with a comparison between the whale as a victim of multiple strikes that leave old harpoons embedded in its hide and the earth as an equally marked quarry, the text is complicated by intertwined notions of earth as organism and quarry and by references to Melville's speculation on the whale's vision seeing two disparate sights—the sea as mother and the human as "messenger of death and change" ("The Invisible Island," *UU,* 148). Through the "fusing of the past and future" in language (as in the whale's vision), the human being has become "an invisible island, as surely as the great whale was an island" (149). An island is a place where extremes—even opposites—can appear, such as dwarf or giant forms, or even "failures" like the fish that became a land creature, can survive. Thus Eiseley speculates on the whimsical side of the unexpected, "as though everything alive had in it a tug of antigravity" (149), which is exemplified by another anecdote of a trickster—this time in the form of a deer mouse that appeared to listen intently to a seminar on the Byzantine Empire, a humbling figure for those who take their roles too seriously.

The "invisible island," then, is at first humankind, and Eiseley pursues the characteristics that make an island and that constitute its uniqueness. "Islands," he writes, "are apt by their seclusion to offer doorways to the unexpected, rents in the living web, opportunities presented to stragglers who might be carrying concealed genetic novelty in their bodies—novelty that might have remained suppressed in a more drastic competitive environment" ("The Invisible Island," *UU*, 159). A "genetic island" requires some "isolating barrier," such as an island in the sea or a mountaintop or a season, which is more important than the overemphasized factor of struggle in the evolutionary process (160). In this context, Eiseley pursues the human future of "selfhood," which emerged from an "invisible island," the human brain, which he calls an "island clouded in a mist of sound" (162). Thus he narrows the "invisible island" to the human brain and specifically to the linguistic ability, which draws together the elements of isolation (the genetic isolation that we are only beginning to explore), anomaly or the "unexpected," and the living screen: "In this way the net of life was once more wrenched aside so that an impalpable shadow quickly wriggled through its strands into a new, unheard-of dimension of existence" (162). The new island seemed without limits, for it was "created by sound vibrating with meaning in the empty air" and based on the most important human tool, which can treat past, present and future, manipulate the absent as well as the present (162). Through language, then, humans can "juxtapose, divide, and rearrange" the world and can "project a phantom domain, the world of culture" (162–63).

Thus in a sense we "belong" to our "invisible" island; we partake of the unexpectedness of the universe, and language itself creates a new world that incorporates awareness of time. Ours is a self-created island whose effects are both blessing and bane. Drawn from "the contingent and the possible" in the human mind and enabling us to separate ourselves from nature ("The Invisible Island," *UU*, 164–65), Eiseley's "island" emphasizes antideterminism, the excitement of the unexpected. For him, the only possibility of return to nature is Thoreau's way, "the power of imaginative insight which has been manifested among a few great naturalists" (165). As a naturalist, Darwin speculated that humankind likely arose in an island environment, but he was ambivalent about whether humans were weaklings or fierce monsters, and he failed to see "the outlines of that invisibly expanding universe which man had unconsciously created out of airy nothing" (166). Eiseley takes the island in a different direction, speculating on humankind as both being and creating an island, a rich and fearful environment for contingent change. The emergence of the human "out of airy nothing" returns the essay to its tutor text, for it homonymously evokes Ariel, mentioned in the beginning of the essay as Shakespeare's opposite to Caliban. In Shakespeare's context, Ariel symbolizes the magician's control over air and fire; in Eiseley's context, Ariel symbolizes the magical power of the "airy nothing" of language to create multiple realities.

Eiseley concludes the essay with a play on the dual potentialities of the self-created island. The island's owner is still Caliban, who remains locked within us. In Shakespeare's text, Prospero speaks of "'This thing of Darknesse,'" which, he concludes, "'I acknowledge mine'" (cited in "The Invisible Island," *UU,* 170). Drawing on the tutor text, Eiseley concludes that although the borders of the invisible, linguistic island are "solely within ourselves," they are "still frequented at midnight by a vengeful Caliban" (171). Yet the island is still "Shakespeare's island of sweet sounds and miraculous voices," "the beginning rent in the curtain" (167) that is, Eiseley always reminds us, only in its beginning stage. Humankind is a "beautiful and terrible . . . island failure," and the invisible island has given us "both triumph and disaster" (169). The earliest sounds—measuring, remembering sounds (as James Bunn notes, the hand is the instrument of cutting and combining, the tongue of the "wordless state" of groping)—created a "shoreless," growing island (170). The island is terrifying because of both the Caliban-like element within us and the "other" dimension of language, its separation of the human from a nature that the primitives recalled in rituals and tales of the bear or the raven. The ability to manipulate language for destructive purposes is, in Eiseley's metaphor, "an island within an island," one that "separated people into many islands" (170). In the spatial/temporal metaphor that is nevertheless "invisible," humankind is positioned in time at the beginning of an island whose dimensions continually expand and whose duality is inescapable. Eiseley's codes remain the standard codes of traditional humanism, but his purpose both embraces humanism and undermines the human-centered view in favor of a nature-centered one. If Ariel, the figure of human potential, still hovers over the island, Caliban, the grim side of humanity, still haunts it.

What becomes of the island and of humankind will not be determined by blind struggle, but by conscious, interactive participation with the forces of nature and language.

The Changing Theater: "The Inner Galaxy"

The world we now know is open-ended, unpredictable.
—Eiseley, "The Inner Galaxy"

"The Inner Galaxy" takes the reader from the invisible linguistic and cultural island to the notion of an "inner galaxy" of human potential, whose domain—Ariel's domain—is vast and seemingly unconfined. The essay focuses on the open-ended human quest for the nature of the self and for the equally open-ended human ability to try to be more than it presently is. Anticipating the theme of the space quest that will be the focus of his next book, Eiseley pursues the human "galaxy" of dreams, which is, in one sense, as unexplored as a new galaxy in the universe. "I remain oppressed by the thought," he writes, "that the venture into space is meaningless unless it coincides with a certain interior expansion, an ever growing universe within, to correspond with the far flight of the galaxies our telescopes follow from without" ("The Inner Galaxy," *UU,* 174).

The central image also anticipates an image that he will pursue in *The Invisible Pyramid,* the image of the universe as a great interactive playhouse in which we make the play as we go. Eiseley tells of attending, with a friend who was a poet, an outdoor performance of an opera. Beneath the bright lights of the tent, in the midst of the performance, "far up, blundering out of the night, a huge Cecropia moth swept past from light to light over the posturings of the actors" ("The Inner Galaxy," *UU,* 175). The poet's insight sets up the central metaphor: "'He doesn't know,' my friend whispered excitedly. 'He's passing through an alien universe brightly lit but invisible to him. He's in another play; he doesn't see us. He doesn't know. Maybe it's happening right now to us. Where are we? Whose is the *real* play?'" (175–76). Thinking of "the universe of the moth and the poet," the narrator is "confounded" (176). He remembers the world of the pharaohs, who constructed boats to use in the afterlife and concludes, "There *was* a real play, but it was a play in which man was destined always to be a searcher, and it would be his true nature he would seek" (176).

In surveying this quest for the "nature" of humankind, Eiseley again asserts the influence of "fashions" of thinking. Quoting Montaigne's statement (which itself undermines the traditional humanistic faith)—" 'The conviction of wisdom is the plague of man' " (cited in "The Inner Galaxy," *UU*, 179)—he introduces human perceptions of the species. In each century, we study ourselves "in the mirror of fashion, and ever the mirror gives back distortions, which for the moment impose themselves" upon whatever the reality may be (179). When we ask what kind of creature fossils show us to be, we hear that we are warring creatures, yet at the same time images of St. Francis and primitive philosophers appear (180). In such self-scrutiny Eiseley finds the recurrent duality of wisdom and danger; the danger is that we may shape our image to the definition that intellectual fashion projects (182–83). Here Eiseley deliberately undermines a major metaphor embedded in our thinking. There is, he contends, "a quality of illusion about all of us" (183), and the nineteenth century's acceptance of Huxley's definition of the human in terms of qualities shared with the ape and the tiger, which "are bad metaphors at best," indicates the danger in self-definition—overlooking the better qualities of the human—for we tend "to visualize our psychological make-up as fixed—as something bestowed upon the first man" (185, 184).

Having spent a century discussing the struggle for survival, we might well, Eiseley argues, look instead at the impulse toward love and compassion ("The Inner Galaxy," *UU*, 179), at the sheer potential that the human being comprises, for the human being contains both the "lights and visions" and the "fearful darknesses" of the future (186). Real as natural selection is, it is still "less 'law' than making its own law from age to age" (187). The world "is open-ended, unpredictable" (186), natural selection having "given way to the creative forces of random mutation" so that "the potential hidden in nature has flowered into a greater variety of behavior" (187). Perhaps the strongest evidence that the human mind has escaped pressure for conformity is language, which makes possible intellectual variety and selective mating, both of which further contribute to the emergent potential (187–88).

Eiseley's conclusion returns anecdotally to the suggestion that we have reason to look to human compassion rather than to human struggle. Based on a simple personal experience, the anecdote is a confession of his identification with another nonhuman creature—an old, tired gull during a time spent on a beach. On this beach, he writes, "I came to know the final phase of love in the mind of man" ("The Inner Galaxy," *UU*, 191). Recalling the young poet's question, "Whose is the real play?" he thinks that the poet's "eye reached farther than the giant lens" of a telescope that he sought as a youth (192). Life, he speculates, will win, whether in human form or in another, because of the ability to love, to project the self beyond the self. In the prehistoric past, as in any generation, some have loved—whether the love was projected onto animals, nature, or other human beings. In drawings on cave walls "—not with the ax, not with the bow—" the human "fumbled at the door of his true kingdom. Here, hidden in times of trouble behind silent brows, against the man with the flint, waited St. Francis of the birds—the lovers . . . who are still forced to walk warily among their kind" (188).

Beyond the Boundaries of Form: "The Innocent Fox" and "The Last Neanderthal"

Beginning with an epigraph from Peter Beagle—"Only to a magician is the world forever fluid, infinitely mutable and eternally new" (cited in *UU,* 194)—"The Innocent Fox" explores the unexpected qualities of infinite mutability and infinite meanings, ending with a renewal that compensates for what he calls "unrealized anticipation" (195) in his own experience of a blurring of the boundaries of life forms, a blurring of the line between human and animal. The experience is one that requires a childlike openness to novelty, the kind of openness to "the unexpected and the beautiful"—the aesthetic qualities that reflective scientists have always associated with good science and the "parsimony" that has become a *topos* in scientific rhetoric. These qualities, Eiseley writes, originally led him into science, yet in a period of despair he felt the need for more than he found there. As a scientist, he expresses distrust of the notion of miracles, but he admits "broad latitudes of definition," for, he admits, his "whole life had been unconsciously a search," which "had not been restricted to the bones and stones of my visible profession" (197).

He defines the term *miracle* as that which is "without continuity," characterized by a "sudden appearance and disappearance within the natural order, although . . . this definition would include each individual person" and adds that "miracles . . . momentarily dissolve the natural order or place themselves in opposition to it" ("The Innocent Fox," *UU,* 200). As an example, he describes driving through a dense forest where he sensed a creature running just beyond the headlights—a creature that sometimes seemed to run upright and sometimes to shift in color. The apparition, he concludes in a realization of the magical shape-shifting of living forms, "was a gliding, leaping mythology," which led him to perceive the always already of disorder among what science perceives as order, to conclude that "what order there might be was far wilder and more formidable than that conjured up by human effort" (202). Though he finally ascertained that the creature was only a spotted dog, he remembers it in terms of the linguistic perception of reality; it was "an illusory succession of forms finally, but momentarily, frozen into the shape 'dog' by me. A word, no more" (203). "We deceive ourselves," he concludes, "if we think our self-drawn categories exist there" (203). Both he and the dog were created by "a nerve net and the lens of an eye" and both would vanish into the unknown (203). Similarly, he realizes that his own mind changed constantly. It is this realization that establishes a context for what he calls his own "miracle" through an encounter, in a similarly isolated setting, with a young fox cub.

The incident occurred on an isolated beach shrouded in mist, which takes on multiple meanings in Eiseley's texts. The mist, like the cloud of dispersed molecules in the conclusion to *Darwin's Century,* signifies the chaos that precedes order, the dispersal from which order can emerge. The mist is a barrier to seeing, but he describes its tactile effect as he sat with closed eyes and felt "the tiny diffused droplets of the fog" on his face (205). He feels himself part of the fog, which can become another fog in another place. Recalling a primitive question of "whether God is a mist or merely a mist maker," Eiseley uses this question as a vehicle to play on the "mist" and his own work "amidst" the artifacts of primitive humans ("The Innocent Fox," *UU,* 205). Recalling a line from Charles Williams—"I am the thing that lives in the midst of the bones"—he plays on the mist, molecular dispersion, his own life as a "bone hunter" (which will be the subject of essays in *The Night Country*), himself as a condensation "from that greater fog to a smaller congelation of droplets," and his thoughts as "vague and smoky wisplets" that were extensions of the self (206). Like the fog moving through the ribs of a broken boat, "the insubstantial substance of memory," he feels himself moving between past and present, moving backward to his father's last gesture, which was to lift and flex his hands as he looked at them "unbelievingly" and dropped them (207).

As he cites the notion that "one can never . . . get around to the front of the universe," but must "see only its far side . . . [and] realize nature only in retreat" ("The Innocent Fox," *UU,* 209), Eiseley sets the context for his own "miracle," which occurs when, in a brief episode with a young fox in the mist, he finds a momentary confrontation with the front of the universe. The experience can come only in the dispersal and disorder of the mist, in an abandonment of self and a blurring of the lines between life forms. At dawn, in the midst of a mist that is both literal and the unformed shifting of thought, he encounters a young fox pup playfully shaking a bone. Here, he finds, is Williams's "thing in the midst of the bones," the movement of the universe so that he can see its face. But that face is small and innocent—"so small that the universe itself was laughing" (210). The episode is, in Eiseley's texts, an abandonment of scientific bearing and the protocol of the writer. "It was not a time for human dignity," he writes. "It was a time only for the careful observance of amenities written behind the stars" (210). Arranging his forepaws and shaking a similar bone in his teeth, he tumbles "for one ecstatic moment" with the creature, in a dispersal of molecules, of order, of forms, of formalities:

> We were the innocent thing in the midst of the bones, born in the egg, born in the den, born in the dark cave with the stone ax close to hand, born at last in human guise to grow coldly remote in the room with the rifle rack upon the wall.
> But I had seen my miracle. I had seen the universe as it begins for all things. It was, in reality, a child's universe, a tiny and laughing universe. (210)

Like his father, who, in death, had come to "the front of things" in looking at his hands, he has discovered the human "miracle" of understanding—that meaning itself is brief and mist-like, quickly dispersed into another semblance of order that deconstructs itself as meaning always does for Eiseley: "There was a meaning and there was not a meaning, and therein lay the agony" (211). In the "mist" that is both order and dispersal he has come to the "meaning," which he finds "all in the beginning, as though time was awry" (211). The effect of the experience, Eiseley asserts, was to reverse time's arrow, however briefly, and it was "the gravest, most meaningful act I shall ever accomplish, but, as Thoreau once remarked of some peculiar errand of his own, there is no use reporting it to the Royal Society" (212).

The Royal Society, perhaps, would not admit Eiseley's experience, his poetic juxtaposition of the *mist* that places him in the *midst* of unexpected new views of a fleeting order of forms and experiences. But the effect of the play on mist and midst, of order in disorder, of role reversal and the blurring of lines between forms is a powerful defamiliarization, in the world of life forms, of the determinism that scientists since the nineteenth century have easily embraced and a celebration, which W. H. Auden links to the notion of "carnival"—of disorder, reversal of hierarchy, and simple revelry (1978, 15–24) and which will recur in *The Night Country*. The mist extends into a new dimension Eiseley's exploration of the "living screen" or the "web" that mediates natural selection but can be torn or evaded or, in this case, dispersed.

In the concluding essay in this collection, "The Last Neanderthal," Eiseley similarly pursues the way that nature's "creatures slip in slow disguise from one shape to another" and explores the oxidation—a similar metaphor of molecular dispersal—that occurs in all life even "in an equally diffuse and mediating web of nerve and sense cells" (*UU*, 213). Although the essay does not equal Eiseley's climactic romp with the fox, it draws on his "own solitary hieroglyphics" to evoke a beach at Curacao and a sea-buried dog, the dizzying movement of lizards beyond the beach, hummingbirds flashing in the underbrush, the apples of a poison tree, and a recollection of a junkman whom he watched in a sixteen-year-old's sudden realization of time. "The junkman," he remembers thinking, "is the symbol of all that is going or is gone" (217). From this disparate assortment, this intellectual junkman's freight of recollections, he realizes the power of an "organization" reminiscent of the organizing force around which much of *The Immense Journey* is structured. This time, however, he refers to the "running down" of the inanimate universe, against which he juxtaposes both organic "organization" and its most remarkable product, the "organization" that the human mind creates:

> Of all the unexpected qualities of an unexpected universe, the sheer organizing power of animal and plant metabolism is one of the most remarkable, but, as in the case of most everyday marvels, we take it for granted. Where it reaches its highest development, in the human mind, we forget it completely. Yet out of it history is made—the junkman on R Street is prevented from departing. (218)

From his intellectual junkman's perspective, Eiseley asks the reader to consider whether we might better understand "humanity" if we could "follow along just a step of the evolutionary pathway in person" ("The Last Neanderthal," *UU*, 219). With this transition, he interweaves a tale of his bone-hunting days in the West— a tale of a baleful girl whose "massive-boned" (222) head and body build made him think that "she is the last Neanderthal, and she does not know what to do" (223). She asked him about his digging but "listened incuriously, as one at the morning of creation might do" (223). Finally she asked, "'Do you have a home?'" (223). As the digging season drew to its end, he remembered an "agonizing, lifelong nostalgia, both personal and, in another sense, transcending the personal," a nostalgia that emerged from "the endurance in a single mind of two stages of man's climb up the energy ladder that may be both his triumph and his doom" (225). Through the girl, he had followed a step along the evolutionary pathway and had come to see himself as "a mental atavism," who, like the girl, would never find home, for it lay far in the past while humankind "was plunging into an uncontrolled future" (226). The human brain has mastered energy and the word, which codes individual energy to endure beyond death, yet Eiseley's awareness is of the "utter homelessness of man" (227).

Drawing on the thema of energy dispersal, the essay undermines the expected unity of a conclusion. In a final image of dispersal, Eiseley speculates on life as production and dissipation of energy, and he speculates on fire—oxidation—as "perhaps the very *essence* of animal" ("The Last Neanderthal," *UU,* 231). Momentarily he thinks that perhaps he is no longer *Homo sapiens,* but only "a pile of autumn leaves seeing smoke wraiths through the haze of my own burning" (232). Concluding that "things get odder on this planet, not less so" (232), he remembers his own journey and the two worlds—the world of nature and the world of the mind—in which he has lived. Instead of closure, then, the book ends with "organization" and dispersal, in nature and in the mind, as Eiseley the narrator senses behind him the molecular dispersal of burning, the mysterious oxidation of life itself.

8
Reexamination of Science II:
The Invisible Pyramid

From his critique of biological determinism in *The Unexpected Universe,* Eiseley turns in *The Invisible Pyramid* to the technologically oriented outlook that biological determinism makes "natural"—the technological orientation that will enable a robot-like political establishment to pursue space exploration whatever the human and environmental costs. Thus *The Invisible Pyramid* continues Eiseley's critique of the embedded mechanistic understanding and of its political consequences. He questions science as an institution, specifically the role of science in the quest for domination of space, and he further explores personal narrative as a vehicle for linking scientific knowledge and sympathy with the natural world. Designations that are accepted as fact in the sciences, particularly in anthropology, Eiseley reexamines in linguistic terms; the metonymic designation of the human being as "tool user," for example, is especially subject to scrutiny. And throughout the text, Eiseley factors back into the understanding of humankind and the world itself elements of contingency, of indeterminism, which he views as essential to life and as elements in which we can consciously participate. On a planet shaped by contingency, we can turn that contingency to the uses of all life forms.

Based on a series of lectures sponsored by the John Danz Fund at the University of Washington in the fall of 1969, *The Invisible Pyramid* is Eiseley's strongest expression of his ecological ethic. The book consciously employs the central metaphor of the pyramid, which Eiseley explores not only as an artistic device but as a spatial form embedded in the human consciousness and exemplified not only in the ancient Egyptian pyramids and their association with death, in the ancient Mayan pyramids and their association with worship, and in the upward thrust of the Gothic arch, but also in the twentieth-century physical and psychological thrust toward space. Structurally and in the metaphors that destabilize accepted metaphorical concepts and embedded notions of the human role, *The Invisible Pyramid* comprises some of Eiseley's most complex textures—textures whose complexity may work to undermine their effectiveness, but which also deserve reexamination because of their literary and rhetorical effectiveness in helping to establish an environmental ethic that grows ever more important in the global society that Eiseley envisioned.

Hinting at the cyclical time that Eiseley as scientist rejects and as poet explores, the epigraph from John G. Heihardt focuses on Halley's comet, the thematic organizing point of the collection:

> Once in a cycle the comet
> Doubles its lonesome track.
> Enriched with the tears of a thousand years,
> Aeschylus wanders back.
>
> (cited in *IP,* ix)

Eiseley's play on cyclical time is also a play on evolutionary time: time is linear, but circular time is implied in the human tendency to track ourselves, to follow our own footsteps, and in the speculative play on life's return in a myriad of forms as if life itself were a single ongoing being or force. This play on time and on multiplicity within the force of life is linked to the ecological ethic, which Eiseley explores through metaphors of the web that represents the interconnection of life forms, through the theme of the returning comet and its symbolic linking of historic time with the present to give a sense of cosmic time, and through the *topos* of burning or oxidation as basic to life but also part of what human beings inflict on the green world.

With the comet's trajectory as a point of reference, Eiseley explores several major themes that undermine established notions of human master-y, separation of the human from other life forms, the human obsession with the machine, the gain and loss that science and technology have brought, and the paradoxical enormity of the human mind enclosed in a variety of "prisons," from the human body to language to the earth itself. In an attempt to undermine the deterministic view of a society dependent on embedded mechanistic concepts, Eiseley takes the Faust myth as a tutor structure—the human being reaching upward for more and more knowledge to be used as power. This reaching upward, the opposite of his recurrent reaching out toward other life forms or other human beings, is metaphorized in the slime mold that, having grown to a point of density at which it can no longer sustain life, thrusts upward a spore-bearing capsule carefully attuned to light, ruptures, and spreads its seed spores into new regions. A personal anecdote of a childhood dice game in an old house becomes an elaborate metaphor for the nonmechanistic factors of contingency in a human playhouse, an interactive theater in which we make and change the plot as we go, constructing in one historical epoch vast pyramids inscribed and associated with ambiguous hieroglyphs (both pyramids and hieroglyphs are linked to death) and constructing in our own historical epoch a similarly death-driven pyramid that is "invisible" because it is epistemological and is embedded in a metaphorical concept. It is the pyramid of devotion to technology, and it requires, Eiseley argues, an almost religious devotion and outpouring of energy.

And in opposition to the single-minded pursuit of power through knowledge, the pursuit that brings Western society to the brink of spreading like slime mold through "capsules" borne by rockets, Eiseley draws on the essential visual philosopheme to meditate on the need for two "eyes," the eye of science and the eye of the poet, in an attempt to find a balance, a dialogue between knowledge and power sought through one "eye" and knowledge and reflection sought through the other eye.

The Dice Game and the Universe as Theater: Prologue,
"The Star Dragon," "The World Eaters," and
"Man in the Autumn Light"

In the Prologue, invoking metaphors that underlie his considerations of the human role, Eiseley chooses for his persona "the ambivalent character of pessimist and optimist . . . because mankind itself plays a similar contradictory role upon the stage of life" (*IP*, 1). He compares modern humankind to John Donne's notion that we live in a prison that is paradoxically precious. From both negative and positive perspectives, Eiseley portrays humankind as a slime mold, but he contrasts that analogy with the analogy of the sunflower forest, "the green world" that is the "sacred center" to which humankind will continue to turn "in moments of sanity" (1). Again he invokes the journey—of knowledge in the form of the heat ladder, of space travel as compared to thistledown drifting to new worlds, of humankind living "in the morning twilight of humanity" and threatening itself with doom before noon (1). He both salutes and fears the future, recognizing the awesome power that humankind now possesses over nature.

The prologue ends with Eiseley's tale of a boyhood adventure—

playing dice alone in an empty house at twilight. At the time, he writes, "I was too young to have known that the old abandoned house in which I played was the universe," and he concludes, "I would play for man more fiercely if the years would take me back" (*IP,* 3). The dice game in the old house is a recurrent theme for Eiseley, one to which he will return in the autobiography. The child playing in the abandoned house made up the roles as he played, and he was too young to recognize the allegory—"that the old abandoned house in which I played was the universe" (3). In retrospect, he declares that he "would play for man more fiercely if the years would take me back" (3).

But the old house represents another level of "playing." In the first essay, "The Star Dragon," Eiseley explores the notion of a cosmic playhouse—a nonmechanistic, textual, cultural analogy for the shifting illusions of the physical world in which the projection of self and interaction with the text actually create the reality. Making the game as he went, the child playing dice in the abandoned house repeated the ongoing human drama, the play whose participants make and alter the "plot" as they go. The "play" takes on multiple meanings: the child playing unaware of his surroundings, the universe as an abandoned house, the universe as a playhouse in which the participants help to shape the outcome, the "play" of chance or contingency, and the participatory universe in which humankind becomes part of the evolution of chance itself. Against the backdrop of the elliptical orbit of Halley's comet, which Eiseley's father held up his three-year-old son to see and which Eiseley hoped to see again before his death, the text marks the interplay of apparent stability—and each species' *attempt* at stability—and inevitable instability. Had Eiseley lived to see the return of Halley's comet, he would undoubtedly have found it appropriate that the comet did not perform exactly as scientists had predicted. In its own way, Halley's comet outperformed science and media hype to confirm the very unexpectedness of the universe that Eiseley explores.

In "The Star Dragon," Halley's comet sets up the dialogic relationship of potential and contingency, which Eiseley explores throughout *The Invisible Pyramid* in terms of human ambiguity and human potential. The epigraph to "The Star Dragon"—Jean Rostand's comment, "'Already at the origin of the species man was equal to what he was destined to become'" (cited in *IP,* 5)—suggests the recurrent Eiseleyan theme of the potential always already in the human species. With his early experience of seeing the comet as a point of transition, Eiseley turns to the cosmic playhouse—

to discoveries of the fossil record that undermined the traditional Judaeo-Christian code of the world as a stage created for the human performers and to the awareness that we might be "intruding upon some gigantic stage not devised for [us]. Among these wastes one felt as though inhuman actors had departed, as though the drama of life had reached an unexpected climax" ("The Star Dragon," *IP*, 13). Recapitulating the drama of humankind, Eiseley traces two revisions of the concept of time—from the notion of six thousand years to "eons of inconceivable antiquity," and, later, from the notion of fixed species to realization of constant change in life forms (14). Time meant, he continues, not just "endless Oriental cycles of civilizations rising and declining" (14), but changes in life itself, for life itself "altered its masks upon the age-old stage. And as the masks were discarded they did not come again as did the lava of the upthrust mountain cores" (15). Both individuals and species died; discarded life forms, like discarded theatrical masks, would not come again.

The role of Darwin's *Origin* in this dramatic awareness is, of course, monumental, for

> the great stage was seen not alone to have been playing to remote, forgotten audiences; the actors themselves still went masked into a future no man could anticipate. Some straggled out and died in the wings. But still the play persisted. As one watched, one could see that the play had one very strange quality about it: the characters, most of them, began in a kind of generous latitude of living space and ended by being pinched out of existence in a grimy corner. ("The Star Dragon," *IP*, 16)

On Eiseley's remotivated stage, the human "hero" is now an evolutionary form that "became a victim of his success and then could not turn backward" (16). What Eiseley sees as "the essential theme that time had dramatized upon the giant stage" (17) is a frightening one for human beings, who in each generation seem to think of the end of their own civilization as the end of time. It is a theme that pursues the ramifications of evolutionary thinking away from the human-centered nineteenth-century view and toward the eventual displacement of the human—the theme that "success too frequently meant specialization, and specialization, ironically, was the beginning of the road to extinction" (17).

Continuing the revised metaphor of the old (play)house, Eiseley refers to bodily remnants of earlier times as "outdated machinery" such as we might find "in the attics of old houses"; such remnants are also "the dropped masks of the beginning of Nature's last great play" ("The Star Dragon," *IP*, 18). This play continues the revisionary approach to the embedded metaphor; it is the "last play" because human beings can conceive of the drama only in human terms. Yet this remotivated play introduces novelty; it is a unique drama because in emerging from the forest we have brought "a new unprophesiable world," the "latent, lurking universe" of language (18). Eiseley links the cosmic playhouse to human language, to the juncture of brain and tongue and hand. The key to this cultural and linguistic world is the highly specialized brain, whose "primary instruments" are "the tongue and the hand, so disproportionately exaggerated in [the human] motor cortex" (19). Thus "brain, hand, and tongue would henceforth evolve together," and language would make possible naming and categorizing of "the vague, ill-defined surroundings of the animal world" (19–20). And beyond language, Eiseley moves to writing, culture, and artifact; for language is the origin of culture, the "first and last source of inexhaustible power," and "with the first hieroglyph, oral tradition would become history" (20). The "new" heredity made possible by the brain is "based not upon the gene but instead upon communication"; it is "a greatly speeded-up social heredity" (20). Because of language, we retain what Eiseley calls "a penumbral rainbow," a quality that A. L. Kroeber called *superorganic* (21). Invoking the visual philosopheme that links language and the eye in a "cloud of ideas, visions, institutions which hover about, indeed constitute human society, but which can be dissected from no single brain," Eiseley observes that the superorganic "rainbow" identifies the human, but it "eludes us and runs onward," though it leaves "no trace . . . in the remnants of an ancient campfire" (21).

The playhouse links the dominant philosophemes; language makes possible the playhouse, which requires consciousness of our seeing and hearing the play and the consciousness that Eiseley links in "The World Eaters" to the emergence of writing. It was writing, which takes us into the past and future, that enabled humankind "to live against the enormous backdrop of the theater," becoming "self-conscious" and "enacting his destiny before posterity" ("The World Eaters," *IP*, 63). To enact is to mime, to "write," and the stage is a blank on which this writing occurs. The consciousness of "performing" for the future, of creating for itself an other identity, made possible a stage existence that echoes what Derrida treats in terms of linguistic *differance*—both different from itself and deferred in time. This rainbow world combines with the code of the journey as Eiseley writes of ancient pathways in the human brain: "As the roots of our phylogenetic tree pierce deep into earth's past, so our human consciousness is similarly embedded in, and in part constructed of, pathways which were laid down before man in his present form existed" (22). In the last section of the essay, Eiseley describes a theoretical solar flare that might account for glasslike areas on moon rocks and "would have seemed from earth like the flame of a dragon's breath" (26). Such a flare, he speculates, might be the origin of dragon myths, the narratives that cloak human memory (28) in a texture woven of the embedded philosophemes of eye and tongue.

The old (play)house continues as a shifting, undermining metaphor for the universe in "Man in the Autumn Light," which Eiseley opens with Jean Cocteau's notion that "'the theatre is a trick factory where truth has no currency, where anything natural has no value, where the only things that convince us are card tricks and sleights of hand of a difficulty unsuspected by the audience'" (cited in "Man in the Autumn Light," *IP,* 119). "The cosmos itself," Eiseley continues, "gives evidence, on an infinitely greater scale, of being just such a trick factory, a set of lights forever changing, and the actors themselves shape shifters" ("Man in the Autumn Light," *IP,* 119). In such a universe, he suggests in a return to his exploration of the code of the "natural," there may be nothing natural in the usual sense of the word, or perhaps "the world is natural only in being unnatural, like some variegated, color-shifting chameleon" (119). "Man in the Autumn Light" explicitly links the epistemological dimension of the dramatic metaphor to modern humankind in Eiseley's forthright questioning of the goals of science, particularly the quest for space. He turns to Marlowe's lines "Thou art still but Faustus / and a man," which, he contends, "epitomize the human tragedy" (cited in p. 134). Through science, modern humankind, like Faustus, seeks knowledge and, through knowledge, power. We are "world eaters and knowledge seekers," yet, like Faustus, we remain human (134). To remain human carries both gain and loss; it means to continue reaching for more, but also to grow weary with each new discovery, in effect to consume all that we touch. The Faustian element in science epitomizes the tension of gain and loss. The Faustian drama contains the magic of illusion and the *hubris* of seeking control; for Eiseley, it signifies both the magic that makes us reach to be more than we are and the diabolical quest to control both human and natural worlds. If the drama is the embedded metaphor of the Judaeo-Christian tradition, Eiseley reminds us that the Faustian drama embodies the best and the worst of human capabilities.

Nietzschean Awareness: "The Cosmic Prison"

From the dialectic of stability and change and the role that time plays in both, Eiseley turns in "The Cosmic Prison" to an exploration of "prisons" of the universe, of form, and of language in an essay that engages in a silent dialogue with the existentialist view that we are "damned to freedom." Opening with Kazantzakis' cryptic statement, "'A name is a prison, God is free,'" Eiseley glosses the tutor text by commenting that, in spite of its value, language nevertheless creates "an invisible prison" because it "implies boundaries" and "fixes" the nature of things so that "henceforth their shapes are, in a sense, our own creation" and "no longer part of the unnamed shifting architecture of the universe" ("The Cosmic Prison," *IP*, 31). Through language, ideas and things can be "transfixed as if by sorcery, frozen into a concept, a word" (31). Perhaps the most important point is that words "are always finally imprisoning because man has constituted himself a prison keeper," not consciously but "because for immediate purposes he has created an unnatural world of his own, which he calls the cultural world, and in which he feels at home" (31–32). Language, then, makes of the universe "a cosmic prison house which is no sooner mapped than man feels its inadequacy and his own" (32). This sense of inadequacy in a world of words is the reason Eiseley posits for the quest for space flight—the ultimate quest for escape. In a senator's reference to us as "masters of the universe," Eiseley notes the dual "comfort of words" and "the covert substitutions and mental projections to which they are subject" (such as the false comfort of considering ourselves "masters" because of a journey in a space capsule) (32). In his political commentary, Eiseley continues to explore the human duality, through the imprisoning aspect and the dreaming aspect of language, both arising from the tendency to believe what we say. Language is the cosmic prison that limits our understanding of whatever reality lies outside our technology. Paradoxically it is, in a sense, a technology that confines us to technological understandings, yet language is also the means of the dream by which we create ourselves as more than we presently are. The universe over which the senator claimed mastery is one of time measured in billions of years, and words are finite. We are, Eiseley writes in an extended analogy, "in the position of the blood cell exploring our body" (36).

Other prisons exist—the prison of smells, for example, that limits the dog's world, or the various social and cultural prisons. Even "the Mendelian pathways are prisons of no return" ("The Cosmic Prison," *IP*, 41). Continents on earth itself—notably Australia and South America—have served as prisons that isolated evolutionary processes so that distinctive forms emerged such as the Australian marsupials and the South American tree-dwelling monkeys (42–44). Meteorological prisons, in the form of circulation of moisture, have made possible emergence of new forms, which themselves are smaller "prisons of form" within the cosmic prison (45). In view of all these prisons, Eiseley attacks the notion that "a walk on the moon means that we have escaped the cosmic prison" (45). He concludes his account of these oppressive multiple "prisons," which anticipate the tutor text of Giambattista Piranesi's etchings the *Carceri* in "The Time Effacers," by saying, "At every turn of thought a lock snaps shut upon us," and "we bow to a given frame of culture—a world view we have received from the past" (46).

Eiseley concludes that the only escape from despair is paradoxically through the very human duality that has created our predicament. Through introspection—"as though we still carried a memory of some light of long ago and the way we had come" ("The Cosmic Prison," *IP,* 46)—we can recognize what the poet Henry Vaughan recognized when he wrote that we see "with an "'Ecclips'd Eye'" and look for understanding "'with Hyeroglyphicks quite dismembered'" (cited in p. 46), or we can turn from the pursuit of space to what the poet Thomas Traherne called "'infinite love [that] . . . must be infinitely expressed in the smallest moment'" (cited in p. 47). Or in scientific terms, we can look to Claude Bernard's observation "that the stability of the inside environment of complex organisms must be maintained before an outer freedom can be achieved from their immediate surroundings" ("The Cosmic Prison," *IP,* 47). Bernard dealt with the evolutionary change that brought warm-blooded animals, but Eiseley compares the insight with that of the seventeenth-century mystic to conclude that the ultimate "discontinuity" is "the separation both of the living creature from the inanimate and of the individual from his kind" (48–49). Eiseley's concern is with the human brain's ability to make a "crossing" of the barriers of form, to love not only one's own kind, but beyond the boundaries of form (49).

The Pyramid and the March of the Machines: "The World Eaters," "The Spore Bearers," and "The Time Effacers"

A central analogy that Eiseley explores in *The Invisible Pyramid* is that of modern humankind and the slime molds, the only two entities on the planet that move from individual pioneering to great aggregations of life from which spring upward clusters of life awaiting dissemination. He compares the "overtoppling spore palaces" of the slime molds to city skyscrapers ("The World Eaters,"

IP, 53). Eiseley continues to examine the implications of metonymic description of human beings as tool users, from primitives in Australia to natives of the northern forests of the United States to primitives in Labrador. Members of these cultures have considerable skills in using tools but do not experiment with their surroundings or see "invention" as a means of "using" the environment; in fact, they see the natural environment "almost as a single tool" in which they survived with little equipment and of which they consider themselves a part (58–59).

What is unique in Western culture, Eiseley contends, is that the human "spore bearers" require consumption of natural resources. And this consumption could not have developed without the "tool" of writing, the vehicle by which humankind manipulates time. Consciously aware of the future, the "record," invention-oriented humans clustered in cities where individualism was replaced by conformity. Writing made possible "a kind of stored thought-energy, an enhanced social brain" ("The World Eaters," *IP,* 67). Now, Eiseley concludes, the Western mentality "respects no space and no thing green or furred as sacred. The march of the machines has entered his blood" (70). And this is the autumn before the winter dispersal. What Eiseley proposes is not abandoning either view, but establishing a dialogue. "Two ways of life are thus arrayed in final opposition," he writes (60). One of these "ways" depends heavily on the primitive perception of nature as text; it "reads deep, if sometimes mistaken, analogies into nature and maintains toward change a reluctant conservatism" (60). The other "way," however "is fiercely analytical," drawing on "sequence and novelty" to see the world as a machine and "to install change itself as progress" (60). Neither view alone will solve contemporary human problems. For Eiseley, the primitive outlook, especially the primitive interactive approach to nature with its emphasis on understanding rather than aggressive materialization of a progress-oriented future, is most closely allied with the poet's insight. He concludes that "a reconciliation of the two views would seem to be necessary if humanity is to survive" (60).

In "The Spore Bearers" Eiseley further pursues the mystique of the machine and the machine as the literal bearer of the human dispersal into space. The epigraph, from Garet Garrett, introduces the mystique of the machine: "Either the machine has a meaning to life that we have not yet been able to interpret in a rational manner, or it is itself a manifestation of life and therefore mysterious" (cited in *IP,* 73). And the machine, we recall, is inseparably tied to the notion of human dominance and manipulation. The essay emerges from an analysis of the curious mechanism of *Pilobolus,* a "fungus which prepares, sights, and fires its spore capsule" in "a curious anticipation of human rocketry" ("The Spore Bearers," *IP,* 75). Growing on cattle dung, the fungus thrusts up a tower containing a light-sensitive cell that aims the spores toward the greatest light so that the explosion of the spore capsule can send spores unimpeded several feet upward. Starting from the analogy of the spore bearer, Eiseley looks at the epistemological implications of the contingencies that have brought us to our spore-bearing stage. The only "scientific civilization in the full sense" (82) is Western civilization, which has separated people from the land and from production of food, contributing further to the human alienation. Eiseley sees in the Gothic arches of the great medieval cathedrals a hint of "the upward surge of the space rocket" (84). Humankind is, for Eiseley, a Faustian figure, "a spokesman of the will," an "embodiment of a restless, exploratory, and anticipating ego" that eagerly creates the future; yet it lacks the ability to turn its view upon itself (85). Yet he adds that science needs the ability of the owl to turn its head 180 degrees.

Exploring the notion of "invention," Eiseley begins with the standard mechanistic notion but adds invention in terms of both the quest for power and the quest for "understanding" ("The Spore Bearers," *IP,* 86). The first invention he considers is the zero, which was invented twice in prehistory, by the Hindus and the Maya. It is, Eiseley notes in an echo of the "domain of absolute zero" from which value is asserted in "The Star Thrower" and in later texts, "a 'no thing,' a 'nothing,' without which Roman mathematics was a heavy, lumbering affair" (86). As a necessity to computer circuitry invented by "an unknown mathematical genius seeking pure abstract understanding," the zero stands as a symbol of the dual bane/blessing of invention and of the "unprophesiable" effects of a single understanding in a contingent world. An epistemological journey such as the scientific journey is at least as costly as other journeys. And the "obsession" with space represents an endeavor that has consumed both economic and creative resources on a scale that he compares with the "public sacrifice" required in building the Great Pyramid at Giza. What is produced today Eiseley calls an "invisible" pyramid that "demands great sacrifice, persistence of purpose across the generations, and an almost religious devotion" (87). This "invisible" pyramid invokes the dual aspect of the epistemological journey, the Faustian quest for knowledge as power that creates the pyramid or the "spore" cities. Science, Eiseley concludes, has become "a giant social institution" in which are invested money, power, and prestige. It is the contemporary replacement of "primitive magic as the solution for all human problems" (90).

Eiseley explores this fearful side of the epistemological journey through the story of the mysterious Mayan culture, which produced its own pyramids during the first millennium before Christ. Science, he reminds us, has solved microscopic mysteries of disease and spectroscopic mysteries of the composition of stars. We are only "dimly" beginning to comprehend, he contends, that we may be our own "ultimate menace," our own "final interior zero" ("The Spore Bearers," *IP,* 92). Like the Maya, he implies, we may have created a wealthy priesthood that our society cannot sustain and that will lead to the society's collapse. Even so, the upward thrust of both life and knowledge, he speculates, seems built into life itself—into the spore capsules of the slime mold, into the thistledown that carries the seed from one crevice to another. It is an impulse toward a kind of hope that we bring from the unknown past of life, an impulse that for Eiseley evokes Bacon's reflection on the new scientific method as "'something touching upon hope'" (cited in p. 94). The "hope" that we as spore bearers carry, however, will require conscious participation, conscious choice of the poet's eye in a dialogue with the eye of science.

A similar dialogue opens "The Time Effacers," as Eiseley focuses on the tension of "two diametrically opposed forces," memory and forgetfulness, in the human being. He begins with his own recollection of a conversation with a coroner about a skull with elaborate and expensive dental work, whose owner had disappeared, apparently in a deliberate attempt to be completely forgotten, without identification. Similarly, he argues, there are times when masses of people attempt to destroy a past: Cromwell's destruction of statuary, Henry VIII's dispersal of the wealth and manuscripts of the abbeys, the Spaniards' destruction of Inca and Aztec civilizations. We, he continues, belong to "evolutionary time," which is "the time of the world-eaters" ("The Time Effacers," *IP,* 103). Returning to the notion of the "track" of the comet, Eiseley pursues Thoreau's notion that each individual "'tracks himself through life'" (cited in p. 103) and compares individual and civilization, each of which, he contends, returns to where it began. In the last three centuries, the Western "personality" has dominated "an increasingly time-conscious, future-oriented society of great technical skill," which has lost the sense of balance with the natural world and has established science, with "the laboratory and its priesthood" increasingly dominating both business and government, as the vehicle of "progress" ("The Time Effacers," *IP*, 104–5). Science is perceived, he argues, as "a kind of twentieth-century substitute for magic" rather than a "social institution" deserving of study (105). The scientist is given a new role like that of an Eastern seer, a role for which most scientists are ill prepared. Human beings are wanderers in "the infinitude of space and time which their science has revealed," and yet "trapped in a world of darkening shadows" (107). For a visual image of the situation, Eiseley invokes as tutor text Giambattista Piranesi's *Carceri*. In describing these eighteenth-century etchings, Eiseley emphasizes the human figure as lost "amidst huge beams and winding stairways ascending or descending into vacancy," among ropes that "hang from spiked machines of unspeakable intent." The image is of yet another prison, the present human world, a "vast maze [that] offers no exit" (107). It resembles Jorge Luis Borghes's winding stairway that leads to nothingness, which, as N. Katherine Hayles points out, leads "not to a door but to vertigo" (1990, 32). Although the notion of "progress" has been read into evolution so that we tend to think in terms of predictable causality, in actuality, the future can only be guessed from multiple possibilities.

In opposition to the thinking that has led us to become "*event-oriented*" and bent upon effacing our relationship to time in our "culmination of the Faustian hunger for experience" ("The Time Effacers," *IP*, 109), Eiseley looks to the simple, magical cultures that "are without causal or novel time"—cultures "where the veil between life and death wavers fitfully at best" and "animals or men, rather easily exchanging shapes, pass to and fro in ways unknown to the sophisticated world" (111). In one such society, that of the Australian aborigines, two concepts of time coexist—ordinary time and "dreamtime," the time of creation and lawmaking that is perceived as ongoing, enduring "like an autumnal light" (the phrase that echoes the autumn of humankind and forms the title of the next essay) in a kind of parallel to ordinary time. It is a time in which the divine beings exist in one form or another and aid the human being (112). Dreamtime comprises past and future; it makes possible a search for timelessness, a dialogue with animal life. In such a society of undisturbed time and interrelationship with nature, humankind is not involved in progress, in "tracking itself," or in effacing the past in order to rush toward a future. For Eiseley, a concept such as the dreamtime provides a useful corrective to the "march of the machines," which has led us to efface time, to devour the world, and to spread beyond the world that we have destroyed.

The Pyramid and the Hieroglyph: "The World Eaters" and "Man in the Autumn Light"

Eiseley's central metaphor of the "invisible pyramid" explores both a geometric form that he speculates may be embedded in the human unconscious and the drive toward death that he finds associated with the pyramid and the hieroglyph. Although the "pyramid" of today is not visible, Eiseley asserts that it "demands great sacrifice, persistence of purpose across the generations, and an almost religious devotion" and that it may someday seem "as strangely antiquated as the sepulchres of the divine pharaohs" ("The Spore Bearers," *IP*, 87). In this evident connection of the contemporary worship of progress and technology that has led to the drive toward space with the ancient Egyptian pyramid, Eiseley clearly implies a link with death—including his assertion that "the invention of a way to pass knowledge through the doorway of the tomb" was the invention of writing ("The World Eaters," *IP*, 63). This play on the pyramid, the hieroglyph, and death is especially evident in "The World Eaters" and "The Last Magician." As noted in chapter 7, for the student of language and of writing, a distinction exists between "hieroglyphic" and "alphabetic" writing. A hieroglyph is a symbol that may be read as ideograph or pictogram, or as a phonetic symbol—hence many of the difficulties in interpreting the Egyptian hieroglyphs, some of which functioned pictographically while others functioned phonetically. A true alphabet requires, ideally, a single symbol for a single sound, or at least (as in the case of the Roman alphabet and the English language) an approximation thereof. The common Western understanding—which Jacques Derrida has been at pains to undermine—is that language is a "container" for ideas, an instrument for communication, and that writing simply extends that instrument.[1] Derrida has noted that writing posses a hieroglyphic quality that does not require sound; in reading, we bypass sound. This "nonphonetic moment" of writing displaces the "voice" and "breath" and "presence" of logos. In his system of "writing" that embodies more than is normally ascribed to the dualistically conceived hierarchy of speech/writing, Derrida finds the hieroglyph "the elementary milieu, the medium and general form of all writing" (1979, 125). And, he continues, "hieroglyphic writing does not surround knowledge

like the detachable form of a container or signifier. It structures
the content of knowledge" (126). Although his project is vastly
different from Derrida's, Eiseley draws on the quality of the hieroglyph that "structures the content," for his "natural" hieroglyph
is, like an emblem, suggestive of a meaning; that meaning is supplied by the individual on the basis of her own experiences, but
the hieroglyph initially suggests. It is not a "container" for a meaning, but a starting point for making meanings.

In earlier cultures, the human being "was so oriented that the
total natural environment occupied his exclusive attention," Eiseley writes ("The World Eaters," *IP*, 59). If the natural environment
did not provide direct help in survival, then parts of it "were often
inserted into magical patterns that did," and primitive humans existed in a nature that "was actually as well read as an alphabet"
and served as "the real 'tool'" that enabled them to survive (59).
Yet that "tool had largely been forged in the human imagination,"
for it represented the way of perceiving the self and a way of
relating to the environment. In this view, "nature was sacred and
contained powers which demanded careful propitiation" (59). Returning to the concept of "natural" as a social construct, Eiseley
contrasts the primitive view with the contemporary perspective
that sees nature as external and manipulable. For the "modern,"
"technology and its vocabulary" create the world, and achieving
whatever is potential is considered necessary (59). Eiseley's contrast of the primitive view of nature that required "reading" nature
as hieroglyph, though the reader actually brought the meaning, and
the contemporary view of human separation from the natural sets
up the link of separation and death, as well as the metaphor of
civilization itself as a hieroglyph.

In "Man in the Autumn Light," Eiseley exemplifies the metaphor
of civilization as a hieroglyph with the Mayan civilization, whose
"remaining hieroglyphs tell us little" ("Man in the Autumn Light,"
IP, 129). In a curious doubling, for Eiseley often refers to the Sanskrit *maya*—which indicates the illusion involved in perceiving
"shapes and structures, things and events around us" as realities
rather than mental constructs (Capra 1977, 78)—he notes that the
Maya Indians could calculate time with an accuracy unheard of in
Europe, and their mathematical accomplishments are the hieroglyphs of their culture. Nevertheless, the hieroglyph points to the
death of the culture as Eiseley notes that their descendants worshipped the upside-down stone tablets of their calculations. Looking for a hieroglyph for modern civilization, he turns to a snow-
covered clearing of fallen trees and abandoned machinery, which

he finds the "pyramid" that modern culture creates, a symbol of "energy beyond anything the world of man had previously known" ("Man in the Autumn Light," *IP*, 133). Clearly linked with the death of nature, which projects the death of culture, the "pyramid" is separated from voice and presence and life. Because the hieroglyph is associated with the pyramids of ancient Egypt—tombs of human beings that have come to emblematize the death of a civilization—modern humans, rather than being the ultimate products of evolution, become the harbingers of decay for a civilization, the ultimate evidence of change and indeterminacy. The snow recalls the recurrent the ice ages, and the energy suggests the human striving that is now out of hand. Fire and ice, death and the striving that characterizes life, are juxtaposed through the search for a "hieroglyphic" figure of twentieth-century culture.

Connoting ambiguity and indeterminacy, phonetic and nonphonetic meaning, life and death, the pyramid and the hieroglyph symbolize both human cultural achievement and inevitable cultural decay. As symbols of the unexpected and unfinished unfolding of ambiguities requiring individual interaction, the hieroglyphs require human probing. And it is the probing that is worth doing, for humankind is by nature "oracular," Eiseley repeatedly contends, seeking messages. To seek these messages through a close relationship with the nature that produced us is, in a sense, to depart from dispassionate science and to engage in dialogue with a primitive mode of observation.

The Eye of the Poet and the Eye of Science: "The Spore Bearers," "Man in the Autumn Light," and "The Last Magician"

In contrast to the world that the "time effacers" have created, Eiseley explores another realm that draws on the kind of imagination that created the "dreamtime." In *The Invisible Pyramid* this realm is effectively created in terms of the visual philosopheme, through the "two eyes" of humankind—the eye of the scientist and the eye of the poet. In "The Spore Bearers," Eiseley introduces two essential eyes of science, the telescope at Mount Palomar and the electron microscope. Together the macro and micro view "are eyes of understanding" that "balance and steady each other" and "give our world perspective" (88). Yet these are technological eyes and thus "subject to their human makers," who make and use such machines and still may have "no true time sense, no tolerance, no genuine awareness of their own history" ("The Spore Bearers," *IP*,

88). What he calls "the balanced eye, the rare true eye of understanding" can encompass history; it can also perceive the link between the "invisible pyramid" of the twentieth century and death,
for this eye can "sense with uncanny accuracy the subtle moment
when a civilization in all its panoply of power turns deathward"
(88).

The "balanced eye" reappears in "Man in the Autumn Light,"
in which Eiseley describes a Brazilian fish with a kind of bifocal
eye that enables it to survey the world of light and air as well as
its own environs. The fish provides a vehicle for exploring the
visual philosopheme, with a focus on the dual vision that Eiseley
contends is necessary for the human. He links the fish's dual vision
to the poetic view as he turns to a comment by Thomas Love
Peacock that "'a poet in our times is a semi-barbarian in a civilized
community. . . . The march of his intellect is like that of a crab,
backward'" (cited in "Man in the Autumn Light," *IP,* 123). Human
culture itself suggests a dual existence, a world superimposed on
the natural and often mistaken for the natural; for "seeing" culture,
Eiseley suggests, we need both the bifocal eye of the fish and the
dual eye of the poet.

Peacock's presumably ironic comment suggests a nineteenth-
and twentieth-century feeling of superiority over "barbarians" and
an assumption that modern, rational, scientific "man" is the end,
the closure, of evolution. Eiseley, identifying with the primitive,
turns the image back on itself, treating the crab's vision as a metaphor for the poet's fortuitous "visual" development. The poet is,
he writes, a "fortunate creation," "born wary and . . . frequently
in retreat because he is a protector of the human spirit" ("Man in
the Autumn Light," *IP,* 124). Poets are marginal figures whom he
imagines "lurking about the edge of all our activities, testing with
a probing eye, if not claw, our thoughts as well as our machines"
(124). The poet thus emphasizes the intellectual groping and probing link of eye and hand that may be ignored by the masses, but
is necessary for a balanced view. Combining this metaphor with
Blake's "double vision of poets," Eiseley argues that "there is no
substitute, in a future-oriented society, for eyes on stalks, or the
ability to move suddenly at right angles from some dimly imminent
catastrophe" (124). This is the kind of "eye" that will enable human
survival—an eye able to see from an odd angle, able to warn
against danger "dimly" perceived. That the poet, crab-like, is able
to see before his contemporaries intertextually evokes Shelley's
notion of poets as "the unacknowledged legislators of the world"
(1971, 513). The metaphor suggests an epistemological exploration

of an "other" way of knowing that is not objective, but mediated with human intuition. In contrast to the space-flight specialists who are heroes today, the poet is a "word-flight specialist," a "spore bearer of thought" whose spores are thoughts and whose flights encompass time and extend beyond the flights of rockets. They are more sensitive than the molecular biologist's electron microscope "because they touch life itself and not its particulate structure" ("Man in the Autumn Light," *IP,* 125). Poets preserve the human being's "venerable, word-loving trait" that enables us to pass on our "eternal hunger," our "yearning for the country of the unchanging autumn light" (125). Only the poet—in this essay personified by a time voyager from a science-fiction story, who, confronted by "deathless machines" (126), programs the machines to produce a machine that would once again incorporate "curiosity and hope"—can foresee that humankind "in the end forgets the message that started him upon his journey" (127).

The need for the "balanced eye" combines with Eiseley's concluding explorations of language in "The Last Magician," which draws together the two "eyes" of science and textuality, the dual "autumnal light" of the "dreamtime" in the previous essay and the autumn of civilization as the time of the exploding spores of the human species as it reaches for space, and the trajectory of the comet, which symbolizes the individual's "tracking" of self or "rounding" toward its natural "home." The opening anecdote is his own recollection of an encounter with his own "last magician." The end to youth, Eiseley posits, is the individual's final confrontation with the person who shaped him. In terms of humankind, the confrontation is with the human "collective brain" ("Man in the Autumn Light," *IP,* 137), which can bring the last magic or the last disaster. His own encounter came with a case of mistaken identity in a train station and his own realization that he had separated himself from "the forest." Similarly, humankind, at the edge of exploring space, has encountered itself in the form of a "looming shadow" that "has pointed backward into the entangled gloom of a forest" (139). Humankind, Eiseley insists, "must learn that, whatever his powers as a magician, he lies under the spell of a greater and a green enchantment which, try as he will, he can never avoid, however far he travels. The spell has been laid on him since the beginning of time—the spell of the natural world from which he sprang" (140).

Plato, Eiseley recalls, suggested that "the mind's eye may be bewildered in two ways, either from advancing suddenly into the light of higher things or descending once more from the light into the shadows" ("The Last Magician," *IP,* 140). The example that he invokes from his own experiences draws on the textual metaphor that he explored in *The Unexpected Universe* and the parallel of textuality and indeterminacy. In addition, Eiseley introduces the notion of civilizations as hieroglyphs that he will explore further in *The Night Country.* Having introduced the "dual vision" of the poet, Eiseley turns again to the primitive notion of "reading" nature, which he has linked to the poetic insight. Recalling a journey of several days on an isolated coast, he writes that he began to see faces looking up from shells. Without human contact, he began "to read again, to read like an illiterate" (141). And we recall his speculation in *Darwin's Century* that "reading" of nature's forms may actually have provided the model for writing itself. His reading on this coast bypassed what Derrida would call the breath of presence, for it "had nothing to do with sound" (141). (Technically, as we read we bypass sound.) Such reading, Eiseley writes, is a mark of the human, an ability "already beyond the threshold of the animal" (141). Through such reading the primitive "had both magnified and contracted his person in a way verging on the uncanny" (141). In the primitive cortex, Eiseley posits a kind of prelinguistic Lacanian visual realm, a proto-Imaginary,[2] which seems to have preceded the apparently swift emergence of language and to have given "biological preparation for its emergence" (142). The primitive human, like primitives today, still lived "in close interdependence with his first world, though already he had developed a philosophy, a kind of oracular 'reading' of its nature" (143). And the primitive saw animals as friends and counselors, remaining inside the natural world, not having "turned it into an instrument or a mere source of materials" (143).

Defining the linguistic term *displacement,* the human ability to verbalize the absent as if it were present, Eiseley refers to it as the means by which humans have survived in nature and the means by which they created the "second world," the world of culture ("The Last Magician," *IP,* 144). Displacement enables human beings "to manipulate time into past and future, transpose objects or abstract ideas in a similar fashion, and make a kind of reality which is not present, or which exists only as potential in the real world" (145). For Eiseley, unity with the natural world and "reading" semiotically are linked, whereas separation from the natural world is linked with semantic reading. As Derrida suggests, "If writing is no longer understood in the narrow sense of linear and phonetic notation, . . . no reality or concept would therefore correspond to the expression 'society without writing'" (1976, 109).

Eiseley also takes the linguistic term *displacement* in a symbolic direction as he asserts that language itself exemplifies the human displacement from and attempt to control the natural world. Language partakes of the dialogue of structure and the flexibility that makes possible evolution itself. Yet language also gives another, more positive ability, the desire to transcend both self and the present. It is this quality that the "eye" of technology can be made to deny. Eiseley exemplifies this "eye" in a reference to a scientist, "a remorseless experimenter," whose face had become distorted from peering into the microscope (144). His image suggests "one bulging eye, the technological scientific eye," which had come to consider both humans and other creatures "in terms of mega-deaths" ("The Last Magician," *IP,* 144). This science, Eiseley argues, has cultivated an "objectivity" that has "become so great as to endanger its master" (144). The same language that makes possible the "bulging eye" of technology also makes possible the "eye" of understanding that emerged powerfully during the time of the great axial religions as manifested by "an intellectual transformation," "a time of questioning," "a turning toward some inner light" (146–47). Plato's notion of emergence into a light that blinds is the vision that Eiseley proposes for this second "eye."

The personal and cultural experience of this "other" understanding is exemplified in Eiseley's own recollection of a sunflower forest of his childhood. If the world itself is "a space ship of limited dimensions" (152) whose journey, like that of the comet, began in prehistory, we now have both an opportunity and a responsibility. Like the comet, we are "both bound and free" ("The Last Magician," *IP,* 155–56). We must now become our own "magician," our own poet (from the Greek *poetas,* "maker"). Like Heidegger (as noted earlier), Eiseley takes the stance that reflection—the realm of both art and *techne* in its original sense—is the power that can save us from the danger of technology, the danger that we will be caught in the Enframing of technology that blinds us to an *other* way of seeing. We can become our own "magician," in Eiseley's view, through the self-reflective abilities of humankind, and these reflective abilities include the self-reflexivity of science. Although we seem driven to pursue space, a "paradox remains"—the paradox that the journey into space has also forced us "to turn and contemplate with renewed intensity the world of the sunflower forest—the ancient world of the body that . . . [we are] doomed to inhabit, the body that completes [our] cosmic prison" (151). Only an incorporation of the sunflower forest of personal and collective youth with the technological vision that has made modern science both bane and blessing will enable us to survive.

9

The Self as a Vehicle for Understanding I: *The Night Country*

How much more we would see, I sometimes think, if the world
were lit solely by lightning flashes from the Elizabethan stage.
What miraculous insights and perceptions might our senses be
trained to receive amidst the alternate crash of thunder and the
hurtling force that give a peculiar and momentary shine to an
old tree on a wet night. Our world might be transformed interi-
orly from its staid arrangement of laws and uniformity of expression into one where the unexpected and blinding
illumination constituted our faith in reality.
—Eiseley, "Strangeness in the Proportion"

The Night Country includes a number of highly personal essays
and marks a further exploration of the personal tale and a greater
departure from the consideration of science, except from the perspective of a life spent in science, than any of the earlier texts. At
first assumed by some to be his autobiography, *The Night Country*
retains the concern with science, but Eiseley's focus increasingly
is on exploring the epistemological possibilities of what he calls
"strangeness." The "night country" is the marginal country between reality and imagination, between today's shifting reality and
the future we create. In the foreword, Eiseley links the two major
themes of wilderness/isolation and running as he writes of the
"many times and places in the wilderness of a single life," a wilderness in which "nights of outer cold and inner darkness" play a
major role, and of the book as "the annals of a long and uncom-
pleted running" (*NC*, xi). The essays in this collection are marked
by Eiseley's personal fascination with the night, a fascination expressed earlier in *The Invisible Pyramid* when he writes that the
human being "has read his way into the future by firelight and by
moonlight," for, in the time of the early humans, "night was the
time for thinking, and for the observation of the stars" ("The Last
Magician," *IP*, 140)—for thought, for observation, and for imagina-

tion. Eiseley continues that he likes to think of the "crossing" into the "second realm," the linguistic realm, as "truly a magical experience" (141). The "running," which recurs throughout his texts, suggests both flight and pursuit; in the autobiography, he will describe himself as "the running man," though he also establishes himself as "every man and no man" (*ASH*, 23). But, as in the earlier texts, this flight/pursuit also incorporates another "crossing"—of the boundaries of form between human and animal, of day and night worlds (the traditional Western day of logic and night of mystery and fear), of the worlds of science and of imagination, and of the human crossing into language and thus into culture, with writing the vehicle of the crossing.

The Night Country also reflects the major themata that have controlled Eiseley's earlier texts: the vastness of time and the human smallness in relation to time, the constant shifting of living forms, the need to create meaning, as primitives have always done, by interacting with nature, and the role of indeterminacy in human thought. Yet these essays are tinged with the melancholy of age and reminiscences of a lost childhood, of a man running (whether from or to something is not always clear, but the running is the human universal). The purpose of these essays is more literary than scientific, with speculation on human creativity a key element in the penultimate essay, "The Mind as Nature," and the author's persona is defined at the outset with an element of grim humor, as that of the disguised fugitive, of the fugitive who wears "the protective coloration of sedate citizenship," which "is a ruse of the fox" ("The Gold Wheel," *NC*, 4). In "The Mind as Nature," Eiseley quotes Thoreau's speculation on his own thinking (a speculation replete with Thoreau's own culturally embedded notion of mechanism and a vague reference to an Emersonian or Shelleyan "greater mind"): "'I catch myself philosophizing most abstractly when first returning to consciousness in the night or morning. I make the truest observations and distinctions then, when the will is yet wholly asleep and the mind works like a machine without friction. . . . As if in sleep our individual fell into the infinite mind, and at the moment of awakening we found ourselves on the confines of the latter'" (*NC*, 217–18). In a sense, the whole of this book is an exploration of this marginal realm between consciousness and unconsciousness, between images and language, between reality and the mental processes that create this reality.

Carnival

Some of the tales in *The Night Country* seem trivial, as Eiseley focuses on a beetle crossing a room, a rat posturing behind a noted speaker, or the "big eyes" of cattle in the night and the "small day-born eyes" of the human being. These tales seem more anecdotal and less related to reflections on science and society than most of his earlier essays. Yet these essays partake of the marginal country between reality and mental process, and they pursue the Eiseleyan *topos* of the human body as vehicle of knowledge and the relationship of that body to the world that produced it, as well as the need for reflection, for a return to the "sunflower forest" that he addressed in the last essay in *The Invisible Pyramid.* These simple personal tales or anecdotes of animals, however, also reflect what W. H. Auden describes in almost Bakhtinian terms as "Carnival" in Eiseley's texts, noting that Eiseley does not overtly describe the human being as the only animal that laughs. Auden defines Carnival in terms of "true laughter (belly laughter)" (1978, 21), tracing the notion to the Middle Ages, where its function is "celebrat[ing] the unity of our human race as mortal creatures, who come into this world and depart from it without our consent" (21) and who live with a sense of ambiguity about the human situation. Auden reminds us of Carnival's suspension of social distinctions, of its reveling in the oddities of the physical body and experience, and he concludes that "the Carnival solution of this ambiguity is to laugh, for laughter is simultaneously a protest and an acceptance" (22). In his own exploration of Carnival, Bakhtin—who approaches language from both cognitive and social perspectives and who looks to the possibilities of dialogue as simultaneously comprising differences (Clark and Holquist 1984, 9), as does Eiseley—views Carnival as a break, a rent in the social fabric, a threat to a dominant ideology (science itself is an ideology in Eiseley's terms) that attempts to present a unified, complete, and stable order (301). Carnival, instead, gives a chance to celebrate the unexpected, to revel in change and diversity. For Bakhtin, Carnival's affront to stasis and its celebration of change provide an expression of opposition to the official position, an expression of the "free laughing aspect of the world, with its unfinished and open character, with the joy of change and renewal" (301).

Though, like Auden, Eiseley could hardly be aligned with the raucous public carnival, several of Eiseley's essays reflect a laughing, reveling, wondering sense of the strange, marginal, sometimes frightening experiences that he associates with the "night country." "The Gold Wheel" begins with a reference to the "curious freedom" that one experiences in the no-man's land between deployed powers, a freedom that still exists "along every civilized road in the West," where civilization encroaches on what remains a vast wild life, a "one dimensional world" that only a "fugitive," "a refugee at heart, a wistful glancer over fences" can experience ("The Gold Wheel," *NC,* 3–4). The reference to a "one dimensional" world recurs in Eiseley's texts to describe a non-linguistic world that simply exists, but whose inhabitants are not conscious of it and do not abstractly discuss it. This marginal country between civilization and wilderness was the country of Eiseley's youth, which he defines both in terms of landscape—in the upland plains subject to tornadoes—and in terms of his own experience— "in the echoing loneliness of a house with no other children; in the silence of a deafened mother; in the child head growing strangely aware of itself as it prattled over immense and solitary games" (4). For such a child, noise became the enemy, the outside, the day; place and circumstance decreed he would explore night and quiet.

The Carnival-like balance comes in the tale of the isolated child's finding a gold wheel and fancifully deciding "to run away upon a pair of them" ("The Gold Wheel," *NC,* 7) by hiding in an old-fashioned tea wagon on its journey into the country to the bishop's house, where he dropped off and hid in the hedge, where he became aware, among lightning and thunder, of hundreds of still and silent birds. The tea wagon having vanished, he thought of the gold wheel that "seemed vaguely linked to my predicament" (10). The next thought of the gold wheel came when, riding in a car behind an antelope across a Western grassland, he saw "a loose golden wheel rolling . . . on the prairie grass" as the front wheels of the car hit the opposite bank of a five-foot gulch that the antelope had cleared with one bound. Although Eiseley writes, "I didn't say I had wished for it," he continues that "you had to be a fugitive" to sense the closeness of another world like that of the birds. He concludes that he wears "the protective coloring of men. It is a ruse of the fox—I learned it long ago" (12). The serious and the Carnival element mingle in the image of the car held by its bumper above the gulch while someone says, "He's gone," and Eiseley, stunned, asks, "The man with the tea wagon?" (12). But the point is that to be in touch with nature one has to be a fugitive from the "real" world of daylight and modern civilization.

"The Places Below" tells of the "Blue Room," a pool of blue water that sometimes, at least in the mind of an elderly prairie homeowner, comes up the steps of the house's cellar. As the obsession of an old man, the blue room is "a part of the places below," whether they are actual or psychological (*NC,* 18). The point, I submit, is that the places in this book are places of the mind— underground places tinged with the eerie quality of night, places that become a comment on the mind's creation of its own reality. The essay includes another childhood story—of a young boy known as the "Rat" and of the two boys' wandering in a storm sewer beneath the street, unknown to their parents, scratching "tribal symbols" on the sewer tiles as the Rat instructed until, one cloudless afternoon, they heard the rumble of water that told them they had to leave. After pushing hard against a manhole cover, they scrambled into the street where a fire hydrant was being tested, only to encounter Eiseley's father, who angrily muttered, "Undisciplined, completely undisciplined" (24). The Rat's unexpected death gives a somber balance to the childhood adventure, but provides the dialogue of dark and quiet with youthful adventure in the western plains.

"Big Eyes and Small Eyes" gives a comic, carnivalesque twist to the skull and the eye, which Eiseley treated with high seriousness in "The Star Thrower." From its opening quotation from Conrad Aiken—"'This is your house'" in reference to the human skull—Eiseley contrasts the "small day-born eyes" (NC, 32) of human beings with the large night eyes of most animals. The anecdotes range from a tale of a man awakened in the middle of the night by a large rat, to a dancing rat that Eiseley saw posturing (like the trickster of primitive cultures) behind a distinguished novelist as he proposed a toast to "man," who "'will turn the whole earth into a garden for his own enjoyment'" (34), to his own awaking from a nap beside a pond on a Western plain to feel the "cold yellow eye" of a blue heron that stood on one foot staring, "like an expert rifleman, down the end of a bill as deadly as an assassin's dagger" (38–39). He concludes an anecdote of his own skipping with a group of toads along a dark road by commenting that "the mind inside us is vaster than the world outside" (42). And the last anecdote concerns his own descent down a mountain side to a spot beside two large guard dogs, where he stayed "for long minutes talking and side-thumping and trying all the dog language I knew" (43). Remembering the moment, he writes that "as one demon to another, . . . I have helped a bat to escape from a university classroom, and I have never told on a frightened owl I once saw perched on the curtain rod above a Pullman berth" (44).

For a writer whose concerns about the intertwining of the human role in the natural world have been passionately serious in all the earlier texts, Eiseley's "bone-hunting" tales—whose tutor structure is the tale of the Old West—introduce a different, but nevertheless carnivalesque, note into The Night Country. "Barbed Wire and Brown Skulls" tells assorted tales of skulls—of the skull that

Uncle Tobias's relatives found among that upstanding lawyer's effects after his death and eagerly donated, anonymously, to Eiseley's collection ("They needn't have been so jittery—that skull had been hundreds of years underground when Uncle Tobias was born" [*NC*, 94]); of the "heads" that he and his grandmother baked many years earlier, modeled after the skulls that he saw at a museum in Lincoln, Nebraska, and abandoned lined up in Hagerty's stable when the family moved; and of Aunt Lucinda's skull, which evokes a real tale of the Old West. "He keeps her in the china closet" (98), said one of the grandsons who summoned Eiseley to convince eighty-year-old Mr. Harney to bury the skull or give it to a museum. Alone in the valley after the long trip from Texas, after his mother died and his father failed to come home, the young boy and Aunt Lucinda were attacked in the morning by the Indians who presumably had killed the father. "One of 'em just picked her off out of the mesquite," the old man says (99). The boy was captured and lived as an Apache until, at age fifteen, his chief sent him back to the white man. Harney went back to the cabin where he found her skull grown over by a prickly pear. Although he settled there, he could not bury the skull of his beautiful Aunt Lucinda in that place; he kept her "safe in the china closet," where sometimes he talked to her (102). Eiseley writes that he would not take the skull because he knew too much about it. Yet as his own collection grew he came to understand the predicament of Uncle Tobias and old Mr. Harney. Despite its grisly subject, a union of life and death, a macabre series of carnivalesque reminiscences, "Barbed Wire and Brown Skulls" draws on carnival's defiance of the human predicament through laughter and extends the tale of the Old West into the living room of the survivor.

Like the other carnivalesque essays, "The Relic Men" draws on personal anecdote and underscores Eiseleyan themes, in this case the imagination of the isolated individual and the creation of order through myth. It is framed with comments from reporters who have come to photograph the archaeologist and an enormous bone out of the Pleistocene. Clearly unimpressed with the rarity of the specimen, the reporters look for "human interest." Then Eiseley recalls the lonesome, change-bringing wind in that country and the old man who claimed to have found a "petrified woman," to which he led the party of bone-hunters. An old bachelor from the East who had settled and grown old alone on the prairie in the early part of this century, Buzby was "a city man dropped, like a seed, by the wind," complete with pince-nez glasses (*NC*, 111). In Buzby's projection of a human form and a history onto what was only a concretion, Eiseley senses "the pathos of a man clinging to order in a world where the wind changed the landscape before morning, and not even a dog could help you contain the loneliness of your days" (112). Seeing the old man's desperation, and eager to get a bison skull that he had hung above his doorway, they agreed to take the concretion, which they dumped into a canyon two days later.

Then there was Old Mullens, a staunch fundamentalist who, red-faced and panting, led the bone hunters over nameless hills to what he called "'a place where all the garbage from the Flood—them big animals and pore human sinners—all got carried along and dropped when the water went back where it came from'" ("The Relic Men," *NC*, 120). Old Mullens led them to a fantastic paleontological find, probably the shallows of a Pliocene lake, but he did not live to see them collected and catalogued. "Old Mullens," Eiseley concludes, "had lived in a small, tight world of marvels, and they had lasted him till the end. . . . Perhaps, in the end, we did not know where we belonged" (123).

"Obituary of a Bone Hunter" tells three tales of Eiseley as a "little bone hunter" who muffed three chances at finding some remnants of ice-age human beings in America. The first chance came in a cave and was foiled by "millions upon millions of daddy-long-legs, packed in until they hung in layers" (*NC*, 184)—innocent spiders that became terrifying as they plopped to the floor of the cave. The second came in a cave with a roof blackened by fire, where he startled an enormous owl, only to find an enormous egg lying on a pile of sticks where the ancient artifacts might have been. And the third was an old man from Missouri who came to Eiseley offering a human jaw in exchange for a statement to the media that the relic represented "the Miocene period—the Golden Age," when "a great civilization existed . . . , far more splendid than this—degenerate time" (189). In each of the three cases there was enough doubt to overcome the desire for a "find." "The egg and the spiders and the madman—in them," Eiseley concludes, "is the obituary of a life dedicated to the folly of doubt, the life of a small bone hunter" (191).

Macbeth *as Tutor Text: "Instruments of Darkness"*

The encounter with the self leads to awareness of the individual's role in creating the future. *The Night Country,* coming after *The Invisible Pyramid* and exploring the darkly imaginative possibilities of the individual human mind, approaches the "magic" of creating the future as a threatening magic. Shakespeare's *Macbeth* is the vehicle for Eiseley's examination of the self-fulfilling prophecy in "Instruments of Darkness." The essay begins with a reading of Macbeth's magical night world of witchcraft, a world in which Macbeth "gains a dubious insight into the unfolding future—a future which we know to be self-created" (*NC*, 47). The power of this scene today Eiseley finds in "its symbolic delineation of the relationship of Macbeth's midnight world to the realm of modern science" (47) in which science and the scientist come to be seen as prophets and employed in the military-industrial complex. Eiseley pursues the "magical" ability of the cultural realm to control the physical by suggesting that the witches are only "an exteriorized portion of ourselves," "smoking wisps of mental vapor that proclaim our subconscious intentions and bolster them with Delphic utterances—half-truths which we consciously accept, and which then take power over us," so that "under the spell of such oracles we create, not a necessary or real future, but a counterfeit drawn from within ourselves, which we then superimpose, through purely human power, upon reality" (48). These exteriorized phantoms lead Macbeth to create the future; similarly, Eiseley argues, every age has its witches who claim omniscience.

Today the phantoms are deterministically oriented scientists often associated with the military-industrial complex, and they "seek control of man's destiny by the evocation of his past" ("Instruments of Darkness," *NC*, 49). These phantoms, claiming to know the future, evoke it. Such "phantoms in military garb" not only evoke the warring elements of evolution, but are all the more misleading because they are associated with both science and authority—"because their spells are woven out of a genuine portion of reality" (49). Ironically, they foretell a future whose "leading characteristic . . . is its fixed, static, inflexible quality" (49). The witchcraft is that they superimpose fixity on what is actually never fixed. Whereas human history includes "transcendence and self-examination," the image of the future that these phantoms project is one of "evoking, rather than exorcising, the stalking ghosts of the past," the ghosts of war and struggle (49–50). These phantoms evoke the human image of bestiality, build on it, and draw a future out of it, rather than leading human beings to examine themselves and to transcend brutality (49–50). For Eiseley, the witchcraft of the modern age is humankind's fulfilling of a violent image of itself. At any point in its evolutionary journey, humankind could have paused and said, "This is man," and could have stopped reaching for more (54). We contain the future, Eiseley contends; we have the capacity to be our own magicians or our own demons.

As an illustration of the organization, the mystery, of life itself and the impossibility of reductionism to explain it, Eiseley uses a fossilized stone bearing the outline of a fish. The fish is extinct, no longer "returnable," but most of the elements that composed it remain. But "the greatest mystery of all" is the personal confrontation with continual change, the tension between the stasis that we desire and the change that we cannot control: "I who write these words on paper," Eiseley adds in a comment on this tension, "cannot establish my own reality" ("Instruments of Darkness," *NC*, 51). What he calls the "entrapment" of modern thinkers in the past is what Emerson called "'man crystallized'" (53). This attitude is partly traceable, he argues, to those evolutionary thinkers who have looked to a real but nevertheless fictionalized past, who have "misread" the past yet "would now venture to produce our own future" (53). Like *Macbeth*'s witches, such thinkers present to us apparent realities—particularly emphasizing the human as a product of warfare—which "take shape in our minds and become the future" (53).

In contrast to the evocation of a future predicated upon a deterministic view of the past as producing a kind of inevitable future, Eiseley argues that "nature . . . is never quite where we see it. It is a becoming as well as a passing, but the becoming is both within and without our power" ("Instruments of Darkness," *NC*, 54). At any point in evolution, someone might have defined "man" with "a magical self-delineating and mind-freezing word," and the magic of self-definition (like the magical spell of Circe to which Eiseley refers in "The Ghost Continent") "could have immobilized us at any step of our journey" (54). His point is that were are never who we think we are—never a defined or definable creature, never determinate products but products of the unexpected, probabilistic universe, yet capable of choosing how we will act. Shakespeare's tutor text pinpoints what Eiseley calls the "terror" of our own age, "our own conception of ourselves," for which he invokes Shakespeare's lines, "'It hath been taught us from the primal state / That he which is was wished until he were'" (cited in p. 55). Self-description as "man," which includes at times bestiality, at others mastery, is for Eiseley our downfall. This is one of those "lightning flashes" from Shakespeare's stage—itself one of Eiseley's figures for the interactive text—that can illuminate the strange human potential that partakes of the "night country" but can also reach beyond itself to be more than it is. Joining metaphors of witchcraft,

magic, and the theater, Eiseley thus displaces reductionist notions
of understanding the human and natural worlds, as well as images
of a fixed, determined future for humanity. Magic, like life, inserts
"indeterminism" into matter, though the choice of magic is ours to
make; and the small choice, as chaos theory tells us, may have
unexpected results.

The Open-Ended Universe: "The Chresmologue"

As in "Instruments of Darkness," the marginalized territory between present reality and future potentiality is the focus of "The
Chresmologue," whose title refers to the ancient Greek "dealers
in crumbling parchment and uncertain prophecy" (NC, 62), with
whose voice Eiseley speaks. After quoting the unknown medieval
author of the Cloud of Unknowing, he argues that the future "is
contained within ourselves" and "drawn from ourselves" (73).
Linking science to the Christian notion of "progress," Eiseley continues the dramatic metaphor, arguing that, though "the great play
has lengthened and become subject to the mysterious contingencies . . . of genetics" ("The Chresmologue," NC, 85), both the
drama and the text are open ended, echoing what he calls "an
open-ended universe" (63).

Through a tutor text, Eiseley continues his undermining of the
embedded and hypostatized dramatic metaphor that has become
associated with a notion of "progress" that is actually static and
stresses the human ability to break from the rigid confines of determinism. The tutor text is a memorable fable of a pair of bettors
and a mysterious bet. Both the game of chance and textuality are
suggested by the "dealers in crumbling parchment and uncertain
prophecy" ("The Chresmologue," NC, 62). Chance and the future
blend allegorically with the journey as Eiseley notes that words of
the chresmologue "may also be spoken upon journeys, for it is
then that humankind in the role of the stranger must constantly
confront reality and decide his pathway" (62). The legend concerns
two English gentlemen riding along a wild English moor near the
coast and encountering a runaway coach that raced by while its
frightened occupants screamed. According to the legend, the two
gentlemen stopped their horses, then galloped to pass the coach
and open the gate while they made a bet that the coach would
lurch over the cliff. After the coach dashed through the gate, no
sound was heard, and the bettors were left wondering about the
outcome, their certainty of knowledge shaken and undermined.

Today, writes Eiseley, outlining the dualism of the allegory, humankind is "both the sporting gentlemen intent on their wager and the terrified occupants of the coach. . . . We are literally enduring a future that has not yet culminated, that has perhaps been hovering in the air since man arose" ("The Chresmologue," *NC*, 61). We are both the bettors and the occupants of the coach, and among us chresmologues, keepers of the dead letters of the parchment, soothsayers, multiply; scientists now are regarded as soothsayers (64). The soothsayer deals not only in dead letters, but in "uncertain prophecy" of the sort suggested by the tale of the bettors in their disregard of the fate of humanity as represented by the passing carriage. The future, however, remains the domain of contingency, as unknown and as surprising as the hush that followed the sporting gentlemen's bet. Though the future may be "hovering," as unknown as the coach's end, Eiseley asserts that we have the power to modify it, though the results of our modifications remain unforeseeable. His answer to the popular penchant for scientific "chresmologues" is another affirmation of humane understanding, for if our concern for the future is totally dependent on "the clever vehicles of science," then that future will be one of "intellectual impoverishment and opportunism" (74). In the absence of true chresmologues, we are left with the unsettling affirmation of epistemological uncertainty represented by the bettors' quandary; if there is an answer, it will come in humane concern for life itself, rather than in the use of science for expansion of governments or struggles for power.

Another ongoing journey underscores the fable as Eiseley tells of a derelict on a train going west from New York. When the conductor asked for his ticket, the man rasped, "Give me . . . a ticket to wherever it is" ("The Chresmologue," *NC*, 63). In that one line, Eiseley writes, the man "personalized the terror of an open-ended universe" (63). After tracing the evolution of the modern concept of time from its Christian origins through the keeping of the sense of time as new and incorporating the notion of "progress" into science, Eiseley ends with the change in conceptions of the human drama, which has acquired "force, direction, and significance beyond the purely episodic" and has led to "intensified" self-examination and the human desire for a single "eternity of which he might be the intellectual master" (67).

Eiseley tells of carrying seeds to mountaintops in the hope of tampering with the world in a small way. The human mind, he analogizes, repeats what is biologically true—the response to the environment, the "defense against satiety" ("The Chresmologue,"

NC, 72). Human thought, using the experimental method, has become "outwardly projected upon time and space" until, without conscious reflection and "inward perception," it may "lose itself, unexamined, in vast distances" (73). The "endurable future," he concludes, cannot come through science alone, but through compassion; like the wagon in the fable, our civilization rushes into the darkness, while we who are both the sporting gentlemen and the inhabitants of the coach await "the crash we have engendered" (74). Like the English bettors, we are left with an unanswered wager—an open-ended universe.

The Archaeological Eye and the Hieroglyph: "Paw Marks and Buried Towns" and "The Creature from the Marsh"

The hieroglyph reappears in *The Night Country* to suggest the ambiguous textuality of nature itself, of civilizations, and of the physical history of humankind. In ruined civilizations, Eiseley sees "marks" of human beings "trying to be human, trying to transcend themselves" in the presence of the vastness of time ("Paw Marks and Buried Towns," *NC*, 80). Like the dog Mickey, whose paw marks in concrete seemed indicative of his desire to be human, none has succeeded, but each civilization leaves "a figurative paw mark—the Shang bronzes, the dreaming stone faces on Easter Island, the Parthenon, the Sphinx, or perhaps only rusted stilettos, chain mail, or a dolmen on some sea-pounded headland" (80). What Eiseley calls "the archaeological eye" is an eye that sees time and change in mundane objects. It is an eye that can "never see quite normally" (81). Such an "eye" is the recompense for studying time and civilizations, for developing a sensitivity to "the melancholy secret of the artifact, the humanly touched thing." (81)

The reader of civilizations as hieroglyphs learns quickly that whatever destroyed past cultures was not ignorance as we think of the ignorance of primitives, for in all the decaying cities there is evidence of "the clever artisan, the engineer devoted to the service of the particular human dream that flourished there" ("Paw Marks," *NC*, 82). Amidst crumbling columns and waterless baths, "still the bold Roman letters" remain over the entrance to a Roman theater, "naming, amid the surrounding ruin, one Annobal, the donor," who has left his "paw mark" and departed (84). What speaks to the "archaeological eye," Eiseley contends in an insight intertextual with Keats's scene in the "Ode on a Grecian Urn," "is the immediacy, across the waste of centuries . . . , the empty seats vacated by a crowd that has just left but is not coming back" (84).

Modern humankind has "turned more stones, listened to more buried voices," than any previous culture, and one would hope that the experience would lead to "a kind of pity that comes with time, when one grows truly conscious and looks behind as well as forward" ("Paw Marks," *NC,* 85). Yet Eiseley finds, beside the sense of pity that knowledge of time brings, a continuing strain of savagery. He closes "Paw Marks" with an anecdote of sitting among fragments of Greek statuary in an enclosed garden at a major university. In contrast to the images as frozen in time as the figures on Keats's Grecian urn, there appear a group of children whose leader, carrying a bow and reminding Eiseley of his own undisciplined childhood, might have represented the savagery of humankind at any stage of its evolution. If, as humans, we create and possess beauty, Eiseley's message to a world passing into Postmodernism is that possession itself is dissolution—on the level of the individual and of the culture. The final reading from his "archaeological eye" is one that emphasizes human interactivity with nature—that no civilization has successfully "sustained beauty without returning it to the earth" (87).

In "The Creature from the Marsh," Eiseley perceives, in the midst of a great city, the ruins of tomorrow, as in "the dead cities of Mexico—the long centuries wavering past with the furious distortion of things seen through deep sea water . . . , past and future being equally resolvable in the curious perspective of the archaeological eye" (*NC,* 154). In the anecdote he looks in the city for "a symbol, something that would stand for us when the time came, something that might be proud after there was no stone upon another" (154). And in a sequence that blends the present city with the impending fragmentation that the archaeological eye perceives, he looks into a shop window, where he sees his symbol:

> And among the bits of glass, a little cluster of feathers, and under a shattered pane, the delicate bones of a woman's hand that, dying, had reached wistfully out, caught there, when the time came.
> "Why not?" I mused. The human hand, the hand is the story. I touched one of the long, graceful bones. It had come the evolutionary way up from far eons and watery abysses only to perish here. (155)

Returning the reader to Eiseley's argument that we as humans need to reflect on the human body and its role in both understanding and participating in nature, the anecdote brings him a sudden awareness of "the terrible *deja vu* of the archaeologist, the memory that scans before and after" (156), as he realizes that the image he has seen is actually his wife's hand, pointing out an object in a shop window.

The essay ends with another anecdote that emphasizes the need to read the physical body as Eiseley tells of discovering that the footprints of the "creature" that he has been tracking in a marsh are his own. Thoreau's notion that we track ourselves reverberates in this penetration into the hieroglyphs of human beings. The "archaeological eye" shows him that "the past of a living species is without memory except as that past has written its physical record in vestigial organs like the appendix or a certain pattern on our molar teeth" ("The Creature from the Marsh," *NC*, 161). The human story, he contends, is not all in the past, for the human body indicates that change is incomplete, and the human mind's dual tendencies indicate that the human being has now only begun to tap its potential, though "it was not by turning back toward the marsh out of which we had come that the truly human kingdom was to be possessed and entered" (165–66).

Finally, Eiseley explores humankind as a palimpsest or a combination of hieroglyphs. In "The Chresmologue," he assumes the title role of a dealer in old parchments and reiterates the uncertainty and ambiguity of the "documents" he studies, for he finds the human being "an indecipherable palimpsest, a walking document initialed and obscured by the scrawled testimony of a hundred ages. Across his features and written into the very texture of his bones are the half-effaced signatures of what he has been, of what he is, or of what he may become" (*NC*, 59–60). As a dealer in such palimpsests, Eiseley emphasizes the presence/absence of man, the presence/absence of what the future may be, and he recalls, as a hieroglyph to be interpreted, the old man on a train buying "a ticket to wherever it is" (63). The palimpsest suggests both the great amount of time that has brought us to where we are and the outlines of previous forms retained in the physical body. For the figure on the train neither time nor destination had meaning; like humankind, he was engaged in a journey whose destination was unknown, and time was only a puzzle. What is important is that he had ceased to ponder the puzzle or his own relationship to it.

The marks of time and the vestiges of the past that the human body carries constitute a form of indeterminate textuality. Such textuality, Eiseley notes, "led Hudson to glimpse eternity in some old men's faces at Land's End" ("Strangeness in the Proportion," *NC*, 148). Such a symbol is the concluding figure of "Strangeness in the Proportion," a man with a dual face, balancing precariously atop a swaying hayrick in the midst of an early evening thunderstorm. The figure physically symbolizes the dual mental nature of humankind, the inner tension that we have not yet resolved, the paradox of death and creation, of destruction and dreaming, the ambiguity of humankind as an unfinished text. We are still writing the text and thus can affect its meaning.

The Dual Visage: "Strangeness in the Proportion"

"Strangeness in the Proportion," originally included in *Francis Bacon and the Modern Dilemma* (1962), is not only a reading of the human hieroglyph, but one of Eiseley's most vivid examinations of the limits of humanism and the limits of science. Opening with Bacon's comment that he was "a stranger in the course of [his] pilgrimage" (*NC,* 127), the essay treats the sense of isolation of any thinking individual and the uniqueness that any individual represents; the "two streams of evolution" (129), biological and cultural, that meet in the human being; the "biform nature" of nature itself (135); and the limitations of a single perspective in the quest for understanding. The essay is an analytical rather than an anecdotal treatment (though it ends with a powerful anecdote) of what I have earlier called Eiseley's "re-nunciation" of his "scientific heritage," drawing on texts from Thoreau, Robert Louis Stevenson, and Bacon, with references to Shakespeare, Van Gogh, and Dostoyevsky.

From the initial anecdote based on Walter de la Mare's story of a traveler who, in a country cemetery, meets a stranger who asks, "Which is the way?" ("Strangeness in the Proportion," *NC,* 128), Eiseley pursues the reader's similar question. For humankind, he argues once again, is like a changeling who has been removed from the safety of nature into a world of "indefinite departure" in which our destination must be partly of our own making. "Far more than the double evolutionary creatures seen floundering on makeshift flippers from one medium to another," Eiseley contends, humankind is "marred, transitory, and imperfect" (129)—dual in nature and uncertain how to deal with this duality. Confronted with the human duality, science has only "exterior and mechanical" powers (129). And the human situation is underscored by human alienation, human inability to compare experiences with other life forms. The scientific project is fraught with contingency, with the ability to move the journey into a new and henceforth unchangeable direction, and again Eiseley asserts that we ascribe to this creation of our own minds "a role of omnipotence not inherent in the invention itself" (130). Attacking the common notion of science as fixed and absolute, he traces the shift from the seventeenth-century view of science and technology as a source for the Baconian New Atlantis—science as a servant of humanity—to the twentieth-century emphasis on power (130–40). Undermining the notion of science as a stable institution, at least after the discovery of that "ill-defined" entity known as the scientific method, he concludes that "there are styles in science just as in other institutions" (139). Although it is a triumph of the individual, science comes to value conformity and to produce "the curious and unappetizing puritanism which attaches itself all too readily to those who, without grace or humor, have found their salvation in 'facts'" (142).

As a counter to the conformity that science seems to breed, Eiseley analogizes from science itself to interpose the element of individual creativity, individual "strangeness." His argument, drawing on theatrical imagery and particle physics, is one of his most vivid expressions of the dialogue of understandings. Drawing on the uncertainty principle of particle physics, he speculates that "something still wild and unpredictable" may underlie the apparently orderly natural world ("Strangeness in the Proportion," *NC*, 136). "It is my contention," he continues, "that this is true, and that the rare freedom of the particle to do what most particles never do is duplicated in the solitary universe of the human mind" (136). This rare freedom, exhibited particularly in the mind of the artist, is what Eiseley proposes as the antidote to conformist science, an antidote in which all can participate through interactive communication (136). *The Tempest,* Dostoyevsky's "midnight world" (137) and Van Gogh's landscapes are what he calls "particle episodes in the human universe . . . , part of that unpredictable newness which keeps the universe from being fully explored" (138). The artist's role is to perpetuate and defend the human flight from uniformity, "just as it is the part of science to seek to impose laws, regularities and certainties" (141). We exist, he contends, in a marginal realm—"half in and half out of nature"—with the mind reaching beyond tools and laws even as genetic indeterminacy seems to reflect subatomic indeterminacy, so that the unexpectedness of the particle plays a role in human variability and the unexpectedness of creativity (141, 138).

Turning again to the writers who have shaped his own sensitivity, Eiseley argues that the parson naturalists such as Gilbert White and Richard Jeffries represented the best of both science and art.

Such writers were able to open human minds and to add to the natural world a new dimension, "something that lies beyond the careful analyses of professional biology" ("Strangeness in the Proportion," *NC*, 143). Drawing on Emerson, he notes that nature has nevertheless been "cast . . . in our image" (143)—a reference that sets up the last essay in the book, "The Mind as Nature." If science has given us the image of nature as a tool, it is the mind of a poet like Thoreau—himself increasingly a scientist—that counsels understanding of nature through both science and "the humane tradition" (146). Yet the limits of humanism are not to be ignored, for the worlds we create from nature bear our own imperfections. What Eiseley calls for "in a conformist age," is a science that would "be wary of its own authority" and an "individual . . . re-created in the light of a revivified humanism which sets the value of man the unique against that vast and ominous shadow of man the composite, the predictable, which is the delight of the machine" (136–37). This "revivified humanism" is a humanism that would recognize and employ the notion of human potential but would nevertheless displace the human being from his assumed centrality and master-y and would engage the human potential for interaction with all life forms.

To draw together the dual nature of human potential, the need to look at nature as a creation of self and of science, and the essential strangeness of humanity, Eiseley ends the essay with a personal anecdote—again a figure riding on a wagon, rushing furiously forward into the future, bearing the mark of human indeterminacy and uncertainty. As a young man, alone on a rural road near nightfall in the midst of a thunderstorm, he encountered a wagon piled high with hay. Sitting high upon his load, the driver lashed and cried out to his horses as they moved furiously through the storm. In an instant, Eiseley writes, he experienced "one of those flame-lit revelations which destroy the natural world forever and replace it with some searing inner vision which accompanies us to the end of our lives" ("Strangeness in the Proportion," *NC*, 147). As the horses and even the rain seemed to pause while he stepped into the road, a flash of lightning illuminated both the scene and the farmer, whose face

> was—by some fantastic biological exaggeration—two faces welded vertically together along the midline, like the riveted iron toys of my childhood. One side was lumpish with swollen and malign excrescences; the other shone in the blue light, pale, ethereal, and remote—a face marked by suffering, yet serene and alien to that visage with which it shared this dreadful mortal frame. (147)

In that moment, he continues, though he could not as a youth explain the experience, he saw "the double face of mankind": "I saw man—all of us—galloping through a torrential landscape, diseased and fungoid, with that pale half-visage of nobility and despair dwarfed but serene upon a twofold countenance. I saw the great horses with the swaying load plunge down the storm-filled track. I saw, and touched a hand to my own face" (148).

Eiseley concludes the essay with the role of the literary naturalist—declared outmoded by some—as interpreter of the natural world of toadstools, of human civilizations, of human faces, or of whales. In Bacon's words, "'There is no Excellent Beauty that hath not some strangeness in the Proportion'" (cited in p. 148). The twentieth-century "modern" individual, "who has not contemplated his otherness, the multiplicity of other possible men who dwell or might have dwelt in him, has not realized the full terror and responsibility of existence" ("Strangeness in the Proportion," *NC*, 148). Through the mind itself humankind passes like the rider on the hay wagon, a creature whose *otherness* is a source of wonder too easily ignored among the conformity of science, a creature who carries "the beast's face and the dream, . . . unable to cast off either or to believe in either. For he is man, the changeling, in whom the sense of goodness has not perished, nor an eye for some supernatural guidepost in the night" (149). In this image lit by lightning Eiseley draws the duality of humankind: the grotesque and the serene, the animal and the human, the two streams of evolution, the human duality and the duality of science, and the need for a dialogue of science and the humanities, a dialogue of understandings. The dual image further incorporates the marginality of modern human existence—participating in a self-created reality, separated from nature and needing to read nature as a hieroglyph whose meaning is partly projected from the reader, incorporating an unexpectedness that is the source of both wonder and terror, incorporating an otherness that has yet to be examined. With a face half-beautiful, half-grotesque, the man is a synecdoche and a symbol for "man, the changeling," who bears a "toppling burden of despair and hope, . . . the beast's face and the dream, but [is] unable to cast off either or to believe in either" (149). In this semiotic figure, in his furious and desperate movement, Eiseley's hieroglyph is readable on both an individual and a universal level.

The Open-Ended Mind: "The Mind as Nature"

What Eiseley describes in "Strangeness in the Proportion" as "the rare freedom of the particle to do what most particles never do" and its repetition in the human mind (136) is the focus of "The Mind as Nature," originally published separately in 1962 as a revision of lectures for the John Dewey Society. He begins with an examination of the "two worlds" of his own childhood—one "dark, hidden and self-examining," characterized by his painful isolation from and futile attempts to communicate with his deaf mother, and the other "external" and "boisterous," characterized by his relationships with others and his father's declamation of the "haunting grandiloquent words" of the Elizabethan stage, which entranced him and made his playmates laugh (*NC*, 195–99). With Dewey's comments on uncertainty in terms of "the *issue* of present experiences," which are troublesome because of our "impatience with suspense" (200), Eiseley develops the sense of uncertainty in the young mind in terms of the positive side of potentiality. His reading of Dewey's statement, "'In the continuous ongoing of life, objects part with something of their final character and become conditions of subsequent experiences'" (cited in p. 201), emphasizes what he has treated earlier in terms of a kind of evolution of chance—an insight based on D'Arcy Wentworth Thompson's observation, in 1897, that chance itself, in an industrial society, is subject to a kind of evolution, for "'change . . . breeds change, and chance—chance. . . . One wave started at the beginning of eternity breaks into component waves, and at once the theory of interference begins to operate'" (cited in "Strangeness in the Proportion," *NC*, 135). Further pursuing this uncanny anticipation of what students of chaos call "sensitive dependence on initial conditions," in which unpredictable changes build upon and enhance each other in a way that is neither totally unconstrained nor simplistically deterministic, Eiseley turns to Dewey's notion of the ongoing changes of chance, a notion that "removes from the genes something of the utter determinacy" that tempts some scientists ("The Mind as Nature," *NC*, 201).

Building on this nondeterministic view, Eiseley examines some literary figures for what Dewey calls certain "statistical constants" that will give insight into the creative process. He notes the isolation that leads to evolutionary experimentation on desert islands, and he analogizes between this isolation and the isolation that leads to human creativity, to the ability of both artist and scientist to "bring out of the dark void, like the mysterious universe itself, the unique, the strange, the unexpected"—the "latent, lurking fertility, not unrelated to the universe from which it sprang" that exists in the human mind ("The Mind as Nature," *NC*, 204). Darwin, he notes, wrote of his isolated pleasure during the voyage of the *Bea-*

gle. Literary figures such as Nathaniel Hawthorne, Hudson, Antoine St. Exupéry, and Herman Melville also worked in isolation.

Yet both the creative artist and the creative scientist, Eiseley continues, may frighten those who seek conformity, as may the teacher (the target audience of the original lecture) who holds up the "mirror" of genius. His point for educators is the need to recognize that education is a lifelong process and to celebrate the "particle episodes" that characterize human insight and creativity. Essential to his argument is the notion that today's faith in science has become so intense that, "though the open-ended and novelty-producing aspect of nature is scientifically recognized in the physics and biology of our time" ("The Mind as Nature," *NC*, 214), awareness of this open-ended quality is acceptable when expressed in "professional jargon" but becomes suspect when expressed in terms (such as his own) of what we do *not* know. This expression of not-knowing is Eiseley's transition to his contention that the universe itself, life, humankind are all process, the making of "something out of nothing, out of the utter void of nonbeing" (214).

In this reiteration of "value arisen from the domain of absolute zero" ("The Star Thrower," *UU*, 87), Eiseley combines his argument for the parallel indeterminacies of the universe and the human brain with his quest to undermine rigid determinacy, reductionism, mechanism, and certainty. The unique human quality that he finds is a linguistic quality—the tendency toward self-examination, whether on the part of poet or scientist. In support of this tendency he cites psychiatrist Lawrence Kubie's speculation that the creative individual draws on "'pre-conscious functions more freely than . . . others who may potentially be equally gifted'" (cited in "The Mind as Nature," *NC*, 218). Eiseley's notion of "preconscious functions," which he will develop more thoroughly in *All the Strange Hours*, is, like Jacques Lacan's prelinguistic realm of images, a realm informed by visual experience. It is a realm to which, he contends, the creative mind seems to have more ready access than other minds, but which most likely exists in all individuals. In Hawthorne's expression "'fragments of a world'" (207)—anticipating the "fragments" that he will work into a mosaic, even a holographic image for his own life, in *All the Strange Hours*—Eiseley embodies the world of preconscious visual potential that he argues is present in all individuals ("The Mind as Nature," *NC*, 212). Like the biological "screen" that precipitates or retards the emergence of a living form, this preconscious realm of the brain exists as potential, and this potential may be affected by culture and experience.

The point, however, is the dialogue between the preconscious realm of images and the culture that retards or promotes its expression and affects the form of expression. Concluding that the task of the educator is to be aware of, to celebrate, the novelty of which the individual is capable, Eiseley expresses this task in terms of the dialogue of understandings: "There are other truths than those contained in laboratory burners, on blackboards, or in test tubes," he argues, recalling the images of his own life's archaeological ruins (an embedded metaphor in the autobiography)—"buttercups, a mastodon tooth, a giant snail, and a rolling Elizabethan line" ("The Mind as Nature," *NC*, 223). Educators are "sculptors in snow," he continues, but they "are sometimes aided by that wild fitfulness which is called 'hazard,' 'contingency,' and the indeterminacy which Dewey labeled 'thinking'" (224). The mind itself, he concludes, "is indigenous and integral to nature itself in its unfolding, and operates in nature's ways and under nature's laws," and it is this operation of the mind as nature that we need to understand and "elaborate" (224).

The conclusion of the essay draws together the interaction of isolation and indeterminacy, the participation of mind in nature, and the quest for knowledge in a line spoken to him in Bimini by a young native woman: "'Those as hunts treasure must go alone, at night, and when they find it they have to leave a little of their blood behind them'" (224). It is an assessment of "the price of wisdom" that maps "the latitude and longitude of a treasure . . . more valuable than all the aptitude tests of this age" ("The Mind as Nature," *NC*, 224). In a concluding line that emphasizes, particularly for an audience of educators, his ongoing dialogue with "measurement" as a product of the reductionist, mechanistic, deterministic, teleological view, Eiseley sets against such "measurement" the wisdom of the young woman who knows the value of the marginal country of night, ambiguity, and indeterminacy and the interaction of sacrifice that she has learned in that realm.

10

The Self as a Vehicle for Understanding II: *All the Strange Hours*

> In this day and age I think a serious writer has to be . . . an ex-suicide, a cipher, naught, zero—which is as it should be, because being a naught is the very condition of making anything.
>
> —Walker Percy, *Signposts in a Strange Land*

In Eiseley's autobiography, *All the Strange Hours,* which at first glance seems only loosely connected, the personal element is woven into reflections on science itself and on a life in science to form a major element in the book. Subtitled *The Excavation of a Life,* the book explores its own fragmented nature. "A biography is always constructed from ruins," Eiseley writes, "but, as any archaeologist will tell you, there is never the means to unearth all the rooms, or follow the buried roads, or dig into every cistern for treasure. You try to see what the ruin meant to whoever inhabited it and, if you are lucky, you see a little way backward into time" (*ASH,* 217). In this "excavation" the dominant themata are time, chance or randomness, and the potential that underlies the real, with continued exploration of the "strangeness" that marked Eiseley's concern in *The Night Country.* As in chaos, however, the theme of randomness and the apparently random recollections in the essays actually mask a carefully structured collection; much is revealed but much also suppressed.

"That the self and its minute adventures may be interesting every essayist from Montaigne to Emerson has intimated," Eiseley writes as he analyzes his transition from a writer out of duty, a writer of scholarship, to popular essayist and writer for the pleasure of writing; he continues that this assumption is true "only if one is utterly, nakedly honest and does not pontificate" (*ASH,* 178). The degree of naked honesty in this autobiography can most likely never be ascertained, though certainly some parallels are evident in the autobiography and the journals (which might be assumed to be nakedly honest, had they not been torn apart and restructured as they were). What is clear is that Eiseley consciously explores identities, tries on and acts out identities as he recalls the past, uncertain of his own identity (248), confident that one plays many roles in a lifetime, always both everyone and no one and always "escaping" before an identity can become fixed (256). It is evident that he has become comfortable with his shifting identities and will deliberately "escape" before any one can be firmly assigned to him.

Despite the dominant images of change and fragmentation, the autobiography blends carefully chosen personae and themata. In the beginning, Eiseley sets up images of excavation of fragments of the past and of a shattered mirror that he recalls from childhood. He divides the book into three sections: Days of a Wanderer, Days of a Thinker, and Days of a Doubter. Through these sections are interwoven the identities—which are not parallel with this division—of fugitive, wanderer, and scholar. But behind the scholar is the trickster, contingency, and the scholar reflects on the Mayan culture and the meaning of the zero, on time's unreturning arrow, and, in a powerful recreation of a dilemma, on the scholar's relationship to his material.

The essays in *All the Strange Hours* are not the elaborately developed essays normally associated with Eiseley's work, but are themselves fragments, reminiscences in which important themes emerge. The larger sections serve as frameworks for these fragments, functioning much as the framework of the "concealed essay" into which, in the earlier texts, Eiseley has woven his reflections, divagations, and tales. What emphatically marks continuity, however, is the ongoing epistemological quest, approached here through both personal and scientific perspectives. Throughout the text, Eiseley describes his quest for knowledge even as he assumes the ultimate skeptic's position—that of the impossibility of knowing. Describing his motivation for writing the book, he reinvokes the earlier image of Odysseus as he says that he is at once "every man and no man," experiencing always "the loneliness of not knowing, of not knowing at all" (*ASH*, 23).

Eiseley portrays himself as a child and as a young man in terms of the "running man"—the fugitive from a household where there was no peace, the child who identifies with three convicts who escaped from prison in the winter that he was five years old, the fugitive from ignorance and irrationality in the form of his deafened mother, the fugitive from poverty, even the fugitive from the West and the remnants of the glacial past that he seems to feel in his blood. From the persona of the fugitive, he turns to that of the thinker, with the epigraph from G. K. Chesterton that links the isolation of the thinker to that of the running man: "'One must somehow find a way of loving the world without trusting it; somehow one must love the world without being worldly'" (*ASH,* 71). His recollections from graduate school are sparse: conversations with and characterization of Frank Speck, his mentor at the University of Pennsylvania; a racial incident involving a Japanese student and two places in Washington, D.C., that belied their status as places of refuge; a recollection of his part in making a hazardous corner on the campus at Penn safer for pedestrians as his chief accomplishment during two years as provost; his recollections of his discovery of Edward Blyth's apparent anticipation of and contribution to Darwin's thinking. These are not normally the stuff of autobiography. The whole is as splintered as the central image of a beautiful, silver-backed Victorian mirror that his mother had shattered in a fit of anger and as holistic as the semiotic realization that each part of the mirror is capable of reflecting the whole rather than just a part, in a manner that anticipates the holographic image of the human brain to which I shall turn later.

Eiseley is not, then, telling his "life story" from poverty to success to disillusionment with clamorous students during the 1960s and with administrative duties that distracted him from what he considered worthwhile. What he does in the book is to censor carefully the personal story and to tell the story of a man with multiple personae, a man who is "every man and no man." This theme of not-knowing, of the loss of knowledge or the lack of knowledge that accompanies the gain in knowledge, has dominated Eiseley's epistemological quest since *Darwin's Century,* and the thema of isolation—not only a perceived personal isolation because of his strange family or his hobo days or his adult sleeplessness, but an isolation that he requires in order to create, to bring the something from nothing, the "value" from the "domain of absolute zero"—dominates the autobiography. The statement complements the book's epigraph from Robert Browning on the changing colors of nature, for this is Eiseley's carefully selected catalogue of how he came to where he was, having turned to the personal essays that he wanted to write as early as high-school days but did not actually begin to write in earnest until he was forty-five. It is a highly personal account of how he chose the persona that he assumes in the last section, titled "Days of a

Doubter." It is an explication of his renunciation of his "scientific heritage"—a renunciation that, as we have seen, was as much a re-announcing as a turning from. It is an exploration of his growing realization of the epistemological uncertainty that accompanies cultural uncertainty and textual indeterminacy. And it is Eiseley's expression of the continuing sense of wonder at the emergence of form from chaos and of value from nothingness.

The Continuing Dice Game

The recurrent image of a dice game in an old house, established in *The Night Country,* reappears as a major theme that subverts cause and effect and reflects the epistemological doubt that dominates Eiseley's third *persona,* that of the scholar. The dice game projects the noncausal, unpredictable element that dominates Eiseley's whole text. "An hour comes when reality," he writes, "the reality we know, gives way to combinations no longer causal or successive in character" (*ASH,* 258–59). Although the object of the game was lost to the child, the man recalls the image that runs as a *leitmotif* through his autobiography. In "The Running Man," he links both his life and his identity to the randomness of a dice game on the western trail (24). Later in the essay he tells of wandering into a ruined farmhouse, where he found only the dice and what looked like school papers bearing, ironically, his own family name. Making up his own game, the child played: "I think I played against the universe as the universe was represented by the wind, stirring papers on the plaster-strewn floor. I played against time. . . . I played for adventure and escape. Then, clutching the dice, but not the paper with my name, I fled frantically down the leaf-sodden unused road, never to return" (*ASH,* 29). Writing— the dead letters of his own name scrawled on a paper found by chance—is linked to time, chance, and, even in childhood, his own identity. For Eiseley, writing is capable of preserving moments in time, though the meaning is as uncertain as the shifting reader, like the meaning of the dice game whose rules the child devised as he played. The dead letters uncannily link the unreturning past with death (earlier I have noted the link of writing with death in ancient hieroglyphs and postmodern theory). In Eiseley's cosmogony teleology has no place; either there is no owner, or the owner will not return. Chance, coupled with probability in the roll of the dice at a given time, determines something as complex as one individual's journey (Serres's *randonee*) or the random path of evolutionary change. As a metaphorical concept, the roll of the dice displaces determinism for probabilism and reinforces Eiseley's notion that life itself inserts "indeterminism" into matter—an indeterminism that makes possible the human participation in shaping the human destiny.

The dice game in the abandoned house of the universe recurs in Part III as Eiseley recalls his major accomplishment as chancellor of the University of Pennsylvania—improving a dangerous street corner. His real concern is with the human relationship with chance: "I seem preoccupied with chance, whether it be the chance that determines life or death upon a street corner, or what it may have been that hovered about me in the ruined farmhouse where, as a child, I threw dice, mimicking a game whose scores I could never possibly determine" (*ASH*, 248). Explaining why he recalls only one small episode in his tenure as a university administrator, he writes of his awareness that daily one plays "harder and more dangerous games" than those in gambling houses (196). Eiseley's answer to an interviewer's question concerns the single episode that may have altered the odds: "I figure that, as on a green table, if the dice are fair, there are some lucky throws. That old short corner here at Walnut gave death a better edge, a percentage. . . . That's what I call the spectral war. It's unseen, but it's everywhere" (204). Although "we all lose eventually" (204), Eiseley's concern is with the human ability to manipulate the probabilities, even if only slightly. Only the "Player," his anthropomorphized figure of chance who blends at times with the Trickster, knows the outcome, "and he plays on all the corners of the world. Watching the percentages. But you can inch him over now and then" (205). The spectral war, on a university campus or in an abandoned house, is the war of contingency, of life and death, of the figures in Darwin's tangled bank. Eiseley's concern is not only for displacing metaphors of fixity, determinism, and teleology, but also for human interactivity in the game.

Eiseley links the dice game to the potentiality that he finds an often-ignored factor in the universe, commenting "that below the existent men of every given generation there lurks an army of *potential* men. . . . In our germ plasm . . . are hidden the freaks, the geniuses, and anomalies of tomorrow" (*ASH*, 120). We think in terms of what *is;* when the hierarchy of actual/potential is reversed, what is potential becomes both the dream and the horror of creative chance. Potential beings appear through life's "gambling machinery," which Eiseley came to know during his college days of working in a hatchery: "I had seen the reality in all its shapes, like dice throws on a green table. Moreover, I had come to know surprises, the one-in-a-million throw. I had learned never to underestimate the potential in favor of the actual" (120). Life as "gambling machinery," rather than the traditional orderly mechanism, subverts reality for potentiality, causality for probability. In *The Lost Notebooks,* he further stresses the probabilistic element in nature, noting that the phrase "survival of the fittest" is itself linked to nineteenth-century concepts of armies clashing. In reality, natural selection is only probabilistic, leaving us "still caught in a chance world of indefinite possibilities, and only possible futures" (*LN,* 173, 211–12).

Finally, for Eiseley the multiple throws of the dice constitute the metaphor of biography itself. He describes himself, invoking the narrative code of the old West, as "an American whose profession, even his life, is no more than a gambler's throw by the firelight of a western wagon" (*ASH,* 24). In the last chapter of the autobiography, he dramatizes his own end in terms of a merging of prehistoric creatures, the westward movement of earlier America, and the rattle of the dice. In a dream sequence that takes him back to an empty boxcar and his days as a drifter during the Great Depression, he sees himself as a young man and as an old man; when the young man jumps from the car, the voice of a third occupant, the Other Player, carries on a dialogue with the old man. The Other Player reminds him of the dice game in the old house. Departing, he tells the old man, "'There is only the one game. . . . That is why the days are counted'" (262). And although "the wisdom . . . was beyond every man," nevertheless, "the counting mattered" (263). The dream of the dice thus introduces the major thema of "counting," of trying to know.

In the dream, the Player controls the dice, but, waking, Eiseley refutes the notion and argues that there is no control of the dice, that the Player is only anthropomorphized chance. Realizing that the dream is gone and that he cannot ask for another throw of the dice, he imagines a last confrontation with the Player in a dramatized blend of the snow of his childhood on the Great Plains and the next ice that he considers inevitable. In the dramatization, time is collapsed, and he is accompanied by the Trickster that he has predicted will posture behind him even in death ("The Star Thrower," *UU,* 78) and by a dog from childhood, as he pursues his quest to erase the boundaries between living things. This dream will take him "further back" to a time when "the dice [are] at last unshaken" and he has crossed the boundary of human and animal forms; he and the dog will be "creatures with no knowledge of contingency or games" (*ASH*, 266). The end he anticipates will be a collapse of time so that he can choose primitive time and simple companions. Though the Player is still present in his dream of collapsed time and blurred boundaries between forms, outside time he is impotent: "The Player could not stop him, for we would be no longer man or dog, but creatures, creatures with no knowledge of contingency or games. All the carefully drawn human lines would be erased between us, the snows deeper, the posse floundering, the dice cup muted in the Player's hand" (266).

The vision of the end is tied to Eiseley's personal past: himself as the young man on the boxcar; his obsession with time and with the impossibility of understanding it, and the importance of attempting to know—of "counting" the marks on the Mayan stelae; the prison escapees with whom as a boy he identified (they were "hunted" but called human); and the image of the Player (curiously like Trickster but a human-like figure rather than the animal form that Trickster mainly assumes). Time remains mystified, as irretrievable as the young man on the boxcar or the prison escape. Knowledge, too (the "counting"), remains mystified.

In the genetic dice, the individual's impromptu dice game in the abandoned house of the universe, and the dream-confrontation with the Other Player, Eiseley's games of chance assert the dominance of contingency and epistemological uncertainty, subvert notions of mechanistic causality or of simple cause and effect, and introduce a probabilistic interpretation of determinism.

Fairy Rings, Quantum Physics, and the Alchemical Meditato: "The Dancers in the Ring" and the Dialogue of Understandings

Probabilistic insights undermine the stability and certainty of science as an institution—a focal point in the autobiography. Appearing near the end of the second section, "Days of a Thinker," the essay "The Dancers in the Ring" describes the scholar's dilemma after years of immersion in a research project and moves to a comparison of scientists to "dancers in the ring," an analogy that emerged in *Francis Bacon and the Modern Dilemma* in terms of folklore surrounding the enchanted "fairy ring" of mushrooms that spring up overnight but is here developed more elaborately in terms that incorporate insights from quantum physics.

Eiseley expresses the scholar's dilemma in the context of his extensive research for *Darwin's Century*. With this research as a vehicle, he treats science, objectivity, creativity, and change. He emphasizes the self-construction of the "treasures" that are to be found in an archaeological dig or in the caverns of a library, for such treasures must be "in the mind that seeks them. Otherwise they are not recognized" (*ASH,* 183). The scholar, however, faces two dangers, the first of which is disorientation to the real as "a given period, or a millennium, . . . become[s] more real than the century he inhabits in the flesh" (183). At this point the researcher may become so lost in her excavations, so obsessed by the incompleteness of her understanding, that she "may never emerge to publish," for no one can assimilate all of the past and the best attempt will be a "relative comprehension" (184). The other danger to the scholar is that, having been lured by research "into a false sense of omnipotence," she "may no longer care to organize this precious knowledge or fix it into a pattern" (184–85). Eiseley recalls the words of Sir Flinders Petrie, who, absorbed in his field research in the Egyptian desert, felt himself absorbed by the past, transported into a wholly other reality: "'I here live and do not scramble to fit myself to the requirements of others. In a narrow tomb with the figure of Nefermat standing on each side of me—as he has stood through all that we know of human history—I have just room for my bed. Behind me is that Great Peace, the Desert'" (cited in p. 185). Like a cavern, Eiseley analogizes, the history of science is filled with "abandoned sinkholes" (185), and he recognizes how close he came to becoming so absorbed with research that he would not "emerge to publish."

The other major realization that Eiseley links to his scholarly work is that, despite the lay public's myth of the progress of science toward some mysterious "truth," the history of science is both a collaborative venture and one "beset by ambiguities, fears, and trends" (*ASH,* 186). In science as in the individual life, "there has to be wandering along bypaths, midnight reading, and sustained effort"; chance plays a major part (186). In the context of post-Kuhnian studies of science, Eiseley speculates on the difficulty of assigning major scientific advances to a single individual "magnified beyond human proportions" (186). And he cites both Rutherford the scientist and Joshua Reynolds the artist on the step-by-step movement of science and the dependence of each thinker on the work of others (187). Yet if science depends upon interwoven insights of many thinkers, it is also subject to a kind of conservatism, an assumption of the rightness of its methods, a reluctance to break from established methods and patterns of thought. Invoking Bacon's image of "those drawn into some powerful circle of thought as 'dancing in little rings like persons bewitched,'" Eiseley insists that "our scientific models do simulate" the fairy ring, the magic circle of mushrooms sprung up overnight within which Celtic folklore says that fairies danced. Once the ring "has encompassed us, it is hard to view objectively," Eiseley concludes. "Within the magic ring . . . may be a truth we come to accept as the whole truth, just as physics became a closed system based upon a substantial, unalterable particle, the atom" (187).

In his own research on Darwin, Eiseley recalls, he spent years "stumbl[ing] back and forth between the fairy ring of fixity and that of organic novelty, as those two circles were beginning to interpenetrate each other in Charles Darwin's young manhood" (*ASH,* 188). The point might have been considered heresy in some scientific circles—that *both* were enchanted circles. Darwin himself once commented that "'to him who convinces belongs all the credit,'" in what Eiseley calls "an unconscious revelation from the very soul of the excavator" (190, 189) and what today we would call an unwitting testimony to the power of rhetoric to create reality. Eiseley closes the essay with a defense of his work on Edward Blyth as anticipator of Darwin as he consciously attempts "to reveal the passions of the excavator, including myself" (190).

Yet this revelation is bound up with the notion that both novelty and fixity were enchanted circles. The probabilistic view is an *other* way of thinking toward which Eiseley has steadily moved—a dialogic view in which nature clings to present forms for stability but incorporates an unpredictable novelty that defies the deterministic, mechanistic explanations that many of Darwin's followers espoused. The image of the dancers in the ring, like the Newtonian model of the atom, is undermined by the "freedom of the particle to do what most particles never do," which, as Eiseley has written earlier, he finds "duplicated in the solitary universe of the human mind" ("Strangeness in the Proportion," *NC,* 136). Paradoxically, this "solitary universe" echoes the old macrocosm/microcosm duality, even as the particle's "freedom" derives from the quantum leap that takes the electron from one shell to another. And this quantum leap is only statistically predictable, not deterministically programmed. Eiseley's image of the dancers in the ring has been expanded to incorporate probabilities, as physics itself has been expanded; reality itself has shifted. In pursuing Blyth and others who worked outside the ring, Eiseley explores the particle-like freedom in the individual mind. Intermingled with personal defensiveness toward his often-maligned defense of Blyth, Eiseley's image of the fairy ring of conventional science is supplemented with an allusion to his own readings in Coleridge, DeQuincey, and

Browne—none of whom found "the ultimate secret," as Eiseley did not find it in science. For all our freedom, he concludes, we track ourselves. Like the comet, "we round back, we return. . . . We are dancers in the ring" (*ASH*, 195).

The last section, Days of a Doubter, refers to the doubt that Eiseley experienced in his own profession and his turn to "wonder" at the risk of criticism from other scientists. He quotes Ruland the Lexicographer in the epigraph to this section, in which he verbalizes his doubts toward stable, professional, institutionalized science and turns to "the Alchemical meditato" as an ancient metaphor for the individual's questioning, a counter to established, "disinterested" science: "'The Alchemical meditato is an inner dialogue with someone who is invisible, as also with God, or with oneself, or with one's good angel'" (cited in *ASH*, 215).

Deriving from the Arabic *al-kimya,* the term *alchemy* refers to the art of transmutation, particularly the notion of transmuting base metals to gold, though its use is now generalized to apply figuratively to any apparently miraculous change or improvement. Alchemy first appeared among the Egyptians, and Hermes was its inventor. In various mythological contexts he appears as the god of science and of commerce, the messenger of Zeus, the conductor of souls to the underworld, the breaker of machines (significant in the context of Eiseley's recurrent criticism of machine metaphors and the social control they imply), the weaver and disconnector of multiple spaces and places of knowledge, the great-grandfather of Ulysses, and the starting point of an epistemology of journeys. Hermes is the link of literature, myth, science, and philosophy in Michel Serres's program, which, like Eiseley's, sees myth in science and science as myth, uses the legend of Odysseus as model of the epistemological journey, and finds in literature the means by which science and myth can be brought together.[1]

Eiseley's first collection of poetry, *Notes of an Alchemist* (1972), establishes the role of the poet as alchemist. As an early attempt at science, alchemy is incomplete, unfinished—a trope for the human incompleteness that Eiseley as a student of ongoing biological and cultural evolution addresses. The Eiseleyan alchemist can "transmute" not lead to gold but details from the natural world to poetic understanding. The poet as alchemist provides a means to a mediated knowledge, to an epistemology in which the mediation of language displaces "pure" observation. And this poetic knowledge transmutes and integrates science so that there is no longer a separation, but a joining of the two kinds of knowledge.

Elsewhere, Eiseley establishes the figurative quality of "alchemy" when he establishes the archaeologist as alchemist because he removes from the earth "those forgotten objects Thoreau called 'fossil thoughts'" ("Thoreau's Unfinished Business," *ST*, 241). Using what Eiseley sometimes calls the "humanly touched thing," the artifact, the archaeologist transmutes an object into thought. Thus archaeology is a model for moving through layers of time, and the archaeological view blends with the poetic, dramatic, and "alchemical" insight into time. The archaeologist as alchemist gives "depth and tragedy and catharsis" to the human drama; the important point is the archaeologist/alchemist as an emblem of the human being confronting time, for only the human being "is capable of comprehending all he was and all that he has failed to be" (241). In comprehending and transmuting himself and his possibilities, the archaeologist/alchemist is capable of little more than any human being. The point is the individual's ability to question understandings, especially to question science as a detached and increasingly destructive institution, and, through a dialogue with himself—a dialogue of understandings—to transmute thought into a new reality, which is itself only process.

Behind Nothing, Before Nothing: The Trickster, the Zero, and the Loneliness of Not Knowing

The mocking figure of the trickster and the mathematical concept of the zero combine with the dice game to form the central epistemological dialogue of the autobiography—the dialogue between knowing and not-knowing that emerges most powerfully in Part III as Eiseley contemplates the emergence of his own doubts as a scientist. "The Coming of the Giant Wasps" narrates the episode that, Eiseley writes, "first aided in implanting some doubts in my mind about the naturalness of nature, or at least nature as she may be interpreted in the laboratory" (*ASH*, 237). The terrifying, mystifying quality of these wasps is their necessary destructiveness (the larvae require the paralyzed form of a cicada for food) coupled with their uncanny ability, in a swift surgical strike that requires exact knowledge of the prey's nerve centers, to render helpless but not to kill a prey larger than themselves. So powerful is the impact of these insects that Eiseley concludes with his statement of ultimate skepticism: "I have come to believe that in the world there is nothing to explain the world. Nothing in nature that can separate the existent from the potential" (238). Between the extremes of metaphysical conclusions based on the individual's "temperamental bent"—whether reductionist or mystical—we as human beings, he writes, "all flounder, choosing to close our eyes to ultimate questions and proceeding, instead, with classification and experiment" (239). Latency, diversity, contingency, and change are the governing forces; as an evolutionist, Eiseley nevertheless repeats that "in the world there is nothing to explain the world. . . . To bring organic novelty into existence, to create pain, injustice, joy, demands more than we can discern in the nature that we analyze so completely" (242). Confronted with this epistemological nothingness, he concludes: "Worship, then, like the Maya, the unknown zero, the procession of the time-bearing gods" (242). In an ongoing dialogue whose tension he cannot resolve, Eiseley contrasts the apparent purposelessness of an endlessly changing genetic code with Emerson's notion of nature as "'a work of ecstasy'" (244). His conclusion is: "I was beyond the country of common belief. . . . I had spent a lifetime exploring questions for which I no longer pretended to have answers, or to fully accept the answers of others" (246). Accepting ignorance, he notes, is the basis of Claude Bernard's statement, "'I put up with ignorance. That is my philosophy'" (cited in p. 247). And his final "metaphysical position," expressed in the context of the sphex wasps, is "I am simply baffled" (*ASH*, 246).

The role of the trickster of primitive tribes prefigures the epistemological dialogue. The trickster, who postures behind seriousness, especially seriousness about one's self, appears from the beginning of the autobiography as the posturing figure of contingency when Eiseley describes a rat that danced behind him at a professional meeting, gaining the audience's attention: "Wasn't it I who had once written that there was a trickster in every culture who humbles what are supposed to be our greatest moments? The trickster who reduces pride, Old Father Coyote who makes and unmakes the world in a long cycle of stories and, incidentally, gets his penis caught in a cleft pine for his pains?" (*ASH*, 11). In this tale, the trickster as a rat has caught the spotlight and captured the applause; he is an ironically humbling figure—a physical figure behind the individual scientist, as Eiseley posits the invisible trickster behind the institution of science. The rat trickster "reduces pride" (12) just as the scientist is most confident. Eiseley comments, "I laughed but the trickster always brings pain," concluding that, in a sense, he "had been the rat that danced" (11–12). In identifying with the dancing rat, the trickster, Eiseley turns the text on itself as he portrays himself as the one who mocks order. The difference in awareness in primitive and modern cultures, he writes, is that "only the old people from the silent cliff houses and the horse people of the plains had known and institutionalized him—the backward dancing man, the caricaturist of order" (12). Lacking a trickster that is now formalized in only a few remote cultures, the scientific institution requires a dancing shadow, a caricaturist of order, and Eiseley sees himself assuming this role.

The trickster returns in a tale Eiseley tells of riding through the Sierra Madres with Manuel, a Latin whose frank *machismo* has led him on a reckless drive accompanied by much discussion of *cojones:*

> Nights later, in a powerful dream by the campfire, a gigantic hooded figure like that of a monk sat on a log opposite me. He made no move. Still he raised his cowled head at my uneasy stare. Beneath the hood there was no face, nothing, merely a chill like the void. I endured the moment unfrightened. The compulsion to look had come from me, had been projected, as it were, from my own darkness. I had been drawn relentlessly to peer beneath the cowl. (*ASH,* 220)

The hooded figure appears in the context of discussions of *la vida,* metaphysics, and *cojones.* Eiseley has concluded privately, amidst the driver's assertion of the importance of *machismo* over life, that *la vida* does indeed matter. The figure, which Eiseley links with time, once again symbolizes uncertainty and indeterminacy as it mocks the rigid order that a system of values imposes—whether the system of values that Manuel represents or the equally rigid system that science itself represents—and shows an existential emptiness lurking behind *la vida.* Eiseley insists that life matters, though he undermines his insistence as he asks, "And why do I speak here toward the end, resolving nothing, seeing nothing but a dreadful duality of power, the good and the evil?" (219).

Like the faceless figure in the Scandinavian myth that formed the tutor text of "The Ghost Continent," the hooded figure in the dream has no true face. This time the figure represents a code by which a society lives, whereas the earlier figure was science itself, "the swirling vapor of an untamed void whose vassals we are—we who fancy ourselves as the priesthood of powers safely contained and to be exhibited as evidences of our own usurping godhood" (*UU,* 20). As always, the trickster undermines the confident set of values, including the determinism and order of institutionalized science. He symbolizes the loss that accompanies gain, the abyss that undermines serious science, the aporia that undermines meaning but is nevertheless necessary for meaningful insight, as confrontation of the abyss is necessary for human creation of meaning out of nothingness.

In the last chapter of *All the Strange Hours*, Eiseley links the mocking trickster, the dice game in the abandoned house, and the zero. The dream sequence with the Other Player, who presents a textual blur between Trickster and anthropomorphized chance, begins with physicist Max Born's comment that "'the brain is a consummate piece of combinatorial mathematics'" and Eiseley's argument, "An hour comes when reality, the reality we know, gives way to combinations no longer causal or successive in character" (*ASH*, 258–59). In the dream sequence on an empty boxcar, the only inhabitants of the car (intertextual with the "car" that carries the scientific figures of the Scandinavian "crossroads religion" in "The Ghost Continent") are the ghostly young man, the old man, and the Other Player, who resembles Trickster as he questions the old man about matters that he cannot understand. Within the dream, the old man awakes on the steps of a Mayan temple, where the Player reminds him of his fascination with the "vertical time" of the Mayas (263), who calculated time in eons, a fascination that Eiseley has expressed often in his earlier texts. The Mayas, Eiseley has noted, never had the wheel, but they had the zero when the Romans were lumbering along with a mathematical system that did not include the zero. The word *maya* itself carries with it an additional suggestion; if, as Heidegger suggests, language indeed "knows" something, the word trails behind it the Sanskrit meaning of "illusion," which resonates in Eiseley's texts, particularly in his poem on the Mayas, which ends with a reference to their ability to manipulate the zero and with the connotations of the zero as "nothing":

> It is all
> a secret of the zeros unfolding.
> Behind, nothing,
> before, nothing.
> Worship it, the zero, and at intervals
> erect the road markers
> the great stelae
> with the graven numbers. What other people
> have had the strength for this
> and quietly
> let their hands fall
> obliterating the secret of the markers
> erasing the constellations
> before their disappearance—
> a blackboard exercise a god might have envied.
> ("The Maya," *AKA*, 23–24, ll. 27–41)

Maya, then, is a signifier with an excess of signifieds; it is both illusion and nothing. Eiseley has invoked the poem in the first essay of the autobiography, as he recalls the rat that danced behind him during a talk in which he linked all the sciences to time. In the stream of consciousness in the first essay, he recalls an old hobo who gave him advice for life: "'The capitalists beat men into line. Okay? The communists beat men into line. Right again?' . . . 'Men beat men, that's all'" (*ASH,* 10). The advice, Eiseley says, "left all my life henceforward free of mobs and movements, free as only very wild things are both solitary and free" (10). But recalling the hobo's line leads him to recall the lines from his own poem, "Before nothing / behind nothing," and to conclude that one either rides time or stops and lets it pass (10).

In the dream, then, the old man sleeps on the steps of the Mayan temple, sleeping "as the temple slept in the timeless Caribbean sun" and understanding finally what "the dots and bars" on the Mayan stelae meant: "The Wisdom could take care of itself. It was beyond me. It was beyond every man. But for all that the counting mattered" (*ASH,* 263). All he can conclude is the need, in any culture, for the "counting," the quest for understanding. The zero is essential to the counting; the counting is playing, participating in the game or the pursuit; and it is the counting, the living, that matters, as Eiseley had tried to explain to Manuel amidst talk of *la vida* and *muerte.* The uncertainty of the dream and the textual blur that occurs when Eiseley introduces "Coyote the trickster" as a figure who will "help" him and the dog (266)—displacing the Other Player as a trickster figure—echo the uncertainty of the dream and of knowledge, the certainty of chance.

The blur at the end of the dream echoes the epistemological blur that Eiseley has explored throughout the text. Ultimately, Eiseley assumes the skeptic's position—that of the inability to know. As "every man and no man," he experiences always "the loneliness of not knowing, of not knowing at all" (*ASH,* 23). He recalls confronting, through an illness as a young man, the mystery of the human body's ability to reassemble itself without conscious direction or outside help. Since that moment, he writes, "I have never been sure what nature is about" (78). The absence of the known is the presence of mystery, of the unknown—not in a metaphysical but in a purely materialist sense (78). Recalling a discussion with his mentor at the University of Pennsylvania, Frank Speck, whose outlook was essentially that of a primitive, Eiseley confesses that he may not have been "a very good scientist, . . . not sufficiently proud, nor confident of my powers, nor of any human powers," and adds that Darwin himself "wobbled" as did he and Speck in their discussion (90). It is as if, Eiseley argues, "the universe were too frighteningly queer to be understood by minds like ours" (90). Science assumes simplicity, but Eiseley's position is that the simplicity of the universe is a kind of trick—not that nature itself tricks the scientists, but "we trick ourselves with our own ingenuity"; and he concludes, "I don't believe in simplicity" (91).

This ultimate confrontation with ignorance, the ultimate humbling brought by the hooded Trickster with no face or by the Other Player who is in waking hours nonexistent, is a powerful irony for one whose life has been devoted to knowing. In the end, knowing is not-knowing. In the next-to-last chapter Eiseley notes his preoccupation with chance and compares his "predicament" in the autumn of his life to the outwash fan of pebbles at the low end of a stream. We are not the pebbles in the fan, however. They are the remnants of the stream, and we are the stream; we are movement, process, change—creatures of multiple identities in our quest for understanding. But in the last chapter, Eiseley's projection of the last dream is further back than the zero and the quest for knowledge, back to a time when taxonomic definitions would be meaningless and only being would be left. In such a dream, the anthropomorphized Player would have no part. Eiseley and the dog Wolf would be "creatures with no knowledge of contingency or games," and "carefully drawn human lines would be erased between us" (*ASH*, 266). The last dream is an image of a last quest for transcendence—not for transcendence as described by the Medieval mystics whom Eiseley has quoted, but for transcendence of the need for knowing, for obliteration of knowledge in simple being. It is a dream of ultimate rejection of knowing, a final existential re-nunciation of knowing for being. Yet if this being is "no longer causal or successive," it nevertheless remains governed by probability, and the surrender to being is a surrender to probability.

But we recall that in the pivotal essays "The Star Thrower" and "The Mind as Nature," the "domain of absolute zero" is the source of meaning projected from the human being onto the text of nature in a process that the Russian statistician and philosopher V. V. Nalimov (who, in his quest for a probabilistic metaphor for human language and human understanding, is also a reader of Eiseley) states in probabilistic and textual terms: "new texts are always a result of free creativity realized on a probabilistic set which may be regarded as an unexposed semantic universe or *nothing,* the semantic vacuum or, metaphorically, an analogue of the physical vacuum," "the concept of *non-existence* (the unobservable state)

as a potential basis for *reality* (the observable state)" (1982, 29, 39).
The zero, then, is the domain of the essential Eiseleyan dilemma; it
is the ultimate not-knowing from which knowing is created by the
individual within a cultural frame. It is the undermining figure for
a stable, fixed, referential, institutionalized science whose "knowing" is often accepted as true and final.

The Mirror Stage and the Holographic Fragment:
Creation of an Identity through Language

As the last stage in a journey of understanding that began with
The Immense Journey, the autobiography reflects Barthes's question "What, after all, is this 'I' who would write himself?" (1983,
98) and presents Eiseley's conscious quest to define and to *write*
the self through science and the personal. The shattered mirror
both symbolizes the fragmentation of personality established in the
first chapter and takes on a holographic dimension in that each
piece reflects the whole. The concluding chapter reinvokes the
opening image—that of "a beautiful silver-backed Victorian hand
mirror" (*ASH,* 3) that his mother had shattered in a fit of anger.
Significantly for Eiseley's ongoing thema of the human duality, the
mirror was a "twin"; while his mother had shattered hers, his aunt
had preserved its twin and left it among her effects. Eiseley remembers the mirror from childhood, recalling that when it "vanished," his own childhood identity vanished. From one perspective
Eiseley's autobiography is as splintered as the image of the shattered mirror. Yet the splintering takes on an almost holographic
quality as one realizes that each fragment of the mirror is capable
of reflecting the whole rather than just a part, in a manner that
anticipates the holographic image of the human brain that Karl
Pribram and David Bohm have explored. Returning to the fragments in the final chapter, Eiseley writes of

> the end of which I spoke in the beginning—the shattered mirror which
> can never be repaired but which lies in bits in the hallways of the mind
> itself. Feet crunch upon the glass as in an abandoned house; sometimes
> a ray of light strikes through a closed shutter and something still glit-
> ters, devastatingly beautiful, upon the floor. Or, similarly, a moonlit
> dream turns the fragments to soft shadows out of which come
> voices. (258)

In analyzing the shattered whole of Eiseley's mirror, I shall first
establish the mirror as a thema in his texts and then explore the

inversion of values from which emerges the almost holographic notion of the shattered mirror in the concluding chapter of the autobiography.

Mirror

As I have noted earlier, traditional philosophy images the mind in terms of a mirror, and the image underlies the conception of "knowledge as accuracy of representation" (Rorty, 1979, 12). Thus the notion of correspondence between the "idea" (from *eidos,* "form") and the world underlies Western philosophical discourse. Following the dominant (scientific) epistemology, we forget the metaphor of the mirror and accept the language of science as transparently referential, a model for the discourse of knowledge. The mirror, however, is also historically associated with the eye as the organ of perception and with the homophone of the "I." Thus the mirror is linked with knowledge, self-knowledge, and identity. But the signifier, as Derrida has argued, can "double back" upon the signified to create a mirror effect in which the signifier is reflected onto the signified, in contradistinction to the traditional notion of the signifier as mirror of the signified, of the vehicle as reflection of the "idea."[2]

The fleeting, constantly shifting quality of the mirror's images makes the mirror an appropriate analogue to the quest for identity in Eiseley's texts, for humans and time and form are treated as fleeting expressions, process. *Mirror* comes from the Old French *miroir,* which derives from the Latin *miror, mirari,* "to wonder at," related to *mirus,* "wonderful"; *admire* derives from the Latin *admirari,* "to wonder at." Etymologically, the mirror is linked to both identity and wonder. But wonder also suggests questioning, "wondering," the intersection of the inexplicable and the unexpected. As Gerber and McFadden note, referring to another derivation from *mirari* in Eiseley's texts, "Miracle is a matter of philosophic seeing. It is an awareness of the vast spaces between subatomic particles and the lurking potentiality of evolutionary change" (1983, 136). Both *wonder* and *wander* are signifiers whose impact is reflected onto the signified.

Throughout Eiseley's texts, the mirror is linked both to identity, which of course is fleeting, and to distortion. The mirror may be a dark window, a simple mirror, or a trick mirror; finally, it is a shattered mirror. Like the fleeting image in the mirror, life, knowledge, and science are not fixed or *present,* but shifting processes. As a metaphor, the mirror helps to displace notions of fixity, of transparent referentiality, of scientific "objective observation," and to link the wanderer/wonderer to the unexpected. Eiseley is concerned, not with the notion of a perfectly clear, unclouded mirror but with "the shifting colors in the enchanted glass of the mind which the extreme Baconians would reduce to pellucid sobriety" (1957, 480). The individual's observation is not to be ignored in the interests of a hypothetical objectivity, though Eiseley would not do away with scientific observation. Exploring the vocabulary that Richard Rorty calls "a seventeenth-century vocabulary largely dependent on visual imagery, especially mirror imagery," Eiseley (like Rorty's "*intentionally* peripheral" figures [1979, 6, 369]), nevertheless attempts to move beyond the seventeenth-century concepts of order and mechanism that have been absorbed into evolutionary thinking. Writing for a popular audience, Eiseley retains the vocabulary, but he undermines the accepted, "stable" referentiality of the vocabulary.

In an early essay, "The Enchanted Glass," which does not appear in any of the collections but essentially defines Eiseley's "literary" texts, the mirror appears as the "enchanted glass of the mind" (1957, 480), which is "colored" by the individual's perception in his definition of the contemplative naturalist's work; but the position remains consistent throughout his texts. Eiseley suggests that the naturalist tells a story from personal experience, blended with science. He contends that the naturalist works in an "indefinable country"—between the human mind and the "natural" (and we remember his subsequent defamiliarization of this term) objects; thus the naturalist actually treats "fragments of a natural history so vast, shifting and impermanent as to confound the strict empiricist" (480). In this view the natural historian is a "magician" who understands the nature of the "glass" intuitively and is able to "create the worlds" that fill an equally magical mirror, that of "the enchanted glass of the social mind" (492). Only such a writer, Eiseley contends—and in the later texts the emphasis is increasingly on this *mysterium* of uncanny or poetic knowledge—can observe the flight of an "instinct-baffled" bird both from a scientific perspective and as part of a larger drama and can perceive the human and the bird as one in the sense of shared evolutionary characteristics as well as the sense of participation in life. The artist, then, is the mediator, the poet who "knows and contains within himself leaf, man and falling bird" (492). The holistic thrust of this early essay anticipates the metaphor of fragmentation/wholeness that is central to Eiseley's autobiography.

Trick mirrors appear, too, as metaphors for the basic concept of the creativity of time and process. In the trick mirror, distortion and creativity subvert fixity and referentiality. In *The Immense Journey,* anticipating the image of the human being as a "palimpsest," he describes the "mirror of time" in terms of "a hall of trick mirrors" in which one sees oneself distorted, even though the distortions seem fixed, at least temporarily. Thus, "if you have come by the mirror that makes men, and somewhere behind you there is a mirror that makes black cats, you can still see the pattern. You and the cat are related; the shreds of the original shape are in your bones and the shreds of primeval thought patterns move in the eyes of both of you and are understood by both" ("Little Men and Flying Saucers," *IJ,* 160). As in chaos theory or in David Bohm's holographic universe (which I shall explore later in this chapter), regularity underlies irregularity, as irregularity underlies apparent regularity.

In *The Unexpected Universe*—his first text overtly concerned with indeterminacy, inexplicability, and the unexpected—Eiseley explores science itself as a mirror, describing the human as "the self-fabricator who came across an ice age to look into the mirrors and the magic of science" ("The Hidden Teacher," *UU,* 55). Like other forms of knowledge, science is fleeting, deceptive, and subject to the whims of fashion, but its magical quality is undeniable. Still, humankind is more than the Narcissistic quest for individual identity: "Surely he did not come to see himself or his wild visage only. He came because he is at heart a listener and a searcher for some transcendent realm beyond himself" (55), a creature who may realize his own identity after seeing in a darkened window the reflection of a face that is his own but not his own—a palimpsest of other faces (66). Like Narcissus gazing into the fountain, like Lacan's child perceiving his own identity before a mirror, the individual identifies herself, but more: she identifies herself in relation to others. The mirror thus becomes a symbol of a mediated vision; it presents an image that is not transparently referential, but an image that requires "reading," interaction, partial creation of the message.

The distorting mirror takes Eiseley's epistemological exploration in the direction of popular and intellectual fashion in "The Inner Galaxy." He suggests that "humanity studies itself in the mirror of fashion, and ever the mirror gives back distortions, which for the moment impose themselves upon man's real image" (*UU,* 179). The mirror of fashion thus is a vehicle for exploring fashions of scientific "truth." The "real image" is not real, but impermanent, constantly shifting. The mirror doubles back upon itself as Eiseley writes of the "real" image, undermining even the present text. Intellectual fashion allows us at one time to perceive "immutable laws" that rule us, at another time to see ourselves as "the product of a meaningless and ever altering chemistry" (179). For example, the mirror that modern writers on prehistory provide, Eiseley contends, may lead us to believe that we are basically animals. Even as the words are pronounced, however, "the picture begins to waver and to change" as other human "reflections" emerge: "St. Francis of the birds broods by the waters; Gilbert White of Selborne putters harmlessly with the old pet tortoise in his garden" (180). We see, then, what intellectual "fashion" tells us to see, and artists' notions of earlier human beings reflect changes in this intellectual mirror. "Fashion"—intellectual and popular—shows the Neanderthal as an "open-mouthed brute" and the Peking human as "neatly groomed," overlooking the fact that both are on the same anatomical level and that the artists' conceptions are also distortions (183).

If we choose to see humanity in terms of the struggle in the tangled bank, then the social mirror reflects that image, though St. Francis of the birds is an equally true image of man; to perceive him we have only to shift our gaze into the mirror. Exploring the social and intellectual mirror, then, Eiseley gives a torque to the traditional metaphor to undermine notions of transparent referentiality in understanding nature and to lead to further displacement of metaphors of fixity and struggle. To move from warring nature to St. Francis's image of peace requires only a shift of perspective. Because "natural selection is real but at the same time it is a shifting chimera, less a 'law' than making its own law from age to age" (187), the fleeting and fluid qualities of the mirror reinforce Eiseley's reexamination and displacement of traditional metaphors.

Fragmented Light: The Kaleidoscope and the Shattered Mirror

In addition to exploring trick mirrors and the mirror as an image of distortion, Eiseley explores the mirror tricks of the kaleidoscope in undermining images of stability and referentiality in the world that science explores. He treats the shifting process that realism and positivism take as reality, the subject of representation, in terms of a kaleidoscope. Epistemologically, the image of the kaleidoscope explores observation, fragmentation, and reassertion of order in a variation of the role of the "piece" in a *collage* or a mosaic, of the individual, when displaced from one context to another.[3] E. H. Gombrich writes of the kaleidoscope as an instrument that "permits us to study [the] contrary effects of fragmentation and integration," suggesting that "it is precisely by draining the individual elements of their identity that the overall order makes them fuse into a large unit which tends to be perceived as an object in its own right" (1979, 157). Because the kaleidoscope depends on repeated symmetries and "exhibits maximal redundance," Gombrich approaches it as a means of exploring "the visual effects of order" (1979, 150). Following Sir Karl Popper's "searchlight theory" of the mind, which asserts the constant activity of any organism in scanning its surroundings, Gombrich notes that such a theory is inevitable in the Darwinian view. Even *paramecia* respond when they bump into impeding objects, and Gombrich posits the continual testing and refuting of hypotheses as part of the learning process of any organism—from the lowly *paramecium* in its encounters with obstacles, to the human eye's search for order in the process of perception, and even to the process of scientific investigation (1979, 1–4). Gombrich's hypothesis and his explication of the kaleidoscope as model in studying fragmentation and integration help to explain the continuing heuristic of Eiseley's visual metaphors. The kaleidoscope is another vehicle for exploring the constantly changing nature of reality that we perceive. The kaleidoscope "shifts," and reality is other than the apparent fixity that we tend to see as "real." The shifting fragments and the mirrored doubling of forms in the kaleidoscope create a model for the way past thought is reorganized into new knowledge or new organizations of reality.

Wolfgang Iser turns to the kaleidoscope as a metaphor for reading—an exploration that is fruitful in the Eiseleyan context of textual metaphors for understanding. The eye's exploration of pattern, dissolution of pattern, and reestablishment of pattern may be compared to the activity of reading, in which, as Iser argues, each sentence gives "a preview of the next and forms a kind of viewfinder for what is to come; and this in turn changes the 'preview' and so becomes a 'viewfinder' for what has been read" (1980, 54). Iser's metaphor for reading language is the kaleidoscope. In viewing the kaleidoscope, the eye "reads" forms; it reads semiotically, looking for pattern, dissolution, and reorganization. In reading language, the eye finds in a sentence a preview, which is changed in turn so that the effect is redoubled.

Eiseley turns frequently to the kaleidoscope as a vehicle for undermining notions of fixity and determinism. In "Strangeness in the Proportion" in *The Night Country,* the kaleidoscope underscores the constantly changing nature of reality in human thinking. Bacon is a model of the "great synthesizer who . . . suddenly produces a kaleidoscopic change in our vision of the world" (131). Such a synthesizer serves both to disturb the pattern and to rearrange it, becoming "a kind of lens or gathering point through which past thought gathers, is reorganized, and radiates outward again into new forms" (131). Like other references to the shifting frame though which science operates, Eiseley's kaleidoscope undermines the stability of scientific paradigms and the referentiality of scientific language, at the same time emphasizing the new unity, the wholeness that comes out of fragmentation to create a new reality.

In the autobiography, Eiseley's kaleidoscope reinforces the important metaphor of the fragmented mirror as it again introduces the wholeness that emerges from fragmentation, the "piece" that looks toward both old and new contexts as it is displaced from one to another. The kaleidoscope serves as a model of how we perceive and of challenges to that perception as Eiseley tells of Herbert Winlock's seeing, through a crevice in an ancient Egyptian tomb, a miniature world with the illusion of movement. In Winlock's illusion that the miniatures prepared to accompany the dead on their journey were actually going about their business, Eiseley finds an analogy for the changing of any individual's world: "Then some fine day, the kaleidoscope through which we peer at life shifts suddenly and everything is reordered" (*ASH*, 99–100). Visual and philosophical perceptions are reordered; the whole is dependent on the shift of the minuscule, so that the image of fragmentation also suggests anti-fragmentation.

Eiseley's most intense sense of the interrelation of parts and whole emerges in the metaphor of the shattered mirror. The notion is intertextual with an ancient Judaeo-Christian tradition of "reality" as a shattering or splintering of a mirror (the model of the transparent referentiality of knowledge was splintered in ancient times, and Shelley calls on a similar code through Neoplatonism when he images life as a "dome of many-coloured glass"), or of light (Schaya 1973, 43). For Eiseley, the mirror of identity is often fragmented. In the autobiography the shattered mirror assembles an identity but also leads to epistemological exploration. Not a standard autobiography in the diachronic sense, *All the Strange Hours* is a fragmented delving into an identity. The shattered mirror is the central metaphor for this "excavation" of the subtitle— for the individual's quest for an identity that is as fleeting as humankind itself or as one's image in a mirror. The metaphor is established in the opening chapter, in which Eiseley describes one of his earliest memories: "a beautiful silver-backed Victorian hand mirror" whose glass his mother had shattered in her deafness and frustration (*ASH*, 3). He remembers "look[ing] into the mirror as a child, admiring the scrollwork on the silver," and he makes clear the mirror's importance as a symbol of identity as he recalls: "Finally it disappeared. The face of a child vanished with it, my own face. Without the mirror I was unaware when it departed" (3).

In Lacanian terms the *stade du miroir* marks the child's passage from the domain of images, the Imaginary, a "pre-verbal register whose logic is essentially visual" and carries a spatial configuration (Jameson 1977, 353, 357), into the domain of language, the Symbolic. The effectiveness of the Lacanian approach for Eiseley's text lies in its emphasis on language as the vehicle of identity. Seeing itself in a mirror, the child perceives its ability to make the image appear and disappear, and its separateness from the mother. Thus the child passes into a linguistic realm in which self and others are identified by names. The child discovers an identity through language, assuming the Name of the Father and entering the linguistic world of Law. Thus occurs the child's "castration" or separation from himself (Lacan, 1981b, 159–77). Submission to the Law of the Father, participating in a world of law that provides him with an identity, separates the individual from the earlier world of images, of shapes, and brings with it the identity of the Father that accompanies the father's Law. Yet the Law is not repression; it is "the order of abstraction," which leads to alienation from the image of the self (Jameson 1977, 362, 373).

The mirror stage thus "marks a fundamental gap between the subject and its own self or *imago* which can never be bridged" (Jameson 1977, 353). The Symbolic order, then, brings an identity through language; but it also brings a fragmentation—"a loss of totality, the fragmentation of the body and the self," not only in Freudian and Lacanian terms of castration, but in "accepting the possibility which language brings of the discontinuity of the self" (Culler 1981, 165). In Lacanian terms the confrontation of Imaginary and Symbolic is not limited to the child, despite the notion of the "mirror stage," but is ongoing. "Pure" experience of either stage is illusory (Jameson 1977, 350). To experience the Real, the individual must also simultaneously experience the other two orders. Indeed, movement into the Symbolic is "a precondition for a full mastery of the Imaginary as well" (360). Though the Real is problematic because it can be experienced only through the mediation of the Imaginary and the Symbolic, these other two, Jameson argues, should not be considered simply in terms of binaries. "The real, or what is perceived as such—," writes Lacan, "is what resists symbolization absolutely" (*Le seminaire,* cited in Jameson 1977, 384). For Lacan, the subject's experience in the mirror stage has "fixed the instance of the ego well before any social determination, in a line of fiction which is forever irreducible for the individual himself—or rather which will rejoin the subject's evolution in asymptotic fashion only" (*Ecrits,* cited in Jameson 1977, 353). The phrase "line of fiction," Jameson argues, emphasizes "the psychic function of narrative and fantasy in the attempts of the subject to reintegrate his or her alienated image" (Jameson 1977, 353).

The concept of fragmentation and discontinuity in the Symbolic is especially significant in Eiseley's context because of his recurrent interest in recognizing discontinuities as essential to understanding and to living in an evolving world whose very particles are free to do "what most particles never do" ("Strangeness in the Proportion," *NC,* 136). Indeed, *All the Strange Hours* may be seen in Lacanian terms as Eiseley's treatment of the *stade du miroir,* his "reflections" on his own passage from the Imaginary into the Symbolic, where, as E. Fred Carlisle suggests, he creates an identity through language (1983, 76). And Eiseley's visual orientation virtually creates an ongoing dialogue between the Imaginary and the Symbolic in the essay titled "Willy." In the passage describing the artist's gallery of the mind (*ASH,* 151) as an explanation of how he writes, Eiseley presents a dialogue of the visual—the realm of the Imaginary—with the Symbolic—the realm of language. His identity as a writer is created from linguistic fragments that synaesthetically suggest the creation of the whole both because of and despite the fragmentation of the shattered mirror, as the individual's identity is created through the interaction of the holistic Imaginary and the fragmented Symbolic realms. The shattered mirror is not only the vehicle for piecing together the individual's identity, but another form of mediated vision: it requires "reading," or interaction of the self and the text in an act of creation/mediation.

Although the mirror departed, Eiseley writes that subconsciously nothing departs, and the mind in age is left "picking endlessly over the splintered glass of a mirror dropped and broken long ago" (*ASH,* 4). In this context the mirror is not only the source of a quest for identity but also the symbol of a life that is "splintered" in the memory. The shattered mirror is the metaphor for the text itself, for the textual "splintering" that is Eiseley's autobiographical method. And the shattered mirror suggests a holographic effect, in that each piece of this splintered glass can reproduce the full image, as the part in a hologram reproduces the whole. In a sense the splintered memories reproduce a life. Thus the shattered mirror is a means of access to an autobiography that is not linear, but a series of fragments, each reflecting a part and, in a sense, the whole of his quest for identity. In the absence of the mirror, he was unaware when his child's face departed, yet while absence equals fragmentation, this fragmentation in turn makes possible a holographic identity.

In the final chapter when Eiseley returns to the shattered mirror, he finds in it a synaesthetic sense of pleasure in the glittering fragment struck by a ray of light that unexpectedly comes "through a closed shutter" or in the dream that "turns the fragments to soft shadows out of which come voices" (*ASH*, 258). Out of the fragments come the unexpected or the beautiful, the realization that reality is not "causal or successive" (259), and the sense that the human brain, which perceives the fragments, "'is a consummate piece of combinatorial mathematics'" (cited in p. 258). The fragmented mirror symbolizes the fragments of an identity that remains shifting while the narrator—as a synecdoche for humankind—reflects what he has discovered through and because of science as a reality that is not "causal or successive" but a matter of the chance combination of fragments. Thus in assembling the fragments of his own identity and the synecdochic identity of humanity, Eiseley also draws the reader into the unstable reality of science and the need for the philosophical reflections that supplement the traditional mirror that "reflects" nature.

Epistemologically, Eiseley's splintered mirror suggests the fragmentation of knowledge, science, vision, culture, individual awareness. In an entry in the *Lost Notebooks* made in 1959, Eiseley suggests that animal minds "are like splinters from a shattered mirror; they give back a very small, particularized reflection of the outer world" (*LN*, 114). The human, he continues, is still a fragment, though a "much greater" one because of the human ability to think abstractly (114). What we know is fragmented, and the fragmentation adds another dimension of distortion to an already fleeting and distorted image. Yet each piece fits, perhaps haphazardly, into the whole, and each piece also reconstructs the whole, in an exploration strikingly similar to the holographic effect that David Bohm suggests and John P. Briggs and F. David Peat further explore as an alternative to the "frag-mentation" of twentieth-century thinking.

The Holographic Principle

Developed in 1947 by Nobel Prize winner Dennis Gabor, the hologram (from the Greek *holos,* "whole," and *gramma,* "picture," "scratch," "written letter") was perfected in the 1960s with the advent of the laser. In the process of holographic photography, laser beams illuminate a subject with light split into two beams by a lens; these two beams are reflected off mirrors, through lenses, and onto a photographic plate in the form of a "reference beam" and an "object beam." The reference beam hits the plate directly, while the object beam is reflected off the object. The result is a pattern of interference that is encoded on the photographic plate. If the hologram is to be viewed in white (normal) light, a special emulsion transforms the interference pattern into tiny ridges cast into a mold and reproduced by a mirror-like aluminum coating, which functions as a mirror to reflect white light through the interference pattern and thus to give the impression of a three-dimensional object (Caulfield 1984, 366–73). The hologram thus appears both natural and not-natural; it is both a reproduction and a creation, devised by a special kind of photography that enhances the illusion of three-dimensionality, which the viewer recognizes as illusion and yet seems compelled to touch.

The holographic principle suggests a movement of information from object to wave storage to image construction through decoding of an interference pattern. As a "text" (in the poststructuralist sense), the hologram suggests wholeness through the multiple perspectives from which it can be viewed. It suggests, further, an enfolding of images on one another, to be decoded by the viewer. And perhaps the most uncanny feature of the holographic plate is that a single piece, rather than reflecting only part of the image, will reflect the entire image, although the smaller the piece, the fuzzier the image will be.

As a model, the hologram has innumerable possibilities. Karl Pribram, for example, contends that the human brain (imaged in the nineteenth century as a telegraph center and in the twentieth as a computer) can best be explained by the holographic model. "The holographic supertheory," writes Marilyn Ferguson, "says that *our brains mathematically construct 'hard' reality by interpreting frequencies from a dimension transcending time and space. The brain is a hologram, interpreting a holographic universe*" (1982, 22; italics in original). Further, suggests Pribram, "the fact that the holographic domain is reciprocally related to the image/object domain . . . implies that mental operations . . . reflect the basic order of the universe" (1982, 34). Because the holographic paradigm depends on reading of coded interference patterns, space-time coordinates cease to be useful, as do our culturally embedded notions of causality; thus "complementarities, synchronicities, symmetries, and dualities must be called upon as explanatory principles" (Pribram 1982, 34). Roger S. Jones, approaching science as a cultural construct, draws on the work of Pribram and Bohm to emphasize that information about the world is processed in all of the brain—again, the part either "contains or implies the whole" (1982, 199–200).

If, as Pribram suggests, human mental operations, which take place in a holographic brain, also reflect a universe that is a hologram, then these mental operations indeed mirror "the basic order of the universe" (1982, 34). If the mental operations of reading and writing take place in a holographic brain, they are both fragmentary and whole, both reenacting the universe and reflecting it through its part. From multiple perspectives, then, the hologram may serve as a model for human mental processes and even for human identity, and its implications reflect major themata of Eiseley's texts.

DETERMINACY/INDETERMINACY

The holographic model (often referred to as the "holographic paradigm") leads us to consider the question of the indeterminacy of the text—itself a concept explored in the sciences and migrating into the (unconscious) thought of structuralists who attempted to construct a "science" of literary studies, into deconstructive analysis, and into reader-oriented theories. Indeed, what I see as the dominant paradigms of science and literature in the twentieth century are, respectively, indeterminacy and textuality. Indeterminacy is traceable to Darwin (though Darwin himself remained caught within the Newtonian mechanistic view) and his reintroduction of "accident" in the spontaneous mutations essential to the evolutionary process. It is confirmed in Heisenberg's uncertainty principle and echoed in Bohr's complementarity principle, which acknowledges determinacy within the specific experiment but accepts the larger indeterminacy that allows light to be both wave and particle.

The holographic model differs from the organic model with which Coleridge and, in the twentieth century, the New Critics were enamored, for the organic model, as M. H. Abrams has demonstrated, takes on a "life" of its own and, in "substitut[ing] the concept of growth for the operation of a mechanism," in effect "exchanges one kind of determinism for another" (1971, 173). The holographic model incorporates both determinacy and indeterminacy, destiny and not-destiny. Like any artist, the holographic artist interacts with her materials in setting up image, mirrors, and light. In this respect, the holographic model suggests both interaction of the artist with her materials and the constraints suggested by the structure—the "grammar"—of the text. Yet the hologram itself suggests indeterminacy in that it is never quite the same from moment to moment because of its variation with the light and with the viewer's angle of vision. The visual effect may be understood in terms of what Jacques Derrida describes as the effect of rhyme:

> Rhyme—which is the general law of textual effects—is the folding-together of an identity and a difference. The raw material for this operation is no longer merely the sound of the end of a word: all "substances" (phonic and graphic) and all "forms" can be linked together at any distance and under any rule in order to produce new versions of "that which in discourse does not speak." (1981, 277)

The impact of this insight is in the linking of "forms" and "substances" in what Derrida metaphorizes as the "hymen"—the text/tissue that is a difference and yet is not present, but whose very absence asserts itself. This text/tissue creates what Gregory Ulmer calls the "moiré effect"—the "marginal" difference in letters and sound which is also a "movement that articulates a strange space *between* speech and writing" (1985, 47). The rhythm of the two meanings and two spellings, in rhyme or in the Derridean homonym *difference/differance* that suggests both similarity and difference, both differing and deferring, is *"like the two overlapping but not quite matching grids that generate the flicker of the moiré effect"* (Ulmer 1985, 47; italics in original). The flicker of the hologram can be understood in terms of this "moiré effect." As they overlap, the reference beam and the object beam form an interference pattern, which is preserved in the hologram. The hologram, then, is viewed as a "flicker" of this grid-like effect.

The hologram suggests wholeness in a way that even Constable and Turner could not suggest with light and shadow, but it remains a form of representation. In the holographic concept, any given angle is not necessarily a mirror, an imitation, but a fragment, yet it is also capable of reproducing the whole. Thus the holographic paradigm suggests a "flicker" between representation and nonrepresentation. It suggests a similar moiré effect involving the constraints imposed by the structure of the medium and the reader's participation in creating meaning. Thus the model undermines the notion of a perfect, one-to-one correspondence between representation and reality. It undermines the positivist epistemology of transparent reality and makes room for a moiré effect or a "flicker" of presence/absence, or of difference/sameness—of the "overlapping but not quite matching grids" of mimesis and nonrepresentation, of reader and text.

Nature/machine

The holographic paradigm combines nature and the machine to create another flicker, like the "wink" of the Derridean hymen—that of the organic and the mechanistic. The hologram incorporates organic concepts in that it is a whole whose parts work together, but it does not, like the organic model, suggest self-generation. The holographic model is technologically conceived, but it suggests a view of humankind, nature, science, and art as artifacts—humanly *made* objects, cultural constructs, "texts." (Roger S. Jones, we recall, approaches physics itself as "metaphor," as a linguistic construct or artifact.)

The psychoanalytic project reinforces such an interaction of nature and *techne* in describing the dream process. In analyzing his dream of his uncle with a yellow beard, Freud illustrates condensation in terms of a photographic process:

> I did not combine the features of one person with those of another and in the process omit from the memory-picture certain features of each of them. What I did was to adopt the procedure by means of which Galton produced family portraits: namely by projecting two images on to a single plate, so that certain features common to both are emphasized, while those which fail to fit in with one another cancel one another out and are indistinct in the picture. In my dream about my uncle the fair beard emerged prominently from a face which belonged to two people and which was consequently blurred. . . . (1965, 328)

Freud argues that such "construction of collective and composite figures" is a major technique in the operation of condensation in dreams (328). Condensation, he contends, works photographically to project two images to create a new one and to enhance relationships of similarity (355). This photographic effect, like the view of the brain as a hologram creating a holographic reality, incorporates both the machine and nature in interpreting the working of the human brain.

Emphasis on the Whole

The holographic paradigm suggests that each part of the surface (or volume) contains all the information in the whole, as one cell contains information about the whole body (Bentov 1982, 136–37). The whole is *coded* almost cybernetically in the part; indeed, the insights of information theory are vital to the holographic theory. Medical researcher John R. Battista, exploring Pribram's belief in the complementarity of the holographic model and the dominant analytical model of the brain's function, concludes that both "are based on a set of holistic assumptions" and that, as a theoretical structure, information theory is "capable of integrating both of them" (1982, 144). Battista finds that the reigning Newtonian, mechanical, thermodynamic paradigm is linked in epistemology, to objectivity; in methodology, to empiricism; in causality, to determinism; in analysis, to reductivism; in dynamics, to entropy (144). The mechanistic model is thus linked to many of the notions that recent developments in philosophy, in the philosophy of science, and in literary theory have attempted to subvert, including (1) the elusive and misleading notions of objective observation and of the transparent referentiality of language; (2) the empirical and reductivist approaches that have created a society that worships science and recognizes only reductivism and empiricism as valid "scientific" methods—a society whose acceptance of this epistemology leads to rejection of the insights of art and failure to see the important link between art and science (C. P. Snow's notion of the "two cultures" indeed is not dead); and (3) the deterministic view that, despite the role of accident in Darwin's thinking, has often dominated evolutionary thinking and has migrated (often unfortunately) into fields such as economics (in the form of Herbert Spencer's "social Darwinism"[4]) and into behaviorist psychology.

If the holographic paradigm runs counter to the mechanistic model, then its close link to cybernetics provides an important extension of the analogy. As Anthony Wilden notes, "Wiener's original definition of cybernetics as the science of control and communication included *living* as well as inanimate systems" (1980, xxv). Western thought, Wilden contends, seems "to have occluded this fact" and to have associated cybernetics with "computer science, information theory, and systems design, depriving cybernetics of its ecological insights" (xxv). The holographic paradigm reunites cybernetics with the notion of the whole that includes physiological, psychological and ecological systems. And because information theory is defined by a relationship between a sender and a receiver, the holographic concept incorporates both artist and reader, both the scientific observer and the observed.

But the "whole" of the holographic paradigm also embraces a decentering of the causal thinking privileged in the West, for it is probabilistic rather than deterministic. In cybernetic systems, change "comes about from shifts in the structural relations, or constraints, of the system"—not "free will" and "determinism," but "goal seeking" and "restraints" (Wilden 1980, xxv), terms that suggest interaction rather than opposition. This "whole" embraces complementarity, contradiction, multiplicity, *differance, pharmakon, hymen.* Opposites can coexist and complement each other; the marginal can create the "moiré effect" that is the flicker of an *other* reality. Because both the analytic and holographic models can be subsumed in an "emerging development of a general holistic theory based on information theory," the holographic model can "complement the analytic model and reveal to us a new means by which information can be generated" (Battista 1982, 50). Information theory introduces probabilities rather than absolutes, and, in Battista's terms, it can "predict *what* is occurring," while analytic and holographic models can "explain *how* something is occurring" (50).

INTERACTIVITY

The holographic model suggests the viewer's active participation as he shifts positions to discover the different views and colors that may be obtained from different angles. The hologram is a text that may be read, an image whose part is also an image of the whole, a source of multiple perspectives, an "enfolding" (Bohm's term) of orders, of wave patterns, that the human brain must decipher, which it does holographically. Viewed from one angle, the hologram may appear as a grey blur, but when the viewer actively participates—that is, through positioning the laser beam or through bringing a white-light hologram into the needed light—the three-dimensional image appears.

This interactive and "enfolded" aspect of the model may be understood in terms of Martin Heidegger's concept of the wholeness of the earth: "All things of earth, and the earth itself as a whole, flow together into a reciprocal accord" that he calls an "unfolding," but "not a blurring of their outlines" (1975b, 47). Earth itself, in Heidegger's view, "unfolds." The "blur" is overcome by the "flicker" of the unfolding. Poetry—all art, for Heidegger, is "essentially poetry"—"unfolds . . . the Open which poetry lets happen" (72). The point is the active *process* of the work of art, which suspends the blurring of outlines and suggests by its very nature an interaction between reader and text.

Jacques Derrida, too, working at the margins of meaning, writes of the "folding together of an identity and a difference" (1981, 277). For Derrida, the "event" of mime is an "unfolding"—a "quasi-tearing" or "dehiscence"; the latter is a botanical term for the opening of closed organs along a seam or fold so that the inner parts can be exposed (215, n. 27). Derrida's "unfolding" is the repetition that is never the same from one time to the next. Although Derrida is caught in the botanical, organic metaphor (which, as we have noted, is intertwined with the mechanistic), so that botanical determinism undermines the indeterminacy that he seeks to express, the "dehiscence," or unfolding, suggests an occurrence that is never quite the same at two different times—a differing and a deferring that suggests the uniqueness of the individual experience of the unfolding of the work, whether of art or of science.

RESPONSIBILITY

A major implication of the holographic paradigm for the individual reader is a greater sense of involvement with, as well as responsibility to, the whole. The reader participates in the *whole* creative process; reading itself, as reader-oriented theorists have insisted, is a creative process that undermines the notion of total autonomy and separateness of the text. The paradigm of textuality stems partly from the New Critics' insistence on the autonomy of the text, though "textuality" as I am considering it is a product of structuralist and poststructuralist concerns with the text that does not become a *text* until a reader actively engages it. The "organic" view of New Criticism was unifocal (univocal), enclosed, self-contained, autonomous; the text was not studied as part of an interrelated network of texts, nor was it viewed as embedded in its culture. The New Critical view was related to Ezra Pound's model of observation, based on his tale of Louis Agassiz's student observing a fish. In the tale, the student is repeatedly sent back to "observe" and record a decomposing fish, to "see" the fish as it "really" is. Pound's version of the tale exemplifies the artist's adoption of a visually oriented epistemology of science that assumes observation alone is enough for understanding. As Robert Scholes notes, Agassiz's model of understanding based on careful observation suggests a kind of "latent Platonism" that underlies any attempt "to see the object as in itself it really is" (1985, 136). What Agassiz's student learned to write, of course, was the discourse of his mentor ("Agassizese"). What the New Critical observer learns is to write the discourse of organicism. As Scholes points out, "we 'see' and 'are' in discourse," and "the way to see the fish and to write the fish is first to see how one's discourse writes the fish" (143–44).

The effectiveness of the holographic model, however, lies in its subversion of organicism, which is the old mechanistic determinism in leafy clothing. The holographic model suggests the interrelationship of the whole not only to its parts, but also to the larger whole—the relationship of the text to the *techne* in which it emerges. Rather than returning to the "unity," "order," and "organic meaning" of the New Criticism, the holographic model reinforces the fragmentation of meanings, the multiple angles of perception, which have occupied theorists since Roland Barthes. The text itself is also a *techne* of multiple perspectives that recalls Heidegger's linking of the *poiesis* of the fine arts with *techne*" ("Question" 34). The holographic model that relies on *techne* but also on the artist's/observer's/reader's interaction with the text may renew the ancient link of *poiesis* and *techne*.

The Brain and the World

If the brain functions holographically in response to a holographic world, our tropology assumes a different focus. The holographic model reinforces Umberto Eco's notion of contiguity as the basis for tropology,[5] for it implies that all things are part of an interlinked web and are actually inseparable. Thus *metonymy* becomes the trope of choice for investigation; the brain, the world, and the text form a web—not a series of linear, separable events whose causality can be discovered, but a whole that we can break into at any point, a linked but constantly shifting whole whose fragmentation results not from its nature, but from a mental intertext, a predetermined way of looking at the relationship of world and text.

The holographic paradigm thus serves as a visual and a conceptual model for an emerging holistic view that partakes of Eastern interrelatedness as well as Western fragmentation. Although some students of this model have inferred an underlying connection between the emerging metaphor from physics and Eastern mysticism, the basis of the holographic model is scientific insight, not an amorphous mysticism. Ken Wilber notes that the mystical view includes a hierarchy of levels of consciousness, with the higher levels transcending, but not equivalent to, the lower; and the physicist works at a single level, not at multiple levels of reality (1982, 158–60, 165). In the holographic metaphor, "it is fundamentally the storage of memory information that is said to occur on the principles of optical holography," whose working can be explained mathematically (180). Further, the notion of the "implicate order" should not be interpreted in terms of a kind of mystical transcendence; the implicate order is seen as *underlying* the explicate, not transcending it (168). The usefulness of both the holographic model and the notion of the enfolded, implicate order is in suggesting an alternative to the particularizing of phenomena rather than in suggesting transcendent states of consciousness. As Wilber argues, the work of the scientists who espouse this paradigm "is too important to be weighed down with wild speculations on mysticism. And mysticism itself is too profound to be hitched to phases of scientific theorizing" (185). For the student of literature and science, the holographic paradigm is a visual and conceptual *supplement*—adding to and yet replacing other models, necessary and yet dangerous.

In this detour into the possibilities of the holographic model, I have noted its undermining of two of the oppositions with which Eiseley is most concerned—of determinacy/indeterminacy and nature/machine; and I have suggested the model's emphasis on the whole and on the interaction of artist and materials, of reader and text, of observer and observed, as well as its implications of the reader's responsibility. In "Tradition and the Individual Talent," T. S. Eliot suggests that not only do previous texts affect the new one, but the new text of literature has an impact on previous texts (1974, 785). This early notion of intertextuality both informs and is informed by the holographic model. What Briggs and Peat call the "flowing movement" (1984, 112) of the holographic model provides a new perspective—one partaking of physics and cognitive psychology, of photography and consciousness—from which to approach the phenomena of nature.

Eiseley's shattered mirror is, in its own context, a virtual hologram, a vehicle for exploring dualities and discontinuities—the dualities and discontinuities of an unexpected universe, the ambiguities of excavating one's own life, the noncausal, nonlinear "poetic" or intuitive knowledge that Eiseley accepts as a supplement to science. Like the mirror that Briggs and Peat employ in *Looking Glass Universe,* Eiseley's mirror is a dual model—of the distortion that science forces on seeing (as opposed to the traditional mirror of transparent referentiality), but also of the fragmented wholeness that undermines reductionist "reality." The shattered mirror, finally, is another form of mediated vision: it requires "reading," or interaction of the self and the text in an act of creation/mediation.

11

Margins of Humanism, Margins of Science: *The Star Thrower* and the Dialogue of Understandings

Throughout this book I have asserted Eiseley's concern with the effects of culturally embedded metaphorical concepts and the *trace* that they leave—the trace effect on thinking of a metaphorical concept that enframes our thinking and, like the framework of Kuhn's "normal science" for the professional, interferes with new conceptualizing, with re-viewing or re-imagining the world and our relationship to it. I have explored, too, Eiseley's attempts to undermine concepts (and the metaphors that reinforce them) that arose out of the Cartesian-Newtonian paradigm—reductionism, fragmentation, mechanism, determinism, fixity—and to remotivate such concepts and metaphors or to displace them so that a holistic, nondeterministic, probabilistic, process-oriented, dialogic understanding of the world, with its own tropology, can lead to an *other* approach to knowledge. Increasingly, especially from *The Unexpected Universe* through the autobiography, Eiseley saw himself working at the margins of science and of humanism, though he had long since quarreled with traditional humanism and argued for a "revitalization" of the humanistic view that would displace human centrism and view the human as part of an interlinked web, a symbiotic network that recognizes the interdependency of the planet itself and all its inhabitants, as opposed to the traditional notion of human mastery. For his last collection, whose title essay is, significantly, the essay that most effectively marks his "re-nunciation"—paradoxically both renouncing its myopic reductionism and determinism and re-announcing his allegiance to his "scientific heritage"—Eiseley selected essays from his earlier collections, from introductions to collections and essays published individually in publications such as *The American Scholar* and *Scientific American,* a few poems, and some unpublished material. The three sections of this last collection—Nature and Autobiography, Early Poems, and Science and Humanism—reflect the epistemological thrust that I find paramount in Eiseley's project, examining nature from the perspective of the reflective scientist, examining the role of the dominant sense organs in knowing, and considering the interrelations of science and art. In examining the essays in this collection, I focus on those not included in the collections and on the unpublished work—notably "The Illusion of the Two Cultures" (1964; originally published in the *American Scholar*) and "Science and the Sense of the Holy" (previously unpublished)—that situate Eiseley's work at the margins of science and the margins of humanism, and I further explore his quest for a dialogue of understandings.

The Reach and the Grasp: "The Long Loneliness," "Man the Firemaker," and "The Lethal Factor"

The essays in *The Star Thrower* recapitulate Eiseley's concern with the dominant philosophemes through which we know, as well as his focus on the duality of the human use of these organs of knowing, both their potential for continuous expansion and exploration of the human capacity for knowledge and their potential diversion to destructive uses. He emphasizes the link of hand, eye, and tongue with writing and with our conceptions of intelligence. In his undermining of the Cartesian-Newtonian deterministic, mechanistic view, Eiseley emphasizes the unfixed nature of human potential, expressed figuratively by the polysemy of the "grasp." Yet even as he explores the making of meaning and the making of self, Eiseley himself never escapes the inner-outer dualism of the Cartesian-Newtonian paradigm that he criticizes.

"The Long Loneliness," originally published in the *American Scholar,* draws on then-current research on the intelligence of the porpoise and uses a tutor structure, the imaginary journey, to explore a new perspective for the human being—the perspective of great intelligence locked in a watery world without written communication. In an essay that is more poetry than science, Eiseley explores an unexpected angle of vision. The essay opens with an assertion of human aloneness: we are alone because we know that we are separated by "a vast gulf of social memory and experiment" from other animals, which lack awareness of past and future; and "only in acts of inarticulate compassion, in rare and sudden moments of communion with nature" can we momentarily break from our isolation ("The Long Loneliness," *ST,* 37). Citing Dr. John Lilly's findings of social organization, gregariousness, even altruism, among porpoises, which live in groups and attempt to help the injured, Eiseley explores the link of the dominant philosophemes, particularly the hand, with intelligence. He contends that because we are locked in a particular kind of intelligence, which "is linked to a prehensile, grasping hand" (38) that gives us power over our environment, we cannot understand the perspective of another highly intelligent creature that lacks our "prehensile" capabilities. In an early profession of the holism with which he concludes the autobiography, he contends that the mysteries of the porpoise's behavior cannot be discovered by reductionism ("the dissector's scalpel"), but "involve the whole nature of the mind and its role in the universe" (39). The question that evokes Eiseley's thought-experiment in this essay is whether intelligence equal to the human can exist without leaving "material monuments" such as we erect (39). We think of intelligence in material terms, he suggests, and we now shape the environment that originally shaped us. We express our intelligence through tools and link tool-using with intelligence in "an unconsciously man-centered way of looking at intelligence" (39).

Using the imaginary journey, Eiseley invites us to descend with the porpoise into a different world and to experience an intelligence not linked to altering its surroundings or recording its deeds. In this journey, he takes us beneath the waters, "naked of possessions" but retaining our social tendencies. In our new form, we see immediately what Eiseley has repeatedly treated in terms of the ancient notion of human duality—the human potential for devastation as well as for social behavior. For having "sacrificed . . . hands for flippers," we realize that "no matter how well we communicate with our fellows through the water medium we will never build drowned empires in the coral; we will never inscribe on palace walls the victorious boasts of porpoise kings" ("The Long Loneliness," *ST,* 40). In a radical shift of intellectual orientation, we find that our new world is "not susceptible to experiment" and our thoughts are limited to immediate experiences (41).

The imaginary journey thus becomes a vehicle for exploring the role of writing in human development. Without writing, Eiseley asserts, we cannot retain a sense of history; for example, we have no records of the four great battles with continental glaciers, not because earlier humans lacked the mental capacity to record these battles, but because writing as a "tool" had not been developed. Yet he retains the view of writing as art, for "only the poet who writes speaks his message across the millennia to other hearts" ("The Long Loneliness," *ST*, 41). Linking writing and printing to "our adaptable many-purposed hands" (42), Eiseley notes that other animals are problem solvers and thus manipulators of their world, but the human capacity to preserve written records sets us apart. The verbs he uses for this human ability partake of the paradoxical duality of humankind: we can *open* (41) the doorway to the past; we can *express* ourselves upon the environment (39); we can *impose* (41) our insights on subsequent generations; we can *wreak* (43) our thought upon the world; we can *foist* (40) upon the world the hand as a symbol of intelligence and thus conceive of "intelligence as geared to things" (42).

In the imaginary journey the eye becomes the eye of reflectiveness rather than the eye that coordinates the hand's actions in toolmaking. "It is as though both man and porpoise," Eiseley speculates, "were each part of some great eye which yearned to look both outward on eternity and inward to the sea's heart" ("The Long Loneliness," *ST*, 43). The point is that perhaps we can learn from the porpoise—and here Eiseley becomes polemic—"from fellow creatures without the ability to drive harpoons through living flesh, or poison with strontium the planetary winds" (43). If we had evolved with flukes instead of hands, he speculates, we might still be able to philosophize, but we would lack "the devastating power to wreak . . . [our] thought upon the body of the world"; like the porpoise, the human would remain "the lonely and curious observer of unknown wreckage falling through the blue light of eternity" (43). Only such "observing"—observing without manipulating—can return us to the childhood of humankind, when we talked with living things and lacked the urge and the power to harm. Only an awareness that reinvokes the primitive attitude toward other creatures can remove from humankind "the long loneliness" that makes us "a frequent terror and abomination even to [ourselves]" (44).

In "Man the Firemaker" (1954; originally published in *Scientific American*), Eiseley turns from the hand-eye link in toolmaking and writing to the human use of fire as a tool. It is a tool that underlies the whole of humankind's "long adventure with knowledge," itself "a climb up the heat ladder" (*ST*, 45), which he traces from early humans' use of fire to cook the meat (making it more easily digested) that provided energy and to expand the grasslands for grazing herds, to the use of fire to crack grain, to baking pottery, to making glass (for flasks, telescopes, microscopes, mirrors), to using metals (not just as tools, but metals in furnaces and machinery), to the "atomic furnaces" and heat-resistant substances that make possible the quest for space. But if fire is essential to the epistemological journey, Eiseley insists that humankind is itself a fire that "has burned through the animal world" in its quest for domination (49). And there is no way back; again, we measure civilization in terms of the number of materials it can manipulate (50). Bearing its dual burden, humankind "partakes of evil and of good," which struggle within us (51). We both use fire and are fire, and this is, Eiseley concludes, what we must learn if we are to survive, for "knowledge without greatness of spirit is not enough" (52).

In "The Lethal Factor" (1963; originally published in the *American Scientist*), Eiseley again explores the human hand. In this essay, however, he uses it to address the quality of all organic matter as contrasted with inorganic—a generalized grasping to be more than it presently is, a grasping that, in humankind, is intellectual as well as physical. This "grasping" carries with it the sense of potential in nature and in humankind whose recognition Eiseley considers essential:

> The organic world, as well as that super-organic state which exists in the realm of thought, is, in truth, prehensile in a way that the inorganic world is not. . . . The creature existing now—this serpent, this bird, this man—has only to leave progeny in order to stretch out a gray, invisible hand into the evolutionary future, into the nonexistent. (*ST*, 252)

In this prehensile reaching forward, one insect-eating creature becomes a bat, while a similar creature "draws pictures in a cave and creates a new prehensile realm where the shadowy fingers of lost ideas reach forward into time to affect our world view and, with it, our future destinies and happiness" (252). The importance of the hand's drawing is not just the representation of what the eye sees, but the hieroglyphic communication of one time with another, the impact of what is "seen" (which is shaped by the seer's way of viewing the world) on what will be seen in the future, the impact of the drawn or "scratched" image in constructing a frame for seeing that extends into the future.

For Eiseley, a metaphorical reaching and "fingering" identify organic life so that, "since the dawn of life on the planet, the past has been figuratively fingering the present" ("The Lethal Factor," *ST*, 252). The implications of this reaching forward and fingering are clear in terms of time: "There is in reality no clearly separable past and future either in the case of nerve and bone or within the less tangible but equally real world of history" (252). This organic "groping" is also a groping of thought toward the future, as the metaphor reverses the sequence and the future reaches to the past as the past reaches toward the future. Even extinct creatures may still be having unknown effects; they have "plucked the great web of life in such a manner that the future still vibrates to their presence," and potential creatures are unimaginable even as we grope in thought toward the future (252).

The probing of space, which is an extension of a probing of time, with the tension of positive and negative, reappears in terms of "tentacular space probes" occasioned by both conflict and intellectual curiosity in ("The Lethal Factor," *ST*, 254). Eiseley emphasizes, throughout history, the dual pull of order and disorder, for "behind every unifying effort . . . there is an opposite tendency to disruption" (258). The one truly "lethal factor" that he identifies is human adaptability, which is a form of epistemological reaching now transferred to space itself. Eiseley treats this groping through the physical hand, which "fumbles" toward adaptation and toward shaping a future that is conceptual as well as physical. On an enormous scale, far out in space, human hands are "already fumbling in the coal-scuttle darkness of a future universe" (264). Human adaptability, then, a part of the figurative "reaching" forward, is the mark of both humanity and the human capability of self-destruction.

The Eyes of Summer and Winter: Thoreau as Tutor Text

Not only the hand and writing, writing and knowledge, but the eye and the text occupy Eiseley's attention in this collection. The blending of the eye as sense organ and the eye as metaphor into the two eyes of poetry and reductionism dominates the two essays that blend criticism and speculation, using Henry David Thoreau's essays as tutor texts. In "Thoreau's Vision of the Natural World," Eiseley begins with Thoreau's remark that "there has been nothing but the sun and the eye since the beginning" (*ST*, 233). (The sun is the eye and the eye is the sun, as Thass-Thienemann has noted, with the Gaelic *suil* used for both [1968, 262].) The "eye" that recurs in Thoreau's text is the poetic eye that defies reductionism. Although the science that intrigued Thoreau would "reduce everything to infinitesimal particles and finally these to a universal vortex of wild energies" ("Thoreau's Vision," *ST*, 233), still Thoreau retained the poetic "eye," no less insistent even though what it experienced was reduced by science to "secondary qualities, the illusions that physics had rejected" (233). Drawing on his own experience among the leaves, Eiseley combines Thoreau's insight with his own: "We were particles but we were also the recording eye that saw the sunlight—that which physics had reduced to cold waves in a cold void. Thoreau's life had been dedicated to the unexplainable eye" (233). The point is that Thoreau, as process philosopher, anticipated what the twentieth century has discovered, the observer's participation in, engagement of, what is seen. This participation of the observer incorporates the "poetic eye," for the individual in Thoreau's world "is forever the eye" (234), participating in a metaphorical "seeing" of process rather than fixed reality.

Eiseley's notion of the function of art is further linked to the "eye" in Walden: Thoreau's Unfinished Business." For Eiseley, "looking is in itself the business of art" (*ST*, 236), no less than of science. Walden Pond is metaphorically two eyes, the eyes of "alternate glazing and reflection"; allegorically, the human being is Walden's "eye of ice and eye of summer" (236–38). One human eye, like Walden Pond, is "gray and wintry and blind," but the summer eye of the poet perceives "another world just tantalizingly visible," though the reductionist dismisses it; this summer eye of poetic or intuitive observation is "the alchemist's touchstone" that Eiseley perceives in Thoreau—an eye that sees the need for balance between the two if we as humans are to "see" in the larger allegorical sense (238). It is only the human mind, "the artist's mind," Eiseley concludes, "that can change the winter" in humankind (238). Eiseley's alchemist's eye does not see transparently, nor does the reductionist's. But the alchemist's eye is aware of its own distortion. The human eye, Eiseley contends, provides an ongoing dialogue with the old paradigm of science, for it "constitutes an awesome crystal whose diffractions are far greater than those of any Newtonian prism"; and the artist helps to shape the "oncoming world . . . by the harsh angles of truth . . . as glimpsed through the terrible crystal of genius" (249–50). Although the "terrible crystal" does not represent perfectly, but distorts, it serves as a counterforce to the equally distorting vision of "objective" science. It is self-reflexive, aware of its own mediation.

Reexploration of the traditional onto-theological metaphor of the sun and the eye thus leads Eiseley into a maze of eye metaphors, but also into a reading of Thoreau that stresses the mediating, distorting power of the eye as opposed to traditional scientific notions of unmediated observation, reality as process and change as opposed to fixity and decidability, and the importance of the unexpected or the inexplicable in nature as opposed to nature as expected, reduced to parts, understood, and controlled. In the *Notebooks* Eiseley writes that the "eye" Thoreau described at Walden is "multitudinous, ineradicable," and that we are "all the eye of the Visitor—the eye whose reason no physics could explain" (*LN,* 231). The eye as metaphor continues Eiseley's project of displacing established metaphors, but provides an opportunity to explore an "other" way of seeing as he explores the relationship between the physical senses and understanding of the human place in the natural world.

In Thoreau's writings Eiseley finds further stimulus for exploration of nature's hieroglyphs. True to the nature of hieroglyphics, which ambiguously sometimes signify either an idea or a sound, an ideogram or a phonetic sound, Thoreau explores the notion of nonphonetic writing when he describes the snow as a kind of tablet on which "were inscribed all the hieroglyphs that the softer seasons concealed" ("Thoreau's Vision," *ST,* 232). Seeing nature's figures as signs of process, Thoreau ultimately found the text self-undermining—like the human text, "as unreadable as it ever was" ("Thoreau's Unfinished," *ST,* 237). For Thoreau, humankind, too, left hieroglyphs, but they were "tiny and brief" (242) in relation to all of nature. Arrowheads were "humanity inscribed on the face of the earth" as "mindprints" or "fossil thoughts" (239).

Taking Thoreau's coinage, *mindprint,* as a unique and simple term for his hieroglyphs, Eiseley draws from the tutor text his own suggestion that both nature and the first and last humans share the leaving of mindprints. Uncannily modern, Thoreau searched nature for its codes as does science today. Nature, for Thoreau, constituted a journal of hieroglyphs, but one whose "script was always changing, like the dancing footprints of the fox on icy Walden Pond" ("Thoreau's Unfinished," *ST,* 242). Textuality is Eiseley's dominant metaphor for achieving knowledge, but to "read" this journal requires interaction with the text, awareness of the "tiny and brief" human role (242), and acceptance of the interdependence of life and nature. Looking at a quartz knife perhaps ten thousand years old, Eiseley adds his own suggestion to Thoreau's reading of the arrowheads or "mindprints": "They were free at last. They had aged out of human history" (243). As a mere extension of nature's hieroglyphs, the arrowhead reflects the role of the artist, who helps to shape the "oncoming world" (250). What is left behind is art, civilization, "mindprint." Thoreau's notion of "mindprint" reinforces Eiseley's image of nature itself as a set of nonphonetic symbols that the individual can read projectively, but it also places human civilizations in perspective as merely part of nature's hieroglyphs, leaving their "tiny interpretable minds" embedded as signs (240).

Darwin, Freud, and Melville's Ship of Questions: "Science and the Sense of the Holy"

> Tomorrow lurks in us, the latency to be all that was not achieved before. This is what led proto-man, five million years ago, to start upon a journey, at a time when night and day were strange and miraculous. . . . It was for this that man adorned his caverns in the morning of time. It was for this that he worshiped the bear. For man had fallen out of the secure world of instinct into a place of wonder. That wonder is still expanding, changing as man's mind keeps pace with it.
>
> —Eiseley, "Man Against the Universe"

Opening with an analogy between the gullies of the Nebraska badlands and the human brain that "contains the fossil memories of its past," Eiseley states in "Science and the Sense of the Holy" (previously unpublished, perhaps because of the criticism that he knew it would evoke in professional circles) that this analogy has preoccupied him and affected his view of the natural world (*ST*, 186). He explains the problem by contrasting the attitude of Charles Darwin at twenty-eight (vividly evoked in the conclusion of *Darwin's Century*) and that of the mature Sigmund Freud. The contrast, he contends, only illustrates a difference in views that has existed throughout the history of science, and he uses it to make the point that he made as early as *The Firmament of Time*— the role of nonobjective, non-"scientific" factors in science itself. For the contrast makes clear "that, in the supposed objective world of science, emotion and temperament may play a role in our selection of the mental tools with which we choose to investigate nature" (187).

Darwin's view is one of sympathy with animals, at a time when the learned viewed animals "as either automatons or creatures created merely for human exploitation"; he speculates that "'animals, our fellow brethren in pain, disease, suffering and famine, . . . may partake of our origin in one common ancestor—we may be all netted together'" (cited in p. 187). The comment, Eiseley contends, "gives every sign of that feeling of awe, of dread of the holy playing upon nature, which characterizes the work of a number of naturalists and physicists down even to the present day," for it "reveals an intuitive sensitivity to the life of other creatures about him," as opposed to the insensitivity of the experimentalist who does not hesitate to inflict pain ("Science and the Sense," *ST,* 187). Eiseley calls Darwin's comment a speculation "that we are in a mystic sense one single diffuse animal," and he argues that it indicates a concern for nature that goes beyond the dissecter's curiosity (187). Freud's attitudes contrast sharply with Darwin's, for he expressed "distrust of that outgoing empathy," relegating "feelings of awe before natural phenomena" to the status of childhood remnants (188). In Freud's statement that he could not find in himself what a friend called the "'oceanic' impulse," Eiseley argues that Freud is "explaining away one of the great feelings characteristic of the best in man" (188).

From the early humans who drew images on cave walls, Eiseley asserts that the human being has looked "to the numinous, to the mystery of being and becoming, to what Goethe very aptly called 'the weird portentous'" ("Science and the Sense," *ST,* 189). This feeling, Eiseley continues, "is part of the human inheritance, the wonder of the world, and nowhere does that wonder press closer to us than in the guise of animals" (189). Eiseley finds two approaches: that of the educated individual who retains "a controlled sense of wonder before the universal mystery" and that of "the extreme reductionist who is so busy stripping things apart that the tremendous mystery has been reduced to a trifle" (190). For the second individual, "the world of the secondary qualities—color, sound, thought—is reduced to illusion" and reality to "the chill void of ever-streaming particles" (190). Troubled by the cruelty that the latter view can bring, Eiseley quotes Blaise Pascal, who wrote in the seventeenth century of "'two equally dangerous extremes, to shut reason out, and to let nothing else in'" (190). Eiseley contrasts Sören Kierkegaard's sense of the "unintelligibility" of existence with Ernst Haeckel's comment in 1877 that the reduction of the cell to matter explains the universe (191). And he quotes Einstein's comment that "'a conviction akin to religious feeling of the rationality or intelligibility of the world lies behind all scientific work of a high order,'" comparing it with Thoreau's question, "'Am I not partly leaves and vegetable mould myself?'" (191).

The second part of the essay describes a modernist painting by Irwin Fleminger (to which I referred in establishing Eiseley's existential worldview). Titled *Laws of Nature,* the painting depicts a waste land with two laths linked by a string from which hung other filaments. The human being, who was "diminished by his absence," had apparently tried "to delineate and bring under natural law an area too big for his comprehension," and the result was that the human "law" "denoted a tiny measure in the midst of an ominous landscape looming away to the horizon" ("Science and the Sense," *ST,* 192). To Eiseley, the painting suggested a sense of the "other," a sense of awe in the presence of waste land. And "perhaps," he suggests, the painting also suggests the "measure" of humankind, who "perhaps . . . has already gone" (192).

If in the reductionist view the numinous is associated with superstition, Eiseley argues that the individual who retains "reverence and compassion" can see beyond "the two slats in the wilderness," whereas the contrasting view produces "the educated vandal without mercy or tolerance," whose "sense of wonder [is] reduced to a crushing series of gears and quantitative formula" ("Science and the Sense," *ST,* 193). It was, in fact, the individual whom Darwin called "speculative" who developed the understanding of evolution—the individual who retained "a sense of marvel, a glimpse of what was happening behind the visible, who saw the whole of the living world as though turning in a child's kaleidoscope" (193). The term *speculative,* as we have seen, carries the multiple senses of sight and of philosophical reflection, but it also suggests childlike or primitive openness to the unexpected as opposed to the positivist nature of the universe. The individual who is open to the shift, who can follow rather than resist the kaleidoscopic shift, is, in Eiseley's reading of Darwin, the one who can solve scientific problems. The kaleidoscope as a child's toy links childlike openness and wonder to the world's great scientific revolutions. As the colors shift, the wonder of "speculative" seeing remains. Induction and division into parts are not enough; the shift comes as the parts are reorganized into a new whole. In Eiseley's evolutionary journey, a major intersection involves a child's toy that helps to displace notions of fixed truth or direct perception of "truth" or "reality" in favor of change, undecidability, and the child's wonder at the unexpected. The whole is not just a reductionist sum of its parts, but also a unity of shifting forms and colors. The child's toy reinforces Eiseley's call for a sense of wonder that seems to decrease as we encounter rigid methodologies—as we become dancers in the ring rather than speculative, imaginative individuals like those "amateurs" who have been responsible for some of the most important scientific insights.

Conceding that Freud may have been right, Eiseley nevertheless asks the reader to look again at the human brain, and he leads us through another consideration of pedomorphism—of the human potential encased in a cranium hardly larger than that of a gorilla at birth, but triple that size after a year of life, of a brain that makes the individual totally dependent at first, but contains the potential for speech and numbers and sympathy. *Potential,* not determined, programmed ascent, explains the human "niche," which is made possible through language and thought. Thus the world itself has become the human niche, but "without the sense of the holy, without compassion," the human brain becomes "a gray stalking horror—the deviser of Belsen" ("Science and the Sense," *ST,* 196). For Eiseley, then, "in science, as in religion, when one has destroyed human wonder and compassion, one has killed man, even if the man in question continues to go about his laboratory tasks" (198).

In the last part of the essay Eiseley turns to Melville's *Moby-Dick* to exemplify in the form of Ishmael and Ahab the confrontation of the two views of life. Melville's fiction, Eiseley asserts, embodies the ancient questions that are still the questions of science: "Here, reduced to the deck of a whaler out of Nantucket, the old immortal questions resound, the questions labeled science in our era. Nothing is to go unchallenged" ("Science and the Sense," *ST,* 198). Ahab pursues the whale because he has an "obsession that lies at the root of much Faustian overdrive in science," for "like Faust he must know, if the knowing kills him" (198–99). Like Haeckel and others, he relies on means that are sane, though his object is mad (199). Ishmael supplies the contrast. He is not only a wanderer, but "the wondering man, the acceptor of all races and their gods" (199). And Ishmael "paints a magnificent picture of the peace that reigned in the giant whale schools of the 1840s," where "the weird, the holy, hangs undisturbed," as Ishmael knows though others do not (199). For Eiseley, though "the tale is not of science, . . . it symbolizes on a gigantic canvas the struggle between two ways of looking at the universe: the magnification by the poet's mind attempting to see all, while disturbing as little as possible, as opposed to the plunging fury of Ahab" (200). And he concludes, "Within our generation we have seen the one view plead for endangered species and reject the despoliation of the earth, the other has left us lingering in the shadow of atomic disaster" (200).

The essay ends with another juxtaposition of disparate journeys—a combination of the archetypal machine of the nineteenth century (associated, perhaps more deeply than we realize, with the Newtonian world machine[1]) and the flower that Eiseley repeatedly associates with what he considers a much-needed return to the primitive, the sunflower. All summer he has watched a sunflower growing on some clods atop a boxcar, only finally to begin a new journey as the boxcar is linked to a train. The sunflower, he knows, will not survive, but the seeds will be scattered abroad. Drawing the reader suddenly to the self-referential nature of the text, Eiseley asserts that he and the sunflower are travelers like Ishmael, escaping, like him, to tell their stories.

From the journey of Ishmael, the wandering/wondering man, to the journey of a flower (we recall Eiseley's early essay on how flowers synecdochically changed the world [*IJ*, 61–77]), the journey of science (Ahab) or of humane or poetic knowledge (Ishmael) is linked to the physical journey of life and to the human journey. Human intervention is omnipresent. The sunflower is growing in an unlikely spot atop a sidetracked boxcar that only in late summer becomes part of a train, but it is growing. Eiseley recalls in the autobiography many hours spent atop a moving train, yet somehow surviving and flourishing. The sunflower is life itself, surviving thus far in spite of human mechanical intervention, wandering—without destiny—because of mechanical intervention, yet able like Ishmael to say, "'I only am escaped to tell thee'" (cited in "Science and the Sense," *ST,* 201).

Reversing the Epistemological Hierarchy: Darwin and Emerson in "Man Against the Universe"

Eiseley's "Man Against the Universe," previously unpublished, takes the code phrase of the title and works toward a definition of both the human and the universe, reflecting on the five hundredth anniversary of Copernicus's birth and the lengthened time period for human and proto-human life. Despite the impact of Copernicus, Eiseley argues that human self-examination long predates the self-reflection that Western thinking experienced after the Copernican revolution. Eiseley sees in Job's confrontation with the whirlwind and ancient Oriental notions of the world as illusion the human sense of alienation and awe at the natural world. Reiterating that no single individual is responsible for a single intellectual achievement, Eiseley continues that the individual speaks not to one audience, for humankind "is not one public" but many, and human messages may not be received until centuries later and, whenever they are received, "are likely to become garbled in transmission" ("Man against the Universe," *ST,* 208–9). Eiseley concludes "that no great act of scientific synthesis is really fixed in the public mind until that public has been prepared to receive it through anticipatory glimpses" (209).

The main point of the essay, however, is an exploration of Darwin and Emerson as tutor texts in which Eiseley reverses the hierarchy of "romantic" and "scientific" thinkers, comparing the two "as opposed yet converging forces in nineteenth-century thought" ("Man Against the Universe," *ST,* 210). Examining the romantic revolt in terms of "assertion of the self against the universe," Eiseley looks at a self that had "escaped" from custom and formality "into a wilder nature of crags and leaping torrents," a nature "full of picturesque revolutionaries and moon-haunted landscapes" (211). In America, Emerson emerges as "a human ego sustained by a creative power greater than itself yet capable of assimilation by the individual" (211). Yet Emerson was well read in science, having read Lyell and having also discussed Ice Age human beings with Louis Agassiz. Emerson would not have written, as did Darwin, about "'the clumsy . . . blundering and horribly cruel works of Nature'" (quoted in p. 213), just as Darwin would not have written of himself as a "transparent eyeball," yet Emerson anticipates Darwin's view of nature as well as the unconscious, "the fauna contained in man's own psyche" (212–13). If Darwin saw the horrors in nature, Emerson also saw "'a crack in everything God has made'" (quoted in p. 213). For the Romantic, Eiseley notes, the scale of nature was deteriorating, and "the existent began to be replaced by process" (214). Emerson turned to stairway metaphors that curiously anticipate Borges's stairway that ends in space, as he wrote: "'We wake and find ourselves on a stair; there are stairs below us which we seem to have ascended, there are stairs above us . . . which go out of sight'" (quoted in p. 214).

Claiming induction but drawn to the "speculative" and the romantic, Darwin was representative of his age, "swept along" in his youth "in the romantic current which included enthusiasm for Odyssean voyages, evidences of past time, and the looming shadow, not just of tomorrow, but of a different tomorrow in some manner derived from today" ("Man against the Universe," *ST,* 214). When Keats wrote about the struggle for existence, about the "'sea, where every maw / The greater on the less feeds evermore,'" Darwin was still a child (cited in p. 214). And Darwin appealed to thinkers who relied on the imagination, the extreme arguments, the enthusiasm of the Romantic age ("Man against the Universe," *ST,* 215). Yet in writing the conclusion of the *Origin,* Darwin submerged the harsh elements of nature in a bland prediction that natural selection works for the good and for a tendency "'to prog-

ress toward perfection'"—that is, the production of "'higher animals'" (quoted in p. 215). This is a view that Eiseley places in the
context of Romanticism and nineteenth-century notions of Northern European *man* as the highest achievement on the planet.

Yet Emerson wrote ironically, "'What is so ungodly as these
polite bows to God in English books?'" (cited in "Man Against the
Universe," *ST*, 215). Instead of such "polite bows," Emerson looks
to process in a nature that "'will not stop to be observed,'" a
nature whose "'smoothness is the smoothness of the pitch of a
cataract,'" whose "'permanence is a perpetual inchoation'" (cited
in p. 215). Emerson's passage, Eiseley contends, would "have
graced Darwin's final paragraphs in the *Origin of Species*," for
Emerson concluded that nature has "'no single end'" and "'that if
man himself be considered as the end, and it be assumed that the
final cause of the world is to make holy or wise or beautiful men,
we see that it has not succeeded'" (cited in pp. 215–16). Nature for
Emerson, has no finality, no will. Adopting Emerson's metaphor,
Eiseley asserts that "crouched midway on that desperate stair
whose steps pass from dark to dark, he spoke as Darwin chose
not to speak," for

> Emerson saw, with a terrible clairvoyance, the downward pull of the
> past. . . . He could sense, not Darwin's automatic trend toward perfec-
> tion, but the weary slipping, the sensed entropy, the ebbing away of
> the human spirit into fox and weasel as it struggled upward while all its
> past tugged upon it from below. ("Man Against the Universe," *ST*, 216)

Though Emerson lacked Darwin's understanding of natural selection, in Eiseley's view he came closer to anticipating later understandings of the human relationship to the world and the human
responsibility for shaping our destiny in an evolutionary world than
did Darwin, who could "speculatively" devise the "key" of natural
selection, but then bowed to cultural orthodoxy (both scientific
and religious) to superimpose on nature a grid of progress toward
perfection for the dominant nation, race, and class.

In the last section of the essay Eiseley turns to his own recollection of a shell in which, as a child, he believed he heard the sea.
His individual experience recapitulates the human experience of
change: "I had suddenly fallen out of the nature I inhabited and
turned, for the first time, to survey her with surprise" ("Man
Against the Universe," *ST*, 217). The point is the nature of the
human being—variously defined by using tools, by laughing or
weeping, by the duality of flesh and spirit. But as Emerson saw,

both science and theology are involved in constructing "nature" by means of a quest for order and for understanding—a quest that Eiseley links to "those mocking questions from the whirlwind" (218). For Eiseley, the modern human dilemma is not, as Freud contended, the product of science but is the human dilemma, expressed in the phrase "the fall out of nature into knowledge" (218). Emerson noted the lack of purpose in nature, yet he focused his work on the very species that "he had so eloquently dismissed" (218). Eiseley's realization is epistemological: if objects and species emanate from one another, "our very thoughts transform us from minute to minute, hour to hour" (219). The shell no longer emblematizes the sea, but the meaning that the human being can create. In Emerson's view, the human being is always departing, always in process. Like Darwin, Emerson experienced both illness and a voyage, as did many nineteenth-century romantics, who brought "something wild and moon-haunted" to both science and art (220). Emerson sensed the latency in humankind, the striving of nature to produce more than its present forms. It was this latency that began the journey of knowledge. Latency is what distinguishes the human, in Eiseley's view, for we have "fallen out of the secure world of instinct into a place of wonder," and the wonder expands as we, like children, hear messages in shells and peek into the cupboards of nature's "old house" (221). Both latency and wonder have made us "inheritor[s] of an echoing and ghost-ridden mansion," with only one mandate, "to grow" (221). The decision of how we grow—in vicious competition or in cooperation with that "emanation" of things—is the burden of such freedom, the burden that made Emerson write about the imperfect human being even in a nonteleological world.

Conclusion: Eiseley's Last Quarrel with the Institution— "The Illusion of the Two Cultures"

One of the last three essays in this collection, Eiseley's "The Illusion of the Two Cultures" (1964; originally published in *The American Scholar*) stands as one of his most powerful "re-nunciations"—a reassertion of his quarrel with the scientific institution that he repeatedly questioned and undermined yet nevertheless embraced. The essay both renounces the particularization and fragmentation associated with the scientific institution and asserts that institution's link with imagination, with the *techne* of ancient times. Focusing on the epistemological quest and again on the self-creation that memory makes possible, Eiseley pursues his quest for a dialogue of understandings. Using an image that recurs throughout the texts and diction that recalls Whitman's "A Noiseless, Patient Spider," Eiseley describes the highest intellectual power in terms of its ability to stretch "an invisible web of gossamer . . . into the past as well as across [living] minds and . . . constantly [to respond] to the vibrations transmitted through these tenuous lines of sympathy" ("The Illusion," *ST*, 267). Yet in Santayana's terms, as visible nature has become "abstracted," we have seemed to try to escape, at least to become ashamed of, imagination (268).

Eiseley responds strongly to C. P. Snow's *The Two Cultures;* although Snow was reporting, not advocating, the bifurcation of the world of learning, his statement became a target for much controversy, and Snow emerged as a spokesperson for an oversimplified view. Against Snow's statement that the world of learning has become divided into "two cultures," with literary intellectuals and scientists representing separate poles, each of which needs to understand the other, Eiseley juxtaposes the ancient link between the fragmenting, quantifying, measuring approach of science and the more holistic, imaginative, "making" approach of art. He declares Snow's "bipolar division" to be "a peculiar aberration of the human mind" that exists on both sides of the controversy, and he argues that Snow's approach was emergent as early as the seventeenth century, noting that scientists themselves attacked early evolutionists such as Lamarck and Chambers for giving rein to the imaginative faculties ("The Illusion," *ST*, 268). In the arguments of those who urge poets to write about molecular structures, Eiseley finds an attempt to reduce even literature to empiricism and to omit "the whole domain of value, which after all constitutes the very nature of man, as without significance" (269).

In "The Illusion" Eiseley undermines the mystification of science and demonstrates the instability of positivist knowledge; at the same time, he underscores the vitality of the poet's insight.[2] Because his concern is always with knowledge and how we acquire it, he contends that awareness of scientific principles need not exclude poetic understanding. Science, he maintains, is not an end, but a means to what Bacon called "the uses of life"—all life, not just human life. Reducing Snow's dichotomizing to "illusion," Eiseley argues that the greatest danger of such dichotomizing is in the human perception of self, for the human being, "the tool user, grows convinced that he is himself only useful as a tool, that fertility except in the use of the scientific imagination is wasteful and without purpose" (*ST*, 269).

Recognizing the irony that his own image as a scientist might call into question his foray into "matters involving literature and science" ("The Illusion," *ST*, 270), Eiseley focuses on a prehistoric flint whose carver embellished his tool with an artist's touch and thus left "an individual trace . . . which speaks to others across the barriers of time and language" (271). Though some would argue that he "had wasted time," this carver clearly "had lingered over his handiwork," for in the carver's dangerous world there was no place "for the delicate and supercilious separation of the arts from the sciences" (271). The carver's stone is a link between utility and art, between *techne* and *poiesis*. The simple "feel of worked flint" (270) serves as a vehicle that links the "two cultures," both as object and as symbol. The individual who worked the stone, Eiseley speculates, must have enjoyed its texture and possessed an appreciation of its beauty. With the stone as vehicle, he argues that artistic creativity and science are inextricably interlinked, disavowing any attempt to focus on individuals but approaching science as a cultural institution that, like any institution, may develop "behavioral rigidities" that are useful to the "mediocre conformist" who takes comfort in "authoritarian dogmas" and in the "impenetrable wall of self-righteousness" ironically spawned by the notion that science is "totally empiric and open-minded by tradition" (272). Eiseley turns, then, to a reexamination of language to explain the link that the carver's stone symbolizes.

Symbols, he contends, are equally necessary in both art and science. Scientific creation requires "a high level of imaginative insight and intuitive perception" ("The Illusion," *ST*, 273). "It is the successful analogy or symbol," he argues, "which frequently allows the scientist to leap from a generalization in one field of thought to a triumphant achievement in another" (274). Progressionism, which carried a spiritual connotation, modeled evolution. But the point is not just the *use* of analogies: these analogies contribute to the thought, shape the thought, and make possible new understandings. For Eiseley, scientific "analogies genuinely resemble the figures and enchantments of great literature, whose meanings similarly can never be totally grasped because of their endless power to ramify in the individual mind" (274). Yet despite the "tension" between attitudes toward literal and figurative meaning (which, in a digression through a passage from John Donne, Eiseley contends are not so easily separable as some believe them to be), tropes provide a means of entry into understanding. To illustrate, he again pursues a tropic reading of nature itself that the prepared mind could apply to natural evidence of the length of time. "Such images drawn from the world of science," he continues, "are every bit as powerful as great literary symbolism and equally demanding upon the individual imagination of the scientist who would fully grasp the extension of meaning which is involved" (275). In a statement that must have rankled many of his contemporaries who had not taken the step of recognizing the role of language in shaping both the observation and the reporting of "facts" in science, Eiseley links the reading of nature and the imaginative perception of tropes in both literature or science: "It is, in fact, one and the same creative act in both domains" (275).

Evolution itself is "such a figurative symbol," Eiseley continues, exploring the implications of this troping for the nonscientific world. Yet "figurative insights" such as evolution or the notion of the expanding universe are able to "escape" from the professionalized world, and once outside the control of the scientific milieu, they may undergo an evolution that anticipates what we know of the chaotic behavior of all phase transitions. They evolve unpredictably, as art evolves, bringing both "enrichment and confusion," but also "something suggestive of the world of artistic endeavor" ("The Illusion," *ST*, 275). Such "figurative insights" partake of the dual human potential, for they can "become grotesquely distorted or glow with added philosophical wisdom" as the understandings go beyond the original thought intended or expressed (275). Thus Eiseley argues that, "in a sense, the 'two cultures' are an illusion, . . . a product of unreasoning fear, professionalism, and misunderstanding"—all of which are more evident in younger disciplines than in older and more "secure" disciplines (276). Yet the great minds transcend such fear and misunderstanding, and Eiseley invokes the names of Leonardo (who preceded science itself), Darwin, Einstein, and Newton, all of whom "show a deep humility and an emotional hunger which is the prerogative of the artist" (276).

Anticipating the dialogue of understandings that has begun to emerge in recent years—particularly in the work of Hayden White, Gerald Holton, Lawrence Prelli, Alan Gross, Donna Harraway, and N. Katherine Hayles—Eiseley further argues that the "sociology of science deserves at least equal consideration with the biographies of the great scientists, for powerful and changing forces are at work upon science, the institution, as contrasted with science as a dream and an ideal of the individual" ("The Illusion," *ST*, 277). Science is, he contends, a "construct" of human beings "and is subject, like other social structures, to human pressures and inescapable distortions" (277). Citing Thomas Hobbes's castigation of poets as members of "'another kingdome, as it were a kingdome of fayries in the dark,'" Eiseley asks the reader to consider this kingdom, concluding that we are "not totally compounded of the nature we profess to understand," for we contain—again in terms that anticipate the predictable/unpredictable elements of chaos—"a lurking unknown future," of which human culture was once a part until we became able to talk and write of what we know (277). Both Bacon and Shakespeare, he continues, perceived the "creativeness" that humans add to nature (278), a creativeness that is a redoubling of nature's own creativity. This notion, expressed by Henri Bergson in terms of the "indetermination" that life brings to matter, is thus not new, but a kind of "intrusion from a realm which can never be completely subject to prophetic analysis by science," just as evolution's pathways cannot be predicted (278). Analyzing the semantic element of his notion of "a nature beyond the nature that we know," Eiseley returns to the point of "How Natural Is 'Natural'?" as he explains that, with repeated reinterpretations, the natural realm has become "as weird as any we have tried, in the past, to exorcise by the brave use of seeming solid words" (278). His conclusion is that we seem to have grown afraid of "mystery and beauty" and of the limits of human power, afraid of thoughts not "clothed in safely sterilized professional speech" (278), which, as he has tried to demonstrate, does not exist.

In concluding, Eiseley asserts that "all talk of the two cultures is an illusion" (279). If, as he has suggested elsewhere, reality itself is illusion, *maya,* it is fitting that what Lord Snow's contingent takes most seriously—a reified, institutionalized approach to knowledge—should be treated as the source of an even greater illusion, the illusion that human culture can be split into separate, describable parts. Only the artifact can tell the human story, Eiseley concludes, for it reveals the "two faces" of humankind, "the artistic and the practical" ("The Illusion," *ST,* 279). If today we are obsessed with the power that science, through technology, brings, the "artistic imagination" is capable of leading us beyond the quest for power (279).

This essay—standing as it does before the reprinted "How Natural Is 'Natural'?" in which Eiseley questions, undermines, and attempts to demystify the twentieth-century notion of the "natural," and "The Inner Galaxy," in which he examines the self-created nature of reality, the feeling of love beyond the boundaries of species form, and the galaxy of potential in life itself—serves as Eiseley's last quarrel with the institution with which he identified and which he was at pains to examine and explain (even, as Christianson has suggested,[3] to represent—in the sense of re-presenting, of creating an image for—a science widely distrusted by the general populace). "The Illusion," then, in the context of the other essays in *The Star Thrower,* brings together the elements of Eiseley's project: recognition of the essential metaphoricity of language; reexamination of the role of language in structuring the scientific view, including the role of metaphor and its "traces" in structuring thought, as exemplified in the role of the Cartesian-Newtonian-Lockean mechanistically oriented view of the world in shaping thought even in the late twentieth century; working within the scientific institution to undermine for the nonscientific reader the positivism, reductionism, determinism, and frag-mentation, and master-y that linger in scientific understanding even after the collapse of the Cartesian-Newtonian paradigm. The essay further recapitulates Eiseley's attempt to re-figure understandings of science in terms of a dialogue of science and the humanities, in terms of an engaging of the intuitive and the rational, and in terms of a holistic, participatory view that displaces the human from the center of the universe and sees life itself as interactive and intertwined, as the young Darwin (before he had absorbed the dominant views of his century) saw the human and the animal "netted together."

"The Illusion" is a fitting conclusion to the project of a thinker who, in spite of his protestations, often did "pontificate" but who repeatedly referred to himself as an "other" thinker or a "humbler" thinker. Like the figures Rorty situates at the "periphery" of modern philosophy—thinkers who doubt notions of "progress" and of "knowledge as accurate representation," who argue that the privileging of a vocabulary derives from its users rather than from its "transparency to the real," and who try to leave room "for the sense of wonder which poets can sometimes cause" (1979, 6, 367–70)—Eiseley works at the margins of disciplines. Believing that the collapse of the Cartesian-Newtonian paradigm as the single model for understanding must be internalized in other minds than the scientific if a new worldview is to be accepted, even devised, among the nonspecialist population, Eiseley not only undermines the figures embedded in the old paradigm but explores the implications of its collapse for human activity. Well before the instability of knowledge had become a fashionable topic, Eiseley painstakingly centers his project on the parallel of the instability of life forms and the instability of human knowledge. As an "intentionally peripheral" and "reactive" (Rorty 1979, 369) figure, he makes room for poetry and carnival, for questioning and quoting, for tales of the old West and tales of city-dump philosophers. Although he never escapes the inner-outer dualism of the Cartesian paradigm, his contribution as a marginal figure questioning the means of knowledge, the stability and certainty of knowledge, and the human role in helping to shape our own unfixed destiny deserves, I believe, the "long second look" that this book has attempted to present.

Notes

Chapter 1. The Scientist as Writer:
Eiseley and the Ironic Imperative

1. For an important recent collection of essays by major thinkers, see *The Interpretive Turn: Philosophy, Science, Culture* (Hiley, Bohman, and Shusterman 1991).

2. Jonathan Culler develops the notion of Barthes's shift from structuralism to poststructuralism (a point that Seamon does not address) in *Roland Barthes* (1983b; see especially pp. 84ff).

3. For details that support this summary of the hieroglyphic and phonetic view, see Irwin 1983, 1–240.

4. For an excellent summary of what she calls the "isomorphism" of indeterminacy in science and in literary and language studies, see Hayles's *Chaos Bound* (1990, xii–xiv).

5. For further details on *S/Z* as both a monument of structuralism and a seminal work of poststructuralism, see Jonathan Culler's *Roland Barthes* (1983b).

6. For a concise analysis of the Vienna Circle's role in the history of science, and of the roles of Feyerabend, Kuhn, Polanyi, and Toulmin in "dismantling" its "foundationalist program," see Prelli 1989, 2–5.

7. For a rhetorically powerful argument about the successes of the fragmenting approach and its limitations, see Capra, *The Turning Point* (1983, 99–262).

8. For explication of the concept of root metaphors or *philosophemes*—"founding ideas of philosophy"—see Ulmer's *Applied Grammatology* (1985, 31–67).

Chapter 2. A Route for Critique: Eiseley's Genre

1. "Poetic knowledge," as I believe Eiseley uses it, indicates intuition, empathy or the ability to identify with other creatures than the self, awareness of the web that interlinks all life, and understanding or insight that cannot be measured or quantified, but is no less real.

2. See Pickering 1982 for a study of these "personas."

3. See Roland Barthes's notion that "we read a text (of pleasure) the way a fly buzzes around a room: with sudden, deceptively decisive turns, fervent and futile" (1975, 31). Though Barthes writes of the text of pleasure, I submit that the reader's response to almost any text is one of stopping and starting—a kind of push-pull, stop-start, insight-aporia that occurs as responses occur. As the reader becomes aware of an insight, there is a contrary pull to stop and play with that insight, a tendency to buzz off into new directions of response. "The double movement of revelation and recoil will always be inherent in the nature of a genuine critical discourse," writes Paul de Man (1985, 289).

4. In the concluding chapter of *The Structure of Scientific Revolutions*, Kuhn,

in rejecting the notion "that changes of paradigm carry scientists and those who learn from them closer and closer to the truth," cites Eiseley's account in *Darwin's Century* of pre-Darwinian attempts to see evolution as fulfilling a goal (1969, 170–72). He does not, however, note Eiseley's references in *Darwin's Century* to science as evolving, but not toward a goal—especially not toward nineteenth-century northern European man as the goal. Nor does Kuhn refer to Eiseley's account of the influence of dominant patterns of thinking—especially Lyell's geological evolution, Bentham's utilitarianism, and Herbert Spencer's emphasis on struggle—on Darwin's shaping of the evolutionary theory, or to Eiseley's debunking of the notion of science as reflecting human "progress." In *Darwin's Century* and later in *The Firmament of Time,* Eiseley develops the notion that Western philosophy, strongly influenced by theology, "caused men to look upon the world around them in a way, or in a frame, that would prepare the Western mind for the final acceptance of evolution" (*DC,* 6). *Darwin's Century* was published in 1958; *The Firmament of Time,* in 1960. Kuhn first published *The Structure of Scientific Revolutions* in 1962.

My point is not that Eiseley influenced Kuhn more than Kuhn was willing to admit, but that the notion of scientific revolutions was, in the twentieth century, what Hayles calls an "isomorphism" and Holton calls a "thematic center" (1987, 251) or a "metaphoric description of the *task* of the sciences" (245), as evolution was in the nineteenth century. The scientist who writes becomes a *textual* and an *intertextual* part of the evolving culture.

5. Gould vividly explores the iconography of these notions in his lectures, using slides for illustration. The text that best illustrates his approach is *The Burgess Shale.*

6. For a further treatment of the hand and the philosopheme of concept, thinking in terms of "having," see Ulmer 1985, 48, 100.

7. For further discussion of the "theoretical" sense of sight, "chemical" senses of taste and smell, and associated philosophemes (in the context of Derrida's program of grammatology), see Ulmer 1985, 34–36.

8. See "Plato's Pharmacy" in *Dissemination* (1981, 61–171).

9. See particularly Gould's Conclusion to *The Mismeasure of Man,* in which he writes of "debunking as positive science" (1981, 321–23).

Chapter 3. The Need for the Hybrid Gene

1. Citing a lecture by Derrida and pointing to a significant portion of Roland Barthes's work, Krauss notes what "simply cannot be called criticism, but . . . cannot, for that matter, be called not-criticism either" (1980, 37). She writes of "the paraliterary space" as "the space of debate, quotation, partisanship, betrayal, reconciliation" (37).

2. Marx traces the theme of intrusion of the machine into the idyllic American setting, beginning with Hawthorne's description of a wooded area near Concord, Massachusetts, a place "like the lap of bounteous Nature" into which bursts "the long shriek, harsh, above all other harshness" of a locomotive that "tells a story of busy men, citizens, from the hot street . . . , men of business" and "of all unquietness" (cited in 1967, 12–13). He traces this theme of mechanistic intrusion into the idyllic quietness from the Jeffersonian garden through Henry Adams's juxtaposition of the dynamo and the Virgin, through Thoreau's sense of the machine's ambiguous relation to nature, and through Fitzgerald's green lawns op-

posed by the valley of the ashes. In British literature he notes the moment when Boswell became aware of technology through Thomson and Schiller. He also cites the "mechanistic habit of mind" (297) that Carlyle found in his contemporaries and the fear of writers such as Carlyle, Ruskin, and Morris that technology might threaten ideals of art or civilization (347).

3. See Carlisle 1974.

4. Here Eiseley emphasizes the four vital factors of vast and linear time, gradual geological change, the "naturalness" of extinction, and the creative role of gradual change in organic and inorganic objects (specifically delineated in *DC*, chaps. 1–8, and *FT*, 70–71).

5. What we have learned can be gathered, however, under several basic headings.

The age of the earth. From *Darwin's Century* onward, Eiseley continually returns to the immense age of the earth. Before Darwin's thesis could be articulated, the immense age of the earth (five billion years is a period of time difficult for the nonspecialist to conceptualize) and the infinitely slow succession of geological forms had to be conceptualized.

Extinction. Time and geological change underlie the "naturalness" of death— extinction—as a concept that had to be established before evolution itself could be articulated.

Continuous change. Form is, Eiseley contends, not reality but illusion. And thus is established the notion of continuous change—constant, indeterminate, nonteleological—that is basic to Eiseley's texts and to his metaphors. Evolution is an ongoing, unfinished process. That this process is continuous, that the human being is not the "roof and crown of things" as Tennyson and other mid-Victorians, including the early Darwinians, conceived him to be (with northern European man the pinnacle of nature's achievements), may be the evolutionary concept most difficult to internalize. The concept is basic to Eiseley's metaphors of change, undecidability, and the unexpected, which help to undermine a notion firmly fixed in the public mind—even in the minds of those who would claim thorough understanding of evolution.

Failures as successes. That evolution creates the paradox of failures that become successes is a concept that undermines even some early evolutionary notions. Eiseley emphasizes that the Devonian fish that struggled ashore was a failure as a fish but an evidence of the tendency of life never to be satisfied with what it is. "There are still things coming ashore" ("The Snout," *IJ,* 54), he suggests. Emphasis on emerging mutations as failures rather than on the warfare in nature, which allows the fittest to survive, is, of course, counter to some fixed notions of the "best" or the "fittest" surviving.

The "community of descent." Eiseley also stresses the "community of descent" of all creatures; again, it is the emphasis, not the fact, that is the subject of his effort. If man persists in seeing himself as nature's highest achievement, he is likely to persist in seeing himself as master of nature, animate and inanimate. Eiseley's emphasis is on the human connection to the long thread of life that runs behind him—exemplified in the narrator's dream-like image of a questioning student attached to a "trunk that stretches monstrously behind him. . . . It writhes, it crawls, it barks and snuffles and roars, and the odor of the swamp exhales from it" ("How Natural Is Natural?" *FT,* 168). The individual is "a many-visaged thing . . . , the weird tree of Igdrasil shaping itself endlessly out of darkness toward the light" (168). The human being may be, in Auden's terms, the quest hero in Eiseley's texts (Introduction, *ST,* 18), but he is such only because he is the product of the "community" that has produced him, of the "reaching" toward more than what it is that characterizes life itself.

The escape from determinism. But if the human being is part of this "community" of life, he has also escaped physical evolution with the emergence of his specialized brain, which makes possible the transfer of such functions as protection against cold and the quest for food to tools conceived by the brain. The new evolution is cultural, and humans can control it—for ill or for good.

Chapter 4. The Genesis of Method I: Darwin's Century

1. Gerber and McFadden isolate six Eiseleyan motifs, which structure their analysis of his texts:

1. Time is immense, linear, and creative. . . .
2. Humanity belongs to the community of descent. . . .
3. The human brain creates a second world. . . .
4. For the evolutionist, the common day has turned marvelous. . . .
5. Guided by Bacon's ideas, science can serve human ends. . . .
6. Scientific knowledge bestows neither freedom nor the capacity for love. . . . (1983, 50–54)

2. Figuratively, Eiseley's journey has been treated as the "journey of an artist" by scholars. See, for example, Carlisle 1974, which stresses the parallel of the journey of exploration and the emergence of science; Haney 1977, chapter 5, which treats Eiseley's conscious experience of the evolutionary journey (188) and the "inward journey back into the psyche" (217); Kassebaum 1979, chapter 1, on *The Immense Journey,* which he considers Eiseley's "most hopeful book" (4); Pickering 1982, which treats of *The Immense Journey* in terms of the *personas* that Eiseley assumes; Schwartz 1977, which parallels Eiseley's focus on the evolutionary journey and his own artistic growth in "The 'Immense Journey' of an Artist" and "The Scientist as Literary Artist"; and Angyal 1983, which considers the journey as a projection of the individual's vision into other life forms, which is one of Eiseley's stated purposes in *The Immense Journey* (Angyal notes the source of Eiseley's title in a comment from Henri Frederic Amiel's *Journey In Time*—a comment that links the immensities of time and space: "It is as though the humanity of our day had, like the migratory birds, an immense journey to make across space" [cited in p. 26]).

3. "Cultural evolution" as the "new evolution" is established in Simpson's landmark study *The Meaning of Evolution* (1949).

4. Earlier, Eiseley has noted the progressionist notion of romantic transcendentalists like Emerson that consciousness might will its way up a unilinear scale, as exemplified in Emerson's lines "And striving to be man the worm / Mounts through all the spires of form" (quoted in *DC,* 52).

5. For insight into the two views of chaos—order out of chaos and chaos out of order—see Hayles 1990, 9–11.

Chapter 6. The Move toward an Ironic View:
The Firmament of Time

1. At times Eiseley comments "almost deconstructively," as John Clifford has noted of his finding loss inextricably interwoven with gain in the autobiography

(1986, 6). Because he is dependent on the language of Western metaphysics, Eiseley cannot be aligned with the attempt to "escape" from metaphysics. Even Derrida has noted that perhaps philosophy cannot escape the metaphysical metaphoricity—inside/outside, light/dark—and equivocality of language: "Philosophical language belongs to a system of language(s). Thereby, its nonspeculative ancestry always brings a certain equivocality into speculation. Since this equivocality is original and irreducible, perhaps philosophy must adopt it, think it and be thought in it, must accommodate duplicity and difference within speculation, within the very purity of philosophical meaning" (1978b, 113).

Chapter 7. Reexamination of Science I:
The Unexpected Universe

1. For the interrelation of order and disorder, of what Eiseley calls "organization" emerging from chaos, see Gleick 1987, 9–31.
2. Derrida uses the term to refer to the basic unit of thought, or root metaphor, parallel to the linguistic *morpheme, phoneme, syntagmeme.* For example, in *Margins* he writes of his own purpose: "To write otherwise. To delimit the space of a closure no longer analogous to what philosophy can represent for itself under this name, according to a straight or circular line enclosing a homogeneous space. To determine, entirely against any philosopheme, the intransigence that prevents it from calculating its margin" (1982, xxiv–xxv). Later, he writes of the "idea" of metaphor—"*eidos* (to be placed before the metaphorical eye)"—as "the 'idealizing' metaphor, which is constitutive of the philosopheme in general" (254). In *Applied Grammatology,* Gregory Ulmer explores Derrida's reexamination of the "idealizing" philosopheme of sight, the "intermediate" philosopheme of hearing, and the "chemical" sense of taste and smell (1985, 34–36).
3. Writing of apostrophe as a genre, Derrida suggests that "the word also speaks of the address to be detoured" (1987, 4). In the Derridean context, "the word" trails meanings throughout the history of Western metaphysics.
4. A significant study of the history of writing is Gelb's *A Study of Writing* (1952). For a consideration of "hieroglyphs" and of "grammatology" in its eighteenth-century inception as "a science of decipherment of nonalphabetic scripts," as well as its contemporary employment by Jacques Derrida, see Ulmer, *Applied Grammatology* (1985, 16–18).

Chapter 8. Reexamination of Science II:
The Invisible Pyramid

1. See Ulmer 1985, especially p. 7.
2. As I shall discuss in greater detail in Chapter 10, Eiseley's prelinguistic realm resembles Jacques Lacan's Imaginary, which is a domain of images that precedes the verbal. In Lacan's Imaginary, the child responds to images, but with the acquisition of language the child moves into the Symbolic, which is the linguistic domain, the world of (linguistic) law and separation from the Imaginary (imagistic) concept of self. Thus in Lacan's scheme the individual enters into a dialectic that can never be resolved, for the Symbolic brings fragmentation and the Real can exist only through simultaneously experience of the other two orders in an

ongoing dialogue of narrative and fantasy. See Jameson 1977, Lacan 1966, and Culler 1983a.

Chapter 10. The Self as a Vehicle for Understanding II: All the Strange Hours

1. See Harari and Bell's Introduction in *Hermes: Literature, Science, Philosophy* (1983, ix–xl).

2. For further discussion of the signifier reflecting onto the signified (as opposed to the standard assumption that the signifier simply reflects the signified), see Derrida 1976, 7, 10–11.

3. For an examination of the effect of the "piece" in *collage,* which, though displaced from its original context, still brings to the new context associations that enrich the new whole, see Ulmer 1985, 59, and Ulmer 1983, 97–107.

4. In his *Philosophy of Style,* Spencer applied a mechanistic determinism to language, stressing the need to reduce the "friction and inertia" of language (1892, 12).

5. For Eco's explication of the notion that "each metaphor can be traced back to a subjacent chain of metonymic connections which constitute the framework of the code and upon which is based the constitution of any semantic field, whether partial or (in theory) global," see *The Role of the Reader* (1984, 68).

Chapter 11. Margins of Humanism, Margins of Science: The Star Thrower *and the Dialogue of Understandings*

1. For the interlinking of machine metaphors and attitudes toward life, see Edge, "Technological Metaphor and Social Control" (1974).

2. A striking parallel, which I have treated elsewhere, exists between this essay and Eiseley's treatment in some of the poems in *Notes of an Alchemist* of the very principle that Snow uses as a touchstone for literary scholars' ignorance of science, the Second Law of Thermodynamics.

3. See Christianson's biography, *Fox at the Wood's Edge* (1991), in which he places Eiseley in the context of post–World War II distrust of science and asserts that Eiseley consciously portrayed himself as a representative of science. The guise is, I believe, another of the identities that Eiseley admits "trying on" as needed.

References

Abrams, M. H. 1971. *The Mirror and the Lamp: Romantic Theory and the Critical Tradition.* New York: Oxford University Press.

Angyal, Andrew. 1983. *Loren Eiseley.* Twayne's United States Authors Series. Boston: Twayne.

Aristotle. 1941. *Rhetoric: Basic Works of Aristotle.* Edited by Richard McKeon. New York: Random.

Aronowitz, Stanley. 1988. *Science as Power: Discourse and Ideology in Modern Society.* Minneapolis: University of Minnesota Press.

Arons, A. B. 1983. "Achieving Wider Scientific Literacy." *Daedalus* 112 (Spring): 91-122.

Auden, W. H. 1978. "Introduction: Concerning the Unpredictable." In *The Star Thrower* by Loren Eiseley, pp. 15–24. New York: Harvest-HBJ.

Bachelard, Gaston. 1940. *L'eau et les reves: Essai sur l'imagination de la matiere.* Paris: Librarie Jose Corti.

Barthes, Roland. 1964. *Essais critiques.* Paris: Seuil.

———. 1968. *Elements of Semiology.* Translated by Annette Lavers and Colin Smith. New York: Hill and Wang.

———. 1972a. "The Structuralist Activity." In *The Structuralists: From Marx to Levi-Strauss.* Edited with an introduction by Richard T. and Fernande M. De George, 148–54. Garden City, N.Y.: Anchor-Doubleday.

———. 1972b. "To Write: An Intransitive Verb?" In *The Structuralists: From Marx to Levi-Strauss.* Edited with an introduction by Richard T. and Fernande M. De George, 155-67. Garden City: Anchor-Doubleday.

———. 1974 [1970]. *S/Z: An Essay.* Translated by Richard Miller. Preface by Richard Howard. New York: Hill and Wang.

———. 1975. *The Pleasure of the Text.* Translated by Richard Miller. New York: Hill and Wang.

———. 1977a. "The Death of the Author." In *Image, Music, Text.* Translated by Stephen Heath, 142–48. New York: Hill and Wang.

———. 1977b. "From Work to Text." In *Image, Music, Text.* Translated by Stephen Heath, 155–64. New York: Hill and Wang.

———. 1982. *Empire of Signs.* Translated by Richard Howard. New York: Hill and Wang.

———. 1983. *A Lover's Discourse: Fragments.* Translated by Richard Howard. New York: Hill and Wang.

Barzun, Jacques. 1981 [1941]. *Darwin, Marx, Wagner: Critique of a Heritage.* 2d ed. Chicago: University of Chicago Press.

Battista, John A. 1982. "The Holographic Model, Holistic Paradigm, Information

Theory and Consciousness." In *The Holographic Paradigm and Other Paradoxes: Exploring the Leading Edge of Science.* Edited by Ken Wilber, 143–50. Boulder, Colo.: Shambhala.

Belenky, Mary, et al. 1986. *Women's Ways of Knowing.* New York: Basic Books.

Bentov, Itzhak. 1982. "Comments on the Holographic View of Reality." In *The Holographic Paradigm and Other Paradoxes: Exploring the Leading Edge of Science.* Edited by Ken Wilber, 136–38. Boulder: Shambhala.

Black, Max. 1962. *Models and Metaphors: Studies in Language and Philosophy.* Ithaca: Cornell University Press.

Blanchot, Maurice. 1982a. "Literary Infinity: The Aleph." In *The Sirens' Song: Selected Essays by Maurice Blanchot.* Edited and with an introduction by Gabriel Josipovici, translated by Sacha Rabinovitch, 222–24. Bloomington: Indiana University Press.

———. 1982b. "The Sirens' Song." *The Sirens' Song: Selected Essays by Maurice Blanchet.* Edited by Gabriel Josipovici, translated by Rabinovitch, 59–65. Bloomington: Indiana University Press.

———. 1982c. "Where Now? Who Now?" *The Sirens' Song: Selected Essays by Maurice Blanchet.* Edited by Gabriel Josipovici, translated by Sacha Rabinovitch, 192–98. Bloomington: Indiana University Press.

Block, Ed. 1991. "Eiseley, Pynchon, and the Malign Magic of the Universe." Society for Literature and Science, Montreal, October.

Boyd, Richard. 1979. "Metaphor and Theory Change: What Is 'Metaphor' a Metaphor For?" In *Metaphor and Thought.* Edited by Andrew Ortony, 357–408. New York: Cambridge University Press.

Briggs, John P., and F. David Peat. 1984. *Looking Glass Universe: The Emerging Science of Wholeness.* New York: Simon-Touchstone.

Bronowski, Jacob. 1971. "Protest and Prospect." In *The Shape of Likelihood: Relevance and the University,* by Loren Eiseley, D. W. Bronk, Jacob Bronowski, and H. M. Jones, 43–62. Preface by Taylor Littleton. University: University of Alabama Press.

Bunn, James H. 1979. *The Dimensionality of Signs, Tools, and Models: An Introduction.* Bloomington: Indiana University Press.

Capra, Fritjof. 1977. *The Tao of Physics: An Exploration of the Parallels between Modern Physics and Eastern Mysticism.* 2d ed. New York: Bantam Books.

———. 1983. *The Turning Point: Science, Society, and the Rising Culture.* New York: Bantam Books.

Carlisle, E. Fred. 1974. "The Heretical Science of Loren Eiseley." *Centennial Review* 18 (Fall): 354–77.

———. 1977. "The Poetic Achievement of Loren Eiseley." *Prairie Schooner* 51 (Spring): 111–29.

———. 1983. *Loren Eiseley: The Development of a Writer.* Urbana: University of Illinois Press.

Cassirer, Ernst. 1971. "Art." In *Critical Theory since Plato.* Edited by Hazard Adams, 994–1013. New York: Harcourt Brace Jovanovich.

Caulfield, H. John. 1984. "The Wonder of Holography." *National Geographic* 165 no. 3 (March): 364–73.

Caws, Mary Ann. 1981. *The Eye in the Text: Essays on Perception, Mannerist to Modern.* Princeton Essays on the Arts. Princeton: Princeton University Press.

Christianson, Gale E. 1991. *Fox at the Wood's Edge: A Biography of Loren Eiseley.* New York: Henry Holt.

Clark, Katerina, and Michael Holquist. 1984. *Mikhail Bakhtin.* Cambridge, Mass.: Belknap Press.

Clifford, John. 1986. "Toward a Reader's Rhetoric of Nonfiction." Division on Literary Nonfiction and Composition Studies. Conference on College Communication and Composition, New Orleans, March.

Culler, Jonathan. 1981. *The Pursuit of Signs: Semiotics, Literature, Deconstruction.* Ithaca: Cornell University Press.

———. 1983a. *On Deconstruction: Theory and Criticism after Structuralism.* Ithaca: Cornell University Press.

———. 1983b. *Roland Barthes.* New York: Oxford University Press.

Curtius, Ernst Robert. 1953. *European Literature and the Latin Middle Ages.* Translated by Willard R. Trask. Bollingen Series XXXVI. Princeton: Princeton University Press.

Dawkins, Richard. 1978. *The Selfish Gene.* New York: Oxford University Press.

D'Amico, Robert. 1989. *Historicism and Knowledge.* New York: Routledge.

de Man, Paul. 1985. *Blindness and Insight: Essays in the Rhetoric of Contemporary Criticism.* 2d ed. Theory and History of Literature, vol. 7. Minneapolis: University of Minnesota Press.

Derrida, Jacques. 1974. "White Mythology." *New Literary History* 6:5–74.

———. 1976. *Of Grammatology.* Translated by Gayatri Spivak. Baltimore: Johns Hopkins University Press.

———. 1978a. "The *Retrait* of Metaphor." *Enclitic* 2 no. 2 (Summer): 5–34.

———. 1978b. *Writing and Difference.* Translated by Alan Bass. Chicago: University of Chicago Press.

———. 1979. "Scribble (writing-power)." *Yale French Studies* 58:116–47.

———. 1981. *Dissemination.* Translated by Barbara Johnson. Chicago: University of Chicago Press.

———. 1982. *Margins of Philosophy.* Translated by Alan Bass. Chicago: University of Chicago Press.

———. 1987. *The Post Card: From Socrates to Freud and Beyond.* Translated by Alan Bass. Chicago: University of Chicago Press.

Eagleton, Terry. 1983. *Literary Theory: An Introduction.* Minneapolis: University of Minnesota Press.

Eco, Umberto. 1984. *The Role of the Reader: Explorations in the Semiotics of Texts.* Bloomington: Indiana University Press.

Edge, David. 1974. "Technological Metaphor and Social Control." *New Literary History* 6 no. 1 (Autumn): 135–47.

Eiseley, Loren. 1957. "The Enchanted Glass." *American Scholar* 26 no. 4 (Autumn): 478–92.

———. 1959 [1957]. *The Immense Journey.* New York: Vintage-Random.

———. 1961 [1958]. *Darwin's Century.* New York: Doubleday.

———. 1962. *Francis Bacon and the Modern Dilemma.* Lincoln: University of Nebraska Press.

———. 1969. *The Unexpected Universe.* New York: Harvest-HBJ.

————. 1970. *The Invisible Pyramid.* New York: Scribner's.

————. 1971a. "Introduction." In *The Shape of Likelihood: Relevance and the University,* by Loren Eiseley, D. W. Bronk, Jacob Bronowski, and H. M. Jones, 3–18. Preface by Taylor Littleton. University: University of Alabama Press.

————. 1971b. *The Night Country.* New York: Scribner's.

————. 1972. *Notes of an Alchemist.* New York: Scribner's.

————. 1973a. *The Innocent Assassins.* New York: Scribner's.

————. 1973b. *The Man Who Saw through Time.* Revised and enlarged edition of *Francis Bacon and the Modern Dilemma.* New York: Scribner's.

————. 1975. *All the Strange Hours: The Excavation of a Life.* New York: Scribner's.

————. 1978. *The Star Thrower.* Introduction by W. H. Auden. New York: Harvest-HBJ.

————. 1979a. *All the Night Wings.* New York: Times Books.

————. 1979b, *Darwin and the Mysterious Mr. X: New Light on the Evolutionists.* New York: Harvest-HBJ.

————. 1980 [1960]. *The Firmament of Time.* New York: Atheneum.

Eliot, T. S. 1974. "Tradition and the Individual Talent." *The Norton Anthology of English Literature.* 3d ed. Edited by M. H. Abrams et al., 2198–2205. New York: W. W. Norton.

Ferguson, Marilyn. 1982. "Karl Pribram's Changing Reality." In *The Holographic Paradigm.* Edited by Ken Wilber, 15–26. Boulder, Colo.: Shambhala.

Foucault, Michel. 1973. *The Order of Things: An Archaeology of the Human Sciences.* New York: Vintage-Random.

Freud, Sigmund. 1965. *The Interpretation of Dreams.* Translated by James Strachey. New York: Avon Books.

————. 1977. *Introductory Lectures on Psychoanalysis.* Translated by James Strachey. New York: Liveright-Norton.

Frye, Northrop. 1968. "Introduction." In *The Psychoanalysis of Fire,* by Gaston Bachelard. Translated by Alan C. M. Ross. Boston: Beacon.

————. 1971 [1957]. *Anatomy of Criticism: Four Essays.* Princeton: Princeton University Press.

Gelb, Ignace J. 1952. *A Study of Writing: The Foundations of Grammatology.* Chicago: University of Chicago Press.

Gerber, Leslie E., and Margaret McFadden. 1983. *Loren Eiseley.* Literature and Life Series. New York: Frederick Ungar.

Gleick, James. 1987. *Chaos: Making a New Science.* New York: Viking.

Gombrich, E. H. 1972. *Art and Illusion: A Study in the Psychology of Pictorial-Representation.* The A. W. Mellon Lectures in the Fine Arts, 1956. Bollingen Series XXXV. Princeton: Princeton University Press.

————. 1979. *The Sense of Order: A Study in the Psychology of Decorative Art.* Ithaca: Cornell University Press.

Gould, Stephen Jay. 1981. *The Mismeasure of Man.* New York: W. W. Norton.

————. 1987a. *Time's Arrow, Time's Cycle: Myth and Metaphor in the Discovery of Geological Time.* Cambridge: Harvard University Press.

————. 1987b. *An Urchin in the Storm: Essays about Books and Ideas.* New York: W. W. Norton.

———. 1989. *Wonderful Life: The Burgess Shale and the Nature of History.* New York: W. W. Norton.

Gross, Alan. 1991. "Does Rhetoric of Science Matter?" *College English* 53 no. 8:933–43.

Gruber, Howard E. 1978. "Darwin's 'Tree of Nature' and Other Images of Wide Scope." In *On Aesthetics in Science.* Edited by Judith Wechsler, 121–40. Cambridge: Massachusetts Institute of Technology Press.

Habermas, Jürgen. 1970. *Toward a Rational Society: Student Protest, Science, and Politics.* Translated by Jeremy J. Shapiro. Boston: Beacon.

———. 1975. *Legitimation Crisis.* Translated by Thomas McCarthy. Boston: Beacon.

Haney, Deanna Kay. 1977. "The Role of the Literary Naturalist in American Culture." Dissertation, University of Michigan.

Harari, Josué V., and David F. Bell. 1983. "Introduction: *Journal à plusiers voies.*" In *Hermes: Literature, Science, Philosophy,* by Michel Serres, ix–xl. Baltimore: Johns Hopkins University Press.

Hayles, N. Katherine. 1990. *Chaos Bound: Orderly Disorder in Science and Literature.* Ithaca: Cornell University Press.

Heidegger, Martin. 1975a. "Language." In *Poetry, Language, Thought.* Translated by Albert Hofstadter, 189–210. New York: Colophon-Harper.

———. 1975b. "The Origin of the Work of Art." In *Poetry, Language, Thought.* Translated by Albert Hofstadter, 15–87. New York: Colophon-Harper.

———. 1975c. " . . . Poetically Man Dwells . . ." In *Poetry, Language, Thought.* Translated by Albert Hofstadter, 211–29. New York: Colophon-Harper.

———. 1975d. "The Thing." In *Poetry, Language, Thought.* Translated by Albert Hofstadter, 165–82. New York: Colophon-Harper.

———. 1977a. "The Age of the World Picture." In *The Question Concerning Technology and Other Essays.* Translated and with an introduction by William Lovitt, 115–54. New York: Colophon-Harper.

———. 1977b. "The Question Concerning Technology." In *The Question Concerning Technology and Other Essays.* Translated and with an Introduction by William Lovitt, 3–35. New York: Colophon-Harper.

———. 1977c. "Science and Reflection." In *The Question Concerning Technology and Other Essays.* Translated and with an introduction by William Lovitt, 155–82. New York: Colophon-Harper.

Heidtmann, Peter. 1991. *Loren Eiseley: A Modern Ishmael.* Hamden, Conn.: Archon-Shoe String Press.

Hesse, Hermann. 1970. *Demian.* Translated by Michael Roloff and Michael Lebeck. New York: Bantam.

Heuer, Kenneth, ed. 1987. *The Lost Notebooks of Loren Eiseley.* New York: Little, Brown.

Hiley, David R., James F. Bohman, and Richard Shusterman, eds. 1991. *The Interpretive Turn: Philosophy, Science, Culture.* Ithaca: Cornell University Press.

Hill, Forbes I. 1983. "The *Rhetoric* of Aristotle." In *A Synoptic History of Classical Rhetoric.* Edited by James J. Murphy. Davis, Calif.: Hermagoras Press. 19–76.

Holman, C. Hugh, and William Harmon. 1986. *A Handbook to Literature*. 5th ed. New York: Macmillan.

Holton, Gerald. 1973. *Thematic Origins of Scientific Thought: Kepler to Einstein*. Cambridge: Harvard University Press.

———. 1987. *The Advancement of Science, and Its Burdens*. Cambridge: Harvard University Press.

Irwin, John T. 1983. *American Hieroglyphics: The Symbol of the Egyptian Hieroglyphics in the American Renaissance*. Baltimore: Johns Hopkins University Press.

Iser, Wolfgang. 1980. "The Reading Process: A Phenomenological Approach." In *Reader-Response Criticism: From Formalism to Post-Structuralism*. Edited by Jane P. Tompkins. Baltimore: Johns Hopkins University Press.

Jakobson, Roman. 1971. "The Metaphoric and Metonymic Poles." In *Critical Theory since Plato*. Edited by Hazard Adams, 1113–16. New York: Harcourt Brace Jovanovich.

Jameson, Fredric. 1974. *The Prison House of Language: A Critical Account of Structuralism and Russian Formalism*. Princeton: Princeton University Press.

———. 1977. "Imaginary and Symbolic in Lacan: Marxism, Psychoanalytic Criticism, and the Problem of the Subject." *Yale French Studies* 55/56:338–95.

Johnson, Nan. 1984. "Ethos and the Aims of Rhetoric." *Essays on Classical Rhetoric and Modern Discourse*. Edited by Robert J. Connors, Lisa S. Ede, and Andrea A. Lunsford, 98–114. Carbondale: Southern Illinois University Press.

Jones, Roger S. 1982. *Physics as Metaphor*. Minneapolis: University of Minnesota Press.

Josipovici, Gabriel. 1982. "Introduction." In *The Sirens' Song: Selected Essays by Maurice Blanchot*. Edited and with an introduction by Gabriel Josipovici. Translated by Sacha Rabinovitch, 1–18. Bloomington: Indiana University Press.

Kassebaum, L. Harvey. 1979. "To Survive Our Century: The Narrative Voice of Loren Eiseley, an Essay in Appreciation." Dissertation, Indiana University of Pennsylvania.

Kneale, J. Douglas. 1986. "Wordsworth's Images of Language: Voice and Letter in *The Prelude*." *PMLA* 101 no. 4:351–61.

Krauss, Rosalind. 1980. "Poststructuralism and the 'Paraliterary.'" *October* 13:36–40.

Kuhn, Thomas S. 1969. *The Structure of Scientific Revolutions*. Chicago: University of Chicago Press.

———. 1981. "Metaphor in Science." In *Philosophical Perspectives on Metaphor*. Edited by Mark Johnson, 63–82. Minneapolis: University of Minnesota Press.

Lacan, Jacques. 1966. *Ecrits*. Paris: Seuil.

———. 1981a. *The Four Fundamental Concepts of Psychoanalysis*. Translated by Alan Sheridan. Edited by Jacques-Alain Miller. New York: W. W. Norton.

———. 1981b [1968]. *Speech and Language in Psychoanalysis*. Translated by Anthony Wilden. Baltimore: Johns Hopkins University Press.

Lakoff, George, and Mark Johnson. 1980. *Metaphors We Live By*. Chicago: University of Chicago Press.

McLuhan, Marshall. 1982. "Sight, Sound, and the Fury." In *The McGraw-Hill Reader*. Edited by Gilbert H. Muller, 339–47. New York: McGraw-Hill.

McMullin, Ernan. 1974. "Two Faces of Science." *Review of Metaphysics* 27:668–70.

Marx, Leo. 1967. *The Machine in the Garden: Technology and the Pastoral Ideal in America.* New York: Oxford University Press.

Mead, Margaret. 1959. "Closing the Gap between the Scientists and Others." *Daedalus* 88:139–46.

Merton, Robert K. 1973. "The Normative Structure of Science." *The Sociology of Science: Theoretical and Empirical Investigations.* Edited by Norman W. Storer, 267–78. Chicago: University of Chicago Press.

Nalimov, V. V. 1982. *Realms of the Unconscious: The Enchanted Frontier.* Edited by Robert G. Colodny. Illustrated by V. S. Gribov. Translated by A. V. Yarkho. Philadelphia: Institute for Scientific Information Press.

————. 1985. *Space, Time, and Life: The Probabilistic Pathways of Evolution.* Edited by Robert G. Colodny. Translated by A. V. Yarkho. Philadelphia: Institute for Scientific Information Press.

Olsen, Everett C., and Jane Ann Robinson. 1975. *Concepts of Evolution.* Columbus: Merrill.

Percy, Walker. 1991. *Signposts in a Strange Land.* Edited by Patrick Samway. New York: Farrar, Straus and Giroux.

Pickering, Deborah Hawkins. 1982. "The Selves of Loren Eiseley: A Stylistic Analysis of the Essays in *The Immense Journey, The Night Country,* and *All the Strange Hours.*" Dissertation, University of Iowa.

Prelli, Lawrence J. 1989. *A Rhetoric of Science: Inventing Scientific Discourse.* Columbia: University of South Carolina Press.

Pribram, Karl H. 1982. "What the Fuss Is All About." In *The Holographic Paradigm.* Edited by Ken Wilber, 27–34. Boulder, Colo.: Shambhala.

Rapaport, Herman. 1983. "Staging: Mont Blanc." In *Displacement: Derrida and After.* Edited by Mark Krupnick, 59–73. Bloomington: Indiana University Press.

Ricoeur, Paul. 1979. *The Rule of Metaphor: Multi-Disciplinary Studies of the Creation of Meaning in Language.* Translated by Robert Czerny, with Kathleen McLaughlin and John Costello. Toronto: University of Toronto Press.

Rorty, Richard. 1976. "Professionalized Philosophy and Transcendentalist Culture." *Georgia Review* 30:757–71.

————. 1979. *Philosophy and the Mirror of Nature.* Princeton: Princeton University Press.

Rosenberg, Charles. 1966. "Science and American Social Thought." In *Science and Society in the United States.* Edited by David D. Van Tassel and Michael G. Hall, 135–62. Homewood, Ill.: Dorsey.

Schaya, Leo. 1973. *The Universal Meaning of the Kabbalah.* Baltimore: Penguin.

Scholes, Robert. 1985. *Textual Power: Literary Theory and the Teaching of English.* New Haven: Yale University Press.

Schwartz, James M. 1977a. "The 'Immense Journey' of an Artist: The Literary Technique and Style of Loren Eiseley." Dissertation, Ohio University.

————. 1977b. "Loren Eiseley: The Scientist as Literary Artist." *Georgia Review* 31:855-71.

Seamon, Roger. 1989. "Poetics against Itself: On the Self-Destruction of Modern Scientific Criticism." *PMLA* 104 no. 3:294–305.

Serres, Michel. 1983. "Michelet: The Soup." In *Hermes: Literature, Science, Philosophy.* Edited by Josué V. Harari and David F. Bell, 29–38. Baltimore: Johns Hopkins University Press.

Shelley, Percy B. 1971. "A Defense of Poetry." *Critical Theory Since Plato.* Edited by Hazard Adams, 499–513. New York: Harcourt Brace Jovanovich.

Shen, Benjamin S. P. 1975. "Science Literacy and the Public Understanding of Science." In *Communication of Scientific Information,* 44–52. Basel Switzerland: Karger.

Simpson, George Gaylord. 1949. *The Meaning of Evolution: A Study of the History of Life and of Its Significance for Man.* New Haven: Yale University Press.

Snow, Charles Percy. 1964. *The Two Cultures: And a Second Look.* 2d ed. Cambridge: Cambridge University Press.

Spencer, Herbert. 1892. *Philosophy of Style.* New York: D. Appleton.

Staton, Shirley, ed. 1987. *Literary Theories in Praxis.* Philadelphia: University of Pennsylvania Press.

Temkin, Owsei. 1949. "Metaphors of Human Biology." In *Science and Civilization.* Edited by Robert C. Stauffer, 167–94. Madison: University of Wisconsin Press.

Thass-Thienemann, Theodore. 1968. *Symbolic Behavior.* New York: Washington Square.

Thomas, Lewis. 1980. *The Medusa and the Snail: More Notes of a Biology Watcher.* New York: Bantam.

Ulmer, Gregory L. 1981. "The Post-Age." *Diacritics* 11 no. 3:39–56.

———. 1983. "The Object of Post-Criticism." *The Anti-Aesthetic: Essays on Postmodern Culture.* Edited and with an Introduction by Hal Foster. Port Townsend, Wash.: Bay Press. 83–110.

———. 1985. *Applied Grammatology: Post(e)-Pedagogy from Jacques Derrida to Joseph Beuys.* Baltimore: Johns Hopkins University Press.

Vickers, Geoffrey. 1978. "Rationality and Intuition." In *On Aesthetics in Science.* Edited by Judith Wechsler, 143–64. Cambridge: Massachusetts Institute of Technology Press.

Watkins, Calvert. 1973. "Indo-European Roots." *American Heritage Dictionary.* Edited by William Morris, 1505–50. Boston: Houghton-Mifflin.

Wechsler, Judith. 1978. "Introduction." In *On Aesthetics in Science.* Edited by Judith Wechsler, 1–5. Cambridge: Massachusetts Institute of Technology Press.

White, Hayden. 1978. *Tropics of Discourse: Essays in Cultural Criticism.* Baltimore: Johns Hopkins University Press.

———. 1987. *Metahistory: The Historical Imagination in Nineteenth-Century Europe.* Baltimore: Johns Hopkins University Press.

Wilber, Ken. 1982. "Physics, Mysticism and the New Holographic Paradigm: A Critical Appraisal." In *The Holographic Paradigm.* Edited by Ken Wilber, 157–86. Boulder, Colo.: Shambhala.

Wilden, Anthony. 1980. "Introduction." In *The Myths of Information: Technology and Postindustrial Culture.* Edited by Kathleen Woodward, xii–xxvi. Madison: Coda.

Wrobel, Arthur, ed. 1987. *Pseudo-Science and Society in 19th-Century America.* Lexington: University Press of Kentucky.

Zinsser, William. 1985. *On Writing Well: An Informal Guide to Writing Nonfiction.* 3d ed. New York: Harper and Row.

Index